The Beginnings of Western Science

THE
BEGINNINGS
OF WESTERN
SCIENCE

THE EUROPEAN SCIENTIFIC TRADITION IN
PHILOSOPHICAL, RELIGIOUS, AND INSTITUTIONAL
CONTEXT, 600 B.C. TO A.D. 1450

David C. Lindberg

The University of Chicago Press / Chicago and London

The University of Chicago Press, Chicago 60637
The University of Chicago Press, Ltd., London
© 1992 by The University of Chicago
All rights reserved. Published 1992
Printed in the United States of America

00 99 98 97 96 95 94 93 92   5 4 3 2

ISBN (cloth): 0-226-48230-8
ISBN (paper): 0-226-48231-6

Library of Congress Cataloging-in-Publication Data
Lindberg, David C.
    The beginnings of Western science : the European scientific
tradition in philosophical, religious, and institutional context,
600 B.C. to A.D. 1450 / David C. Lindberg.
        p.   cm.
    Includes bibliographical references and index.
    1. Science, Ancient—History.   2. Science, Medieval—History.
I. Title.
Q124.95.L55   1992
509.4—dc20   91-37741

⊗ The paper used in this publication meets the minimum requirements of the American
National Standard for Information Sciences—Permanence of Paper for Printed Library
Materials, ANSI Z39.48-1984.

To
Marshall Clagett
and
Edward Grant
who set an example

Contents

	List of Illustrations	xi
	Preface	xv
1	Science and Its Origins	1
	What Is Science?　1	
	Prehistoric Attitudes toward Nature　4	
	Babylonian and Egyptian Science　13	
2	The Greeks and the Cosmos	21
	The World of Homer and Hesiod　21	
	The First Greek Philosophers　25	
	The Milesians and the Question of Ultimate Reality　27	
	The Problem of Change　32	
	The Problem of Knowledge　34	
	Plato's World of Forms　35	
	Plato's Cosmology　39	
	The Achievement of Early Greek Philosophy　44	
3	Aristotle's Philosophy of Nature	46
	Life and Works　46	
	Metaphysics and Epistemology　48	
	Nature and Change　51	
	Cosmology　54	
	Motion, Terrestrial and Celestial　58	
	Aristotle as a Biologist　62	
	Aristotle's Achievement　67	
4	Hellenistic Natural Philosophy	69
	Schools and Education　69	
	The Lyceum after Aristotle　74	
	Epicureans and Stoics　77	
5	The Mathematical Sciences in Antiquity	85
	The Application of Mathematics to Nature　85	
	Greek Mathematics　86	
	Early Greek Astronomy　89	

Cosmological Developments   97
Hellenistic Planetary Astronomy   98
The Science of Optics   105
The Science of Weights   108

6   Greek and Roman Medicine   111
Early Greek Medicine   111
Hippocratic Medicine   113
Hellenistic Anatomy and Physiology   119
Hellenistic Medical Sects   122
Galen and the Culmination of Hellenistic Medicine   125

7   Roman and Early Medieval Science   135
Greeks and Romans   135
Popularizers and Encyclopedists   137
Translations   147
The Role of Christianity   149
Roman and Early Medieval Education   151
Two Early Medieval Natural Philosophers   158

8   Science in Islam   161
Learning and Science in Byzantium   161
The Eastward Diffusion of Greek Science   163
The Birth, Expansion, and Hellenization of Islam   166
Translation of Greek Science into Arabic   168
The Islamic Response to Greek Science   170
The Islamic Scientific Achievement   175
The Decline of Islamic Science   180

9   The Revival of Learning in the West   183
The Middle Ages   183
Carolingian Reforms   184
The Schools of the Eleventh and Twelfth Centuries   190
Natural Philosophy in the Twelfth-Century Schools   197
The Translation Movement   203
The Rise of Universities   206

10   The Recovery and Assimilation of Greek and
Islamic Science   215
The New Learning   215
Aristotle in the University Curriculum   216
Points of Conflict   218
Resolution: Science as Handmaiden   223
Radical Aristotelianism and the Condemnations of 1270
and 1277   234
The Relations of Philosophy and Theology after 1277   240

11   The Medieval Cosmos   245
The Structure of the Cosmos   245

The Heavens   248
The Terrestrial Region   252
The Greek and Islamic Background to Western
   Astronomy   261
Astronomy in the West   267
Astrology   274

12   The Physics of the Sublunar Region   281
Matter, Form, and Substance   282
Combination and Mixture   285
Alchemy   287
Change and Motion   290
The Nature of Motion   292
The Mathematical Description of Motion   294
The Dynamics of Local Motion   301
The Quantification of Dynamics   304
The Science of Optics   307

13   Medieval Medicine and Natural History   317
The Medical Tradition of the Early Middle Ages   317
The Transformation of Western Medicine   325
Medical Practitioners   327
Medicine in the Universities   329
Disease, Diagnosis, Prognosis, and Therapy   332
Anatomy and Surgery   339
Development of the Hospital   345
Natural History   348

14   The Legacy of Ancient and Medieval Science   355
The Continuity Debate   355
The Medieval Scientific Achievement   360

Notes   369
Bibliography   407
Index   441

Illustrations

MAPS

1	The Greek world about 450 B.C.	28
2	Alexander the Great's Empire	70
3	The Roman Empire	134
4	Islamic expansion	167
5	Carolingian Empire about 814	186
6	Medieval universities	207

FIGURES

1.1.	Babylonian clay tablet	15
1.2.	Egyptian medical papyrus	19
2.1.	Statue of Zeus	22
2.2.	Shrine at Delphi	24
2.3.	Ruins of ancient Ephesus	30
2.4.	Plato	36
2.5.	The five Platonic solids	41
2.6.	The celestial sphere according to Plato	43
3.1.	Aristotle	48
3.2.	Square of opposition of the elements	55
3.3.	The Aristotelian cosmos	57
4.1.	Parthenon	72
4.2.	The schools of Hellenistic Athens	72
4.3.	Epicurus	78
5.1.	Incommensurability of the side and diagonal of a square	87
5.2.	The method of "exhaustion"	88
5.3.	Two-sphere model of the cosmos	90

5.4.	Retrograde motion of Mars	91
5.5.	Eudoxan spheres	92
5.6.	Eudoxan spheres and the hippopede	93
5.7.	Aristotelian nested spheres	96
5.8.	Method of calculating distances to the sun and moon	98
5.9.	Ptolemy's eccentric model	100
5.10.	Ptolemy's epicycle on deferent model	101
5.11.	Ptolemy's epicycle on deferent model (planet on the inside of the epicycle)	101
5.12.	The retrograde motion of a planet explained	101
5.13.	Ptolemy's equant model	102
5.14.	Ptolemy's model for the superior planets	104
5.15.	Geometry of vision	106
5.16.	Vision by reflected rays	107
5.17.	Ptolemy's theory of refraction	108
5.18.	Apparatus for measuring angles of incidence and refraction	108
5.19.	The balance beam	109
5.20.	Dynamic explanation of the balance beam	109
5.21.	Archimedes' proof of the law of the lever	110
6.1.	Asclepius	112
6.2.	Theater at Epidaurus	114
6.3.	Hippocrates	114
6.4.	Greek physician	123
7.1.	Roman Forum	135
7.2.	Cicero	139
7.3.	Pliny's monstrous races	142
7.4.	Macrobius on rainfall	143
7.5.	Martianus Capella on the motions of Venus and Mercury	147
7.6.	A monk in his study	154
7.7.	A medieval scribe	157
8.1.	Ḥunayn ibn Isḥāq on the anatomy of the eye	170
8.2.	Ibn Ṭūlūn Mosque	172
8.3.	Motion of Mercury according to Ibn ash-Shāṭir	178
8.4.	The eyes and visual system according to Ibn al-Haytham	179
8.5.	The Great Mosque of Cordoba	181
9.1.	Personification of the quadrivium	187
9.2.	Cloister, Santa Maria de Ripoll	189
9.3.	A grammar school scene	192
9.4.	Chartres Cathedral	194
9.5.	Chained library	194

9.6.	Hugh of St. Victor	196
9.7.	God as architect of the universe	199
9.8.	Merton College, Oxford	210
9.9.	The medieval schools, Oxford	210
10.1.	Page from Avicenna's *Physics*	219
10.2.	Basilica of St. Francis, Assisi	222
10.3.	The skeleton of Robert Grosseteste	225
10.4.	Albert the Great	228
10.5.	Cathedral of Notre Dame, Paris	236
11.1.	Simplified Aristotelian cosmology	249
11.2.	Theodoric of Freiberg's theory of the rainbow	254
11.3.	A T-O map	255
11.4.	A modified T-O map	256
11.5.	A portolan chart	257
11.6.	Nicole Oresme	259
11.7.	Astrolabe	265
11.8.	An "exploded" view of the astrolabe	265
11.9.	Stereographic projection of astrolabe plates	266
11.10.	Ibn al-Haytham's solid-sphere model	266
11.11.	Quadrant of Profatius Judaeus	269
11.12.	An astronomer observing with an astrolabe	270
11.13.	Model for the superior planets	271
11.14.	The Alfonsine Tables	273
11.15.	Albumasar	278
12.1.	Alchemical apparatus	289
12.2.	Use of a line segment to represent the intensity of a quality	297
12.3.	Distribution of temperatures in a rod	297
12.4.	Nicole Oresme's system of geometrical representation	297
12.5.	Distribution of velocities in a rotating rod	298
12.6.	Velocity as a function of time	298
12.7.	Representation of various motions	299
12.8.	Geometrical proof of the Merton rule	300
12.9.	Incoherent radiation	310
12.10.	Mixing of light rays in the eye	311
12.11.	The visual cone and the eye	312
12.12.	Page from an optical treatise	314
13.1.	Page from a Greek manuscript of Dioscorides	319
13.2.	Miraculous healing of a leg	322
13.3.	Arabic surgical instruments	324
13.4.	Constantine the African	326

13.5.	Fetuses in the womb	328
13.6.	Trotula	328
13.7.	Medical instruction	331
13.8.	An apothecary shop	334
13.9.	Urine color chart	336
13.10.	Diagnosis by pulse	337
13.11.	A physician's girdle book	338
13.12.	Operation for cataract and nasal polyps	340
13.13.	Operation for scrotal hernia	340
13.14.	Human anatomy	343
13.15.	Human dissection	344
13.16.	A medieval hospital	347
13.17.	Page from an herbal	349
13.18.	Page from a medieval bestiary	352

Preface

The decades since World War II have seen an explosion of research on the history of ancient and medieval science. Much of this research is stunning in quality, and it has greatly enriched our understanding of early Western science. However, the broad synthetic and interpretive efforts needed to communicate the fruits of this research to a larger readership have been surprisingly meager. Indeed, while the curve of research publications continues to rise, the production of books broad in coverage and addressed to the general educated reader and scholars specializing in other fields seems to be declining.

A brief survey of the available literature on ancient and medieval science may help to make the point. The first substantial, knowledgeable, postwar account of ancient and medieval science appeared in a book by E. J. Dijksterhuis, published originally in Dutch as *Mechanisering van het Wereldbeeld* (1950), subsequently translated into English as *The Mechanization of the World Picture* (1961). By the time the English translation of Dijksterhuis became available, Alistair Crombie's *Augustine to Galileo* (1952) had already been around for nearly a decade, helping to fuel a growing sense of purpose and excitement among historians of medieval science. Perhaps Crombie's success frightened off the competition. Whatever the reason, nearly twenty years passed before the appearance of another synthetic work on medieval science: Edward Grant's brief *Physical Science in the Middle Ages* (1971), followed three years later by Olaf Pedersen and Mogens Pihl's *Early Physics and Astronomy: A Historical Introduction* (1974), both books restricted (as their titles indicate) to the physical sciences. Nothing has appeared since Pedersen and Pihl except the collection prepared under my editorship, *Science in the Middle Ages* (1978), which marshaled the talents of sixteen distinguished historians of medieval science to interpret the current state of the field for a relatively advanced audience. Although many of the essays contained in *Science in the Middle Ages* retain their authoritative status, the book as a whole suffers from a lack of unity, gaps in coverage, and (increasingly) old age.

The only books thus far mentioned that deal more than incidentally with ancient science are those of Dijksterhuis and of Pedersen and Pihl. For better or for worse, ancient science and medieval science have developed separate identities and separate interpretive literatures. Benjamin Farrington's *Greek Science* (two parts, published in 1944 and 1949, respectively) led the way on the history of Greek science. This was soon supplanted by the authoritative book of Marshall Clagett, *Greek Science in Antiquity* (1957). Giorgio de Santillana's *The Origins of Scientific Thought* followed in 1961. Roman science received separate treatment in William H. Stahl's *Roman Science* (1962). And in the early 70s G. E. R. Lloyd produced two widely praised volumes—*Early Greek Science: Thales to Aristotle* (1970) and *Greek Science after Aristotle* (1973)—which have held the field unchallenged for the past two decades.

Twenty years after Lloyd and forty years after Crombie (the last author to offer broad coverage of medieval science) does not seem too soon for a new attempt. The present book is a product of that conviction. I do not expect this book to replace its predecessors, particularly Lloyd's excellent volumes, but to achieve certain other goals. First, I have endeavored to take into account a considerable body of research that was not available to my predecessors. (For example, about two-thirds of the items in the bibliography at the back of this book were not available to Lloyd and Grant in the early 70s.) Second, by conjoining ancient and medieval science in a single volume, I have gained the opportunity to examine questions of continuity between ancient and medieval science in ways discouraged by separate treatment of the two halves, and also to raise problems of transmission that are otherwise apt to disappear into the crack.

Third, as the subtitle of this book is meant to suggest, I believe that I have more persistently attempted to place ancient and medieval science in philosophical, religious, and institutional (largely educational) context than have the authors of previous surveys. I am hardly the first to keep an eye on philosophical context. But I do not believe there is any other modern survey that has taken seriously the religious context and done so without embarrassment and without an apologetic or polemical agenda. If I have made any original contribution, this is probably where it lies.

My aim in this book is synthetic rather than encyclopedic. I endeavor to be broad in reach, confronting the major themes in the history of ancient and medieval science, while supplying enough reliable factual data to meet the needs of the reader who at the outset knows nothing about the subject. I have, of course, built on the accumulated scholarship of the past, but I have not hesitated to offer new interpretations and fresh judgments on old historical disputes. I have doubtless been more dependent on existing interpretive traditions for ancient science (where, in all honesty, I am an interested outsider) than for medieval science (my home ground). And, of course, I do not claim to have gotten all of it "right"—or even to have asked all of the right questions—on either the ancient or the medieval period; my hope is that this book will be received as a contribution to a continuing dialogue on the subjects it addresses.

I have written with a diverse audience in mind. Passages in which I lecture the reader on the proper ways of doing history and warn against a variety of perils (one reader of the manuscript has twitted me for the frequency of "anti-Whig inoculation") will be immediately recognizable as the products of long classroom experience; and it is my hope that this book will indeed prove suitable for classroom use. I am also hopeful that it will serve the general educated reader and historians who do not specialize in the history of ancient and medieval science.

Finally, two remarks about endnotes and bibliography. First, I have used the notes not only for purposes of documentation and acknowledgment of scholarly debt, but also as an opportunity for a running bibliographical commentary, in which I suggest sources (frequently at an advanced level) where the subject at hand may be fruitfully pursued. Second, in both the notes and the bibliography I have (with the student audience and the general reader in mind) heavily emphasized English-language literature. Sources in foreign languages are included only where it seems to me there is nothing comparable in English.

Nobody covers a subject as large as this one without a great deal of help, and I am profoundly indebted to friends and colleagues who have done their best to instruct me in the intricacies of their various specialties and rescue me from confusion and error. I have not always been an apt pupil, and some will still find in this book interpretations that they do not like.

Every chapter has been read and commented upon by colleagues knowledgeable in its subject matter. My greatest debt is to four people who read the manuscript from beginning to end and helped me to see its most glaring deficiencies: Michael H. Shank, Bruce S. Eastwood, Robert J. Richards, and Albert Van Helden. Others who read one or more chapters in their areas of expertise are: Thomas H. Broman, Frank M. Clover, Harold J. Cook, William J. Courtenay, Faye M. Getz, Owen Gingerich, Edward Grant, R.

Stephen Humphreys, James Lattis, Fannie J. LeMoine, James Longrigg, Peter Losin, A. G. Molland, William R. Newman, Franz Rosenthal, A. I. Sabra, George Saliba, John Scarborough, Margaret Schabas, Nancy G. Siraisi, Peter Sobol, Edith D. Sylla, the late Victor E. Thoren, Sabetai Unguru, Heinrich von Staden, and David A. Woodward. The manuscript was evaluated for classroom use, or given actual test runs in the classroom, by several scholars; for their feedback, I wish to thank Edward B. Davis, Frederick Gregory, Edward J. Larson, Alan J. Rocke, and Peter Ramberg. For help in identifying and obtaining illustrations I am indebted to Bruce S. Eastwood, Owen Gingerich, Edward Grant, John E. Murdoch, and David A. Woodward. And for maps, I thank the University of Wisconsin Cartographic Laboratory. If this list is remarkable for its length, I can only explain that I needed all the help I could get.

The idea for this book emerged from discussions at the University of Florida in the spring of 1986 about a possible history of science textbook project; for inspiration and encouragement I would like to thank Frederick Gregory (the moving force behind the meeting) and other discussants, including William B. Ashworth, Richard Burkhardt, Thomas L. Hankins, and Frederic L. Holmes. The book was written during my term as director of the Institute for Research in the Humanities at the University of Wisconsin. It is unlikely that the project would have remained on course without the unfailing efficiency of my administrative assistant, Loretta Freiling, and the steadfast encouragement and support of colleagues in the Humanities Institute and the Department of the History of Science. The book was finished during a month's residency at the Bellagio Study and Conference Center of the Rockefeller Foundation; I am indebted to the foundation and to the directors of the Bellagio Center, Francis and Jackie Sutton, for providing an incomparable setting in which to reflect and write. And finally, I am deeply grateful for the forbearance of my wife, Greta, and my son, Erik, who served as unpaid consultants on prose style and know this book as a series of disconnected fragments, not in any particular sequence.

ONE

Science and Its Origins

WHAT IS SCIENCE?

The nature of science has been the subject of vigorous debate for centuries—a debate conducted by scientists, philosophers, historians, and other interested parties. Although no general consensus has emerged, several conceptions of science have attracted powerful support. (1) One view holds science to be the pattern of behavior by which humans have gained control over their environment. Science is thus associated with craft traditions and technology, and prehistoric people are regarded as having contributed to the growth of science when they learned how to work metals or engage in successful agriculture. (2) An alternative opinion *distinguishes* between science and technology, viewing science as a body of theoretical knowledge, technology as the application of theoretical knowledge to the solution of practical problems. On this view, the technology of automobile design and construction is to be distinguished from theoretical mechanics, aerodynamics, and the other theoretical disciplines that guide it; and only the theoretical disciplines are to count as "sciences."

Those who adopt this second approach, viewing science as theoretical knowledge, do not generally wish to concede that all theories (regardless of their character or content) are scientific; and for such people the task of definition has just begun. If they wish to exclude certain kinds of theories, they must propose criteria by which to judge one theory scientific and another unscientific. (3) It has become quite popular, therefore, to define science by the form of its statements—universal, law-like statements, preferably expressed in the language of mathematics. Thus Boyle's law (formulated by Robert Boyle in the seventeenth century) states that the pressure in a gas is inversely proportional to its volume if everything else remains constant. (4) If this seems too restrictive a criterion, science can be defined instead by its methodology. Science is thus associated with a particular set of procedures, usually experimental, for exploring nature's secrets and confirming or disconfirming theories about her behavior. A claim is therefore scientific if and only if it has an experimental foundation. (5) Such a definition, in turn, yields easily to attempts to define science by its epistemological status (that is, the kind of warrant its claims are held to possess) or even the tenacity with which its practitioners hold its doctrines. Thus Bertrand Russell has argued that "it is not *what* the man of science believes that distinguishes him, but *how* and *why* he believes it. His beliefs are tentative, not dogmatic; they are based on evidence, not on authority or intuition."[1] Science on this view is a privileged way of knowing and of justifying one's knowledge.

(6) In many contexts science is defined not by its methodology or epistemological status, but by its content. Science is thus a particular set of beliefs about nature—more or less the current teachings of physics, chemistry, biology, geology, and the like. By this test, belief in alchemy, astrology, and parapsychology is unscientific. (7) The terms "science" and "scientific" are often applied to any procedure or belief characterized by rigor, precision, or objectivity. Sherlock Holmes, according to this usage, adopted a scientific approach to the investigation of crime. (8) And finally, "science" and "scientific" are often simply employed as general terms of approval—epithets that we attach to whatever we wish to applaud.

What this brief and incomplete survey demonstrates is something that should perhaps have been obvious from the beginning—namely, that many words (including most of the interesting ones) have multiple meanings, varying with the particular context of usage. These meanings are sometimes mutually compatible and complementary, sometimes not. Moreover, it seems futile to attempt to eliminate diversity of usage. After all, language is not a set of rules grounded in the nature of the universe, but a set of conventions adopted by a group of people; and every meaning of the term "science" discussed above is a convention accepted by a sizable community, which is unlikely to relinquish its favored usage without a fight. Or to put the point in a slightly different way, lexicography must be pursued as a descriptive, rather than a prescriptive, art. We must acknowledge, therefore, that the term "science" has diverse meanings, each of them legitimate.

Even if we could find a definition of modern science that would satisfy everybody, the historian would still face a difficult problem. If the historian of science were to investigate past practices and beliefs only insofar as those practices and beliefs resemble modern science, the result would be a distorted picture. Distortion would be inevitable because science has changed in content, form, method, and function; and therefore the historian would not be responding to the past as it existed, but looking at the past through a grid that does not exactly fit. If we wish to do justice to the historical enterprise, we must take the past for what it was. And that means that we must resist the temptation to scour the past for examples or precursors of modern science. We must respect the way earlier generations approached nature, acknowledging that although it may differ from the modern way, it is nonetheless of interest because it is part of our intellectual ancestry. This is the only suitable way of understanding how we became what we are. The historian, then, requires a very broad definition of "science"—one that will permit investigation of the vast range of practices and beliefs that lie behind, and help us to understand, the modern scientific enterprise. We need to be broad and inclusive, rather than narrow and exclusive; and we should expect that the farther back we go, the broader we will need to be.[2]

This admonition is particularly important for anybody embarking on a study of the ancient and medieval worlds. If we were to restrict our attention to anticipations of modern science, we would be focusing on a very narrow range of activity, no doubt distorting it in the process, and overlooking many of the very beliefs and practices of ancient and medieval culture that should be the object of our study—those that will help us to understand the development, much later, of modern science.

I will do my best to heed my own advice in the pages that follow, adopting a definition of science as broad as that of the historical actors whose intellectual efforts we are attempting to understand. This does not mean, of course, that all distinctions are forbidden. I will distinguish between the craft and theoretical sides of science—a distinction that many ancient and medieval scholars would themselves have insisted upon—and I will focus my attention on the latter.[3] The exclusion of technology and the crafts from this narrative is not meant as a commentary on their relative importance, but rather as an acknowledgment of the magnitude of the problems confronting the history of technology and its status as a distinct historical specialty having its own skilled practitioners. My concern will be with the beginnings of scientific *thought,* and that will prove quite a sufficient challenge.

A final word about terminology. Up to now I have consistently employed the term "science." The time has come, however, to introduce the alternative expressions "natural philosophy" and "philosophy of nature," which will also appear frequently in this book. Why are these new expressions needed, and what are they meant to convey? The term "science" has connotations, both ancient and modern, that differentiate it somewhat (and in some contexts) from the subjects to which our investigation is directed. The modern term has all of the ambiguity sketched above, and the ancient terms (*scientia* in latin, *epistēmē* in Greek) were applicable to any system of belief characterized by rigor and certainty, whether or not it had anything to do with nature. It was thus common in the Middle Ages to refer to theology as a science (*scientia*). This book will explore ancient and medieval attempts to investigate *nature,* and the least ambiguous name for that enterprise was and is "natural philosophy" or "philosophy of nature."

But it is important that we not construe the use of these latter expressions as a demotion of the medieval investigation of nature from "scientific" to some lesser status. We would do well to remember that natural philosophy was the intellectual venture to which so important a scientific luminary as Isaac Newton (late seventeenth century) assigned his own work, entitling his great book on mechanics and gravitational theory *The Mathematical Principles of Natural Philosophy.* Natural philosophy, the investigation of nature, was conceived by Newton, as by his ancient and medieval predecessors, to be an integral part of the larger philosophical exploration of the total reality that humans confront.

In this book I will employ a diverse vocabulary, making practical concessions to the heterogeneity of customary usage. I will make regular use of the expression "natural philosophy," either to denote the scientific enterprise as a whole or to signify its more philosophical side. The term "science" will also be employed, most often as a synonym for "natural philosophy," sometimes to designate the more technical aspects of natural philosophy, and occasionally simply because conventional usage calls for that term in a certain context. There will be ample talk simply of "philosophy," for there is no hope of understanding natural philosophy if we ignore the larger enterprise to which it belonged. And, of course, I will make frequent reference to the subdisciplines of natural philosophy, the specific sciences: mathematics, astronomy, physics, optics, medicine, natural history, and the like. Careful attention to context should make the meaning clear in every case.

PREHISTORIC ATTITUDES TOWARD NATURE

From the beginning, the survival of the human race has depended on its ability to cope with the natural environment. Prehistoric people developed impressive technologies for obtaining the necessities of life. They learned how to make tools, start fires, obtain shelter, hunt, fish, and gather fruits and vegetables. Successful hunting and food-gathering (and, after about 7000 or 8000 B.C., settled agriculture) required a substantial knowledge of animal behavior and the characteristics of plants. At a more advanced level, prehistoric people learned to distinguish between poisonous and therapeutic herbs. They developed a variety of crafts, including pottery, weaving, and metalworking. By 3500 they had invented the wheel. They were aware of the seasons and perceived the connection between the seasons and certain celestial phenomena. In short, they knew a great deal about their environment.

But the word "know," seemingly so clear and simple, is almost as tricky as the term "science"; indeed, it brings us back to the distinction between technology and theoretical science, discussed above. It is one thing to know how to do things, another to know why they behave as they do. One can engage in successful and sophisticated carpentry, for example, without any theoretical knowledge of stresses in the timbers one employs. An electrician with only the most rudimentary knowledge of electrical theory can successfully wire a house. It is possible to differentiate between poisonous and therapeutic herbs without possessing any biochemical knowledge that would explain poisonous or therapeutic properties. The point is simply that practical rules of thumb can be effectively employed even in the face of total ignorance of the theoretical principles that lie behind them. You can have "know-how" without theoretical knowledge.

It should be clear, then, that in practical or technological terms, the knowledge of prehistoric humans was great and growing. But what about theoretical knowledge? What did prehistoric people "know" or believe about the origins of the world in which they lived, its nature, and the causes of its numerous and diverse phenomena? Did they have any awareness of general laws or principles that governed the particular case? Did they even ask such questions? We have very little evidence on the subject. Prehistoric culture is by definition oral culture; and oral cultures, as long as they remain exclusively oral, leave no written remains. However, an examination of the findings of anthropologists studying preliterate tribes in the nineteenth and twentieth centuries, along with careful attention to remnants of prehistoric thought carried over into the earliest written records, will allow us to formulate a few tentative generalizations.

Critical to the investigation of intellectual culture in a preliterate society is an understanding of the process of communication. In the absence of writing, the only form of verbal communication is the spoken word; and the only storehouses of knowledge are the memories of individual members of the community. The transmission of ideas and beliefs in such a culture occurs only in face-to-face encounter, through a process that has been characterized as "a long chain of interlocking conversations" between its members. The portion of these conversations considered important enough to remember and pass on to succeeding generations forms the basis of an oral tradition, which serves as the principal repository for the collective experience and the general beliefs, attitudes, and values of the community.[4]

There is an important feature of oral tradition that demands our attention—namely, its fluidity. Oral tradition is typically in a continuous state of evolution, as it absorbs new experiences and adjusts to new conditions and needs within the community. Now this fluidity of oral tradition would be extremely perplexing if the function of oral tradition were conceived to be the communication of abstract historical or scientific data—the oral equivalent of a historical archive or a scientific report. But an oral culture, lacking the ability to write, certainly cannot create archives or reports; indeed, an oral culture lacks even the *idea* of writing and must therefore lack even the *idea* of a historical archive or a scientific report.[5] The primary function of oral tradition is the very practical one of explaining, and thereby justifying, the present state and structure of the community, supplying the community with a continuously evolving "social charter." For example, an account of past events may be employed to legitimate current leadership roles, property rights, or the present distribution of privileges and obligations. And in order to serve this function effectively, oral tradition must be capable of adjusting itself fairly rapidly to changes in social structure.[6]

But here we are principally interested in the *content* of oral traditions, especially those portions of the content that deal with the nature of the universe—the portions, that is, which might be thought of as the ingredients of a worldview or a cosmology. Such ingredients exist within every oral tradition, but often beneath the surface, seldom articulated, and almost never assembled into a coherent whole. It follows that we must be extremely reluctant to articulate the worldview of preliterate people on their behalf, for this cannot be done without *our* supplying the elements of coherence and system, thereby distorting the very conceptions we are attempting to portray. We may, nonetheless, formulate certain conclusions about the ingredients or elements of worldview within preliterate oral traditions. (The conclusions that follow are based on a mixture of evidence from prehistoric cultures and contemporary preliterate societies and, unless there is explicit warning to the contrary, should be taken to apply to both settings.)

It is clear that preliterate people, no less than those of us who live in a modern scientific culture, have a need for explanatory principles capable of bringing order, unity, and especially meaning to the apparently random and chaotic flow of events. But we should not expect the explanatory principles accepted by preliterate people to resemble ours: lacking any conception of "laws of nature" or deterministic causal mechanisms, their ideas of causation extend well beyond the sort of mechanical or physical action acknowledged by modern science. It is natural that in the search for meaning they should proceed within the framework of their own experience, projecting human or biological traits onto objects and events that seem to us devoid not only of humanity but also of life. Thus the beginning of the universe is typically described in terms of birth, and cosmic events may be interpreted as the outcome of struggle between opposing forces, one good and the other evil. There is an inclination in preliterate cultures not only to personalize but also to individualize causes, to suppose that things happen as they do because they have been willed to do so. This tendency has been described by H. and H. A. Frankfort:

> Our view of causality . . . would not satisfy primitive man because of the impersonal character of its explanations. It would not satisfy him, moreover, because of its generality. We understand phenomena, not by what makes them peculiar, but by what makes them manifestations of general laws. But a general law cannot do justice to the individual character of each event. And the individual character of the event is precisely what early man experiences most strongly. We may explain that certain physiological processes cause a man's death. Primitive man asks: Why should *this* man die *thus* at *this* moment? We can only say that, given these conditions, death will always occur. He wants to find a cause as specific and individual as the event which it must explain. The event . . . is experienced in its complexity and individuality, and these are matched by equally individual causes.[7]

Oral traditions typically portray the universe as consisting of sky and earth, and perhaps also an underworld. An African myth describes the earth as a mat that has been unrolled but remains tilted, thereby explaining upstream and downstream—an illustration of the general tendency to describe the universe in terms of familiar objects and processes. Deity is an omnipresent reality within the world of oral traditions, though in general no clear distinction is drawn between the natural, the supernatural, and the human; the gods do not transcend the universe but are rooted in it and subject to its principles. Belief in the existence of ghosts of the dead, spirits, and a variety of invisible powers, which magical ritual allows one to control, is another universal feature of oral tradition. Reincarnation (the idea that after death the soul returns in another body, either human or animal) is widely believed in. Conceptions of space and time are not (like those of modern physics) abstract and mathematical, but are invested with meaning and value drawn from the experience of the community. For example the cardinal directions for a community whose existence is closely connected to a river might be "upstream" and "downstream," rather than north, south, east, and west. Some oral cultures have difficulty conceiving of more than a very shallow past: an African tribe, the Tio, for example, cannot situate anybody farther back in time than two generations.[8]

There is a strong tendency within oral traditions to identify causes with beginnings, so that to explain something is to identify its historical origins. Within such a conceptual framework, the distinction that we make between scientific and historical understanding cannot be sharply drawn and may be nonexistent. Thus when we look for the features of oral tradition that count for worldview or cosmology, they will almost always include an account of origins—the beginning of the world, the appearance of the first humans, the origin of animals, plants, and other important objects, and finally the formation of the community. Related to the account of origins is often a genealogy of gods, kings, or other heroic figures in the community's past, accompanied by stories about their heroic deeds. It is important to note that in such historical accounts the past is portrayed not as a chain of causes and effects that produce gradual change, but as a series of decisive, isolated events, by which the present order came into existence.[9]

These tendencies can be illustrated with examples from both ancient and contemporary oral cultures. According to the twentieth-century Kuba of equatorial Africa,

> Mboom or the original water had nine children, all called
> Woot, who in turn created the world. They were, apparently
> in order of appearance: Woot the ocean; Woot the digger,
> who dug riverbeds and trenches and threw up hills; Woot the
> flowing, who made rivers flow; Woot who created the woods
> and savannas; Woot who created the leaves; Woot who created
> the stones; Woot the sculptor, who made people out of

wooden balls; Woot the inventor of prickly things such as fish, thorns, and paddles; and Woot the sharpener, who first gave an edge to pointed things. Death came to the world when a quarrel between the last two Woots led to the demise of one of them by the use of a sharpened point.[10]

Notice how this tale not only accounts for the origin of the human race and the major topographical features of the Kuba world, but also explains the invention of what the Kuba clearly considered a critically important tool—the sharpened object.

Similar themes abound in early Egyptian and Babylonian creation myths. According to one Egyptian account, in the beginning the sun-god, Atum, spat out Shu, the god of air, and Tefnut, the goddess of moisture. Thereafter,

Shu and Tefnut, air and moisture, gave birth to earth and sky, the earth-god Geb and the sky-goddess Nut. . . . Then in their turn Geb and Nut, earth and sky, mated and produced two couples, the god Osiris and his consort Isis, the god Seth and his consort Nephthys. These represent the creatures of this world, whether human, divine, or cosmic.[11]

A Babylonian myth attributes the origin of the world to the sexual activity of Enki, god of the waters. Enki impregnated the goddess of the earth or soil, Ninhursag. This union of water and earth gave rise to vegetation, represented by the birth of the goddess of plants, Ninsar. Enki subsequently mated first with his daughter, then with his granddaughter, to produce various specific plants and plant products. Ninhursag, angered when Enki devoured eight of the new plants before she had the opportunity to name them, pronounced a curse on him. Fearing the consequences of Enki's demise (apparently the drying up of the waters), the other gods prevailed on Ninhursag to withdraw the curse and heal Enki of the various ailments induced by the curse, which she did by giving birth to eight healing deities, each associated with a part of the body—thus accounting for the origin of the healing arts.[12]

It will be convenient to pause for a moment on the healing arts, which can serve to illustrate some of the characteristics of oral cultures. There can be no doubt that healing practices were extremely important in ancient oral cultures, where primitive conditions made disease and injury everyday realities.[13] Minor medical problems, such as wounds and lesions, were no doubt treated by family members. More dramatic ailments—

major wounds, broken bones, severe and unexpected illness—might require assistance from somebody with more advanced knowledge and skill. A certain amount of medical specialization thus came into existence: some members of the tribe or the village became known for their herb-gathering ability, their proficiency in the setting of bones or the treatment of wounds, or their experience in assisting at childbirth.

But so described, the primitive medicine practiced in preliterate societies sounds remarkably like a rudimentary version of modern medicine. A more careful look reveals the healing arts within oral cultures to be inseparable and indistinguishable from religion and magic. The wise woman or the "medicine man" was valued not simply for pharmaceutical or surgical skill, but also for knowledge of the divine and demonic causes of disease and the magical and religious rituals by which it could be treated. If the problem was a splinter, a wound, a familiar rash, a digestive complaint, or a broken bone, the healer responded in the obvious way—by removing the splinter, binding the wound, applying a substance (if one were known) that would counteract the rash, issuing dietary prohibitions, and setting and splinting the broken limb. But if a family member became mysteriously and gravely ill, one might suspect sorcery or invasion of the body by an alien spirit. In such cases, more dramatic remedies would be called for—exorcism, divination, purification, songs, incantations, and other ritualistic activities.

There is one last feature of belief in oral cultures (both ancient and contemporary) that demands our attention—namely, the simultaneous acceptance of what seem to us incompatible alternatives, without any apparent awareness that such behavior could present a problem. Examples are innumerable, but it may suffice to note that the story of the nine Woots related above is one of seven (or more) myths of origin that circulate among the Kuba, while the Egyptians had a variety of alternatives to the story of Atum, Shu, Tefnut, and their offspring; and nobody seems (or seemed) to notice, or else to care, that all of them could not be true. Add to this the seemingly "fanciful" nature of many of the beliefs described above, and we inevitably raise the question of "primitive mentality": do the members of preliterate societies possess a mentality that is prelogical or mystical or in some other way different from our own; and, if so, exactly how is this mentality to be described and explained?[14]

This is an extremely complex and difficult problem, which has been hotly debated by anthropologists and others for the better part of the twentieth century, and I am not likely to resolve it here. But I can at least offer a word of methodological advice: namely, that it is wasted effort, contributing absolutely nothing to the cause of understanding, to spend time wishing that preliterate people would employ (or had employed) a conception and criteria of knowledge that they have (or had) never encountered—a conception, in the case of prehistoric people, that was not invented until centuries later. We make no progress by assuming that preliterate people were trying, but failing, to live up to our conceptions of knowledge and truth. It requires only a moment of reflection to realize that they must have been operating within quite a different linguistic and conceptual world, and with different purposes; and it is in the light of these that their achievements must be judged.

The stories embodied in oral traditions are intended to convey and reinforce the values and attitudes of the community, to offer satisfying explanations of the major features of the world as experienced by the community, and to legitimate the current social structure; stories enter the oral tradition (the collective memory) because of their effectiveness in achieving those ends, and as long as they continue to do so there is no reason to question them. There are no rewards for skepticism in such a social setting and few resources to facilitate challenge. Indeed, our highly developed conceptions of truth and the criteria that a claim must satisfy in order to be judged true (internal coherence, for example, or correspondence with an external reality) do not generally exist in oral cultures and, if explained to a member of an oral culture, would probably seem quite useless. Rather, the operative principle among preliterates is that of sanctioned belief—the sanction in question emerging from community consensus.[15]

Finally, if we are to understand the development of science in antiquity and the Middle Ages, we must ask how the preliterate patterns of belief that we have been examining yielded to, or were supplemented by, a new conception of knowledge and truth (represented most clearly in the principles of Aristotelian logic and the philosophical tradition it spawned). The decisive development seems to have been the invention of writing, which occurred in a series of steps. First there were pictographs, in which the written sign stood for the object itself. Around 3000 B.C. a system of word signs (or logograms) appeared, in which signs were created for the important words, as in Egyptian hieroglyphics. But in hieroglyphic writing, signs could also stand for sounds or syllables—the beginnings of syllabic writing. The development of fully syllabic systems about 1500 B.C. (that is, systems in which all nonsyllabic signs were discarded) made it possible and, indeed, reasonably easy for people to write down everything they could say. And finally fully alphabetic writing, which has a sign for each sound (both consonants and vowels), made its appearance in Greece about 800 B.C. and became widely disseminated in Greek culture in the sixth and fifth centuries.[16]

One of the critical contributions of writing, especially alphabetic writing, was to provide a means for the recording of oral traditions, thereby freezing what had hitherto been fluid, translating fleeting audible signals into enduring visible objects.[17] Writing thus served a storage function, replacing memory as the principal repository of knowledge. This had the revolutionary effect of opening knowledge claims to the possibility of inspection, comparison, and criticism. Presented with a written account of events, we can compare it with other (including older) written accounts of the same events, to a degree unthinkable within an exclusively oral culture. Such comparison encourages skepticism and, in antiquity, helped to create the distinction between truth, on the one hand, and myth or legend, on the other; that distinction, in turn, called for the formulation of criteria by which truthfulness could be ascertained; and out of the effort to formulate suitable criteria emerged rules of reasoning, which offered a foundation for serious philosophical activity.[18]

But giving permanent form to the spoken word does not merely encourage inspection and criticism. It also makes possible new kinds of intellectual activity that have no counterparts (or only weak ones) in an oral culture. Jack Goody has argued convincingly that early literate cultures produced large quantities of written inventories and other kinds of lists (mostly for administrative purposes), far more elaborate than anything an oral culture could conceivably produce; and, moreover, that these lists made possible new kinds of inspection and called for new thought processes or new ways of organizing thoughts. For one thing, the items in a list are removed from the context that gives them meaning in the world of oral discourse, and in that sense they have become abstractions. And in this abstract form they can be separated, sorted, and classified according to a variety of criteria, thereby giving rise to innumerable questions not likely to be raised in an oral culture. To give a single example, the lists of precise celestial observations assembled by early Babylonians could never have been collected and transmitted in oral form; their existence in writing, which allowed them to be minutely examined and compared, made possible the discovery of intricate patterns in the motions of the celestial bodies which we associate with the beginnings of mathematical astronomy and astrology.[19]

Two conclusions may be drawn from this argument. First, the invention of writing was a prerequisite for the development of philosophy and science in the ancient world. Second, the degree to which philosophy and science flourished in the ancient world was, to a very significant degree, a function of the efficiency of the system of writing (alphabetic writing having a great advantage over all of the alternatives) and the breadth of its diffusion among the people. We see the earliest benefits of the use of word signs or logograms in Egypt and Mesopotamia, beginning about 3000 B.C. However, the difficulty and inefficiency of logographic writing inevitably limited its diffusion and made it the property of a small scholarly elite. In sixth- and fifth-century Greece, by contrast, the wide dissemination of alphabetic writing contributed to the spectacular development of philosophy and science. We must not imagine that literacy was sufficient of itself to produce the "Greek miracle" of the sixth and fifth centuries; other factors no doubt contributed, including prosperity, new principles of social and political organization, contact with Eastern cultures, and the introduction of a competitive style into Greek intellectual life. But surely the most important element in the mix was the emergence of Greece as the world's first widely literate culture.[20]

THE BEGINNINGS OF SCIENCE IN EGYPT AND MESOPOTAMIA

I will turn to the Greek world in the next chapter. Before doing so, I must touch briefly on pre-Greek developments in Egypt and Mesopotamia (the region between the Tigris and Euphrates rivers, site of ancient Babylonia and Assyria and of modern Iraq—see map 2). I have said enough about creation myths in the preceding section to reveal key features of Egyptian and Mesopotamian cosmological and cosmogonical speculation. Here I will restrict myself to the Egyptian and Mesopotamian contribution to several other subjects or disciplines that subsequently found a place within Greek and medieval European science: mathematics, astronomy, and medicine. The evidence is scanty but sufficient to convey a general picture.

The Greeks themselves believed that mathematics originated in Egypt and Mesopotamia. Herodotus (fifth century B.C.) reported that Pythagoras traveled to Egypt, where he was introduced by priests to the mysteries of Egyptian mathematics. From there, according to ancient tradition, he was carried captive to Babylon, where he came into contact with Babylonian mathematics. Eventually he made his way home to the island of Samos, bearing Egyptian and Babylonian mathematical treasure to the Greeks. Whether this and similar tales regarding other mathematicians are historically accurate or legendary is less important than the larger truth they convey—namely, that the Greeks were (and knew they were) the recipients of Egyptian and Babylonian mathematical knowledge.

By 3000 about B.C., the Egyptians developed a number system that was decimal in character, employing a different symbol for each power of 10 (1, 10, 100, and so forth). These symbols could be lined up, as in Roman numerals, to form any desired number. Thus if I represented 1, and ∩ represented 10, then the number 34 could be expressed as IIII∩∩∩. By about 1800 B.C. additional symbols had been devised for other numbers, so that, for example, 7 could be represented by a sickle (ʔ) rather than by seven vertical strokes. Addition and subtraction were simple operations in Egyptian arithmetic, performed as with Roman numerals, but multiplication and division were extremely clumsy; and the generalized concept of a fraction was unknown, the general rule allowing only unit fractions (fractions with a numerator of 1). Elementary problems of the following type could be solved: if one-seventh of a quantity is added to the quantity, and the sum equals 16, how large is the quantity?[21]

Egyptian geometrical knowledge seems to have been oriented toward practical problems, perhaps those of surveyors and builders. Egyptians were able to calculate the areas of simple plane figures, such as the triangle and the rectangle, and the volumes of simple solids, such as the pyramid. For example, to find the area of a triangle they took one-half the length of the base times the altitude; and, to find the volume of a pyramid, one-third the area of the base times the altitude. For calculating the area of a circle, the Egyptians worked out rules that correspond to a value for π of about 3.17. Finally, in one of the most obvious areas of applied mathematics, the Egyptians devised an official calendar consisting of twelve months of thirty days each, plus an additional five days at the end of the year—a calendar substantially simpler, because of its fixed character, than contemporary Babylonian calendars and those of the early Greek city-states, which attempted to take into account the lunar, as well as the solar, cycle.[22]

The contemporary mathematical achievement in Mesopotamia was an order of magnitude superior to that of the Egyptians. Clay tablets (see fig. 1.1) recovered in large quantities reveal a Babylonian number system, fully developed by about 2000 B.C., that was simultaneously decimal (based on the number 10) and sexagesimal (based on the number 60). We retain sexagesimal numbers today in our system for measuring time (sixty minutes to an hour) and angles (sixty minutes in a degree and 360 degrees in a circle). The Babylonians had separate symbols for 1 (▼) and 10 (◀); these could be combined like Roman numerals to form numbers up to 59. The number 32, for example, would be expressed by three of the tens symbol plus two of the units symbol, as in table 1.1.

But beyond 59 an important difference appears. Instead of forming the

Fig. 1.1. A Babylonian clay tablet (ca. 1900–1600 B.C.), containing a mathematical problem text dealing with bricks, their volumes, and their coverage. Yale Babylonian Collection, YBC 4607. The text is translated and discussed in O. Neugebauer and A. Sachs, eds., *Mathematical Cuneiform Texts*, pp. 91–97.

Table 1.1. Five Babylonian Sexagesimal Numbers and Their Hindu-Arabic equivalents.

	60^3	60^2	60	1	$1/60$	$1/60^2$	Modern Hindu-Arabic Equivalent
(1)				◀◀◀▼▼			32
(2)			▼▼	◀▼▼▼			$2 \times 60 + 16 = 136$
(3)		▼	◀▼▼	◀◀▼▼▼▼			$1 \times 3600 + 12 \times 60 + 23 = 4.343$
(4)	▼▼	◀◀▼▼					$2 \times 216000 + 22 \times 3600 = 511.200$
(5)					◀◀	◀▼▼	$2 \times 1/60 + 12 \times 1/3600 = 1/30 + 1/300 = 11/300$

▼ = 1 ◀ = 10

number 60 by lining up six symbols for 10, the Babylonians used a place system similar to our own. In our number 234, the numeral 4 (situated in the "units column") signifies simply the number 4; the numeral 3, situated in the tens column represents the number 30; while the numeral 2, situated in the hundreds column stands for the number 200. Thus 234 is 200 + 30 + 4. The Babylonian place system worked similarly, except that successive columns represent powers of 60 rather than powers of 10. Thus in the second example in table 1.1, the two unit symbols in the sixties column represent not 2, but $2 \times 60 = 120$; and in the third example the unit symbol in the 60^2 column represents not 1 but $1 \times 60^2 = 3,600$. There was no equivalent of the decimal point by which to locate the units column, and this information would therefore have to be inferred from context. Multiplication tablets, tables of reciprocals, and tables of powers and roots were used to facilitate calculation. One of the great advantages of the sexagesimal system was the ease with which calculations could be performed using fractions.[23]

The full superiority of Babylonian mathematics over its Egyptian counterpart is evident when we turn to more difficult problems, which *we* would solve algebraically. Historians of mathematics sometimes refer to these problems as "algebra," useful shorthand for this aspect of the Babylonian mathematical enterprise, perhaps, but dangerous if it is taken to mean that they practiced genuine algebra—that is, that they had a generalized algebraic notation or an understanding of what we consider algebraic rules. What we can safely say is that Babylonian mathematicians used arithmetical operations to solve problems for which *we* would employ a quadratic equation. For example, we find many Babylonian tablets, including teaching texts, demonstrating how to solve problems such as the following: given the product of two numbers and their sum or difference, find the two numbers.[24]

One of the areas to which Babylonians applied their mathematical techniques was astronomy. The stars have been objects of investigation and speculation since earliest times. Some of our oldest written records, going back more than 4,000 years, are astronomical in character. There were several reasons for this interest in the heavens. One was agricultural, for it was obvious even from fairly casual observation that the agricultural seasons—the times for planting and harvesting—correspond to the motion of the sun and the position of certain stars and constellations in relation to the sun. A second reason was religious, for the heavens, especially the sun and moon, were usually associated with divinity. A third was astrological. And a fourth was calendric.

Some of the earliest effort was devoted to mapping the heavens—identifying and naming prominent stars and constellations, observing their relationships, and linking their visibility to the seasons. In Mesopotamia, systematic astronomical observation came to be practiced in the temples, for religious, astrological, and calendric purposes. Temple priests not only mapped the fixed stars but also identified the "wandering stars" or planets—the planets now named Mercury, Venus, Mars, Jupiter, and Saturn. (The sun and moon were also considered planets, because they too moved in relation to the fixed stars.) These seven planets were observed to move slowly through the heavens within the narrow band of the zodiac. By about 500 B.C. Babylonian priests had defined this band and identified the constellations that mark it off into twelve segments of thirty degrees each, thus giving us the signs of the zodiac. Once defined, the zodiac could function as a handy measuring system for charting the exact motions of the sun, moon, and remaining planets, and as a source of astrological predictions.[25]

The astrological aspect of Babylonian astronomy requires brief mention (on astrology, see also below, chap. 11). It is clear that astrological needs were a major motivation for the development of Babylonian mathematical astronomy. Out of astral religion—the association of the stars (specifically the wandering stars) with the gods—and the obvious fact that celestial events were connected with the seasons and the weather, there developed a system of judicial astrology, the attempt to make short-term predictions affecting the king and the kingdom on the basis of the current celestial configuration. It is also possible that horoscopic astrology, which predicts the course of a person's life from the celestial configuration at the time of birth, developed in the late Babylonian period. What is important is that both kinds of astrology called for detailed knowledge of solar, lunar, and the other planetary motions. Babylonian astrology was transmitted to the Greeks, who further developed it and passed it on to the Middle Ages, the early modern period, and ultimately the twentieth century. Through most of that long history, it should be noted, the astronomical and astrological traditions have been intimately linked.[26]

We do not have space to examine the development of Babylonian mathematical astronomy in detail. What is important is that in the period 500–300 B.C., the Babylonian astronomer-priest developed his art to the point where he could manipulate large quantities of astronomical data and make a variety of astronomical predictions. He had numerical models, in the form of arithmetic progressions, that enabled him to chart the daily motions of the sun and moon through the zodiac. From such data he could predict the first appearance of the new moon (important for the calendar, since the new moon signified the beginning of a new month); he could also predict lunar eclipses and the possibility or impossibility of solar eclipses. We must stress that he did this not by the use of geometrical models, as Greek astronomers would do, but simply through the use of numerical methods that extrapolated past observations into the future.[27]

The final area of Egyptian and Mesopotamian achievement to be considered is medical. A number of Egyptian medical papyri (written in the period 2500–1200 B.C.) have survived, and these offer us a fragmentary picture of the healing arts in ancient Egypt. From several of the papyri it becomes clear that a principal cause of disease was thought to be invasion of the body by evil forces or spirits. Relief was to be gained through rituals designed to appease or frighten the spirits—exorcism, incantation, purification, or the wearing of an appropriate amulet. The gods could be appealed to for protection: a prayer to the god Horus, found in the Leyden papyrus, reads in part, "Hail to thee, Horus . . . . I come to thee, I praise thy beauty: destroy thou the evil that is in my limbs."[28] Certain gods came especially to be associated with healing functions or healing cults: Thoth, Horus, Isis, and Imhotep. The view that each bodily organ was ruled by a specific god, who could be invoked for healing of that organ, seems to have been widespread. And all of this ritual, of course, required the assistance of an expert who was of acknowledged purity, who knew the required incantations, and who could assure that the ritual was properly performed to the smallest detail; this was the priest-healer.

Healing therapies in ancient Egypt were not limited to prayer, incantation, and ritual. Pharmacological remedies, prepared from animal, vegetable, or mineral substances, were also widespread—though their effectiveness was believed to be conditional on having been prepared and administered under appropriate ritual conditions. The Ebers papyrus (written about 1600 B.C., but containing material copied from much older texts) contains medical recipes for dealing with diseases of the skin, eyes, mouth, extremities, digestive and reproductive systems, and other internal organs; for treating wounds, burns, abscesses, ulcers, tumors, headaches, swollen glands, and bad breath.[29]

Surgery is dealt with in another papyrus, known as the Edwin Smith papyrus (written about the same time as the Ebers papyrus), which contains a surgical manual that systematically describes the treatment of wounds, fractures, and dislocations (see fig. 1.2).[30] One of the notable features of the Ebers and Edwin Smith papyri is the careful arrangement of case studies, beginning with a description of the problem and proceeding to diagnosis, verdict (as to whether or not the ailment is treatable), and treatment.

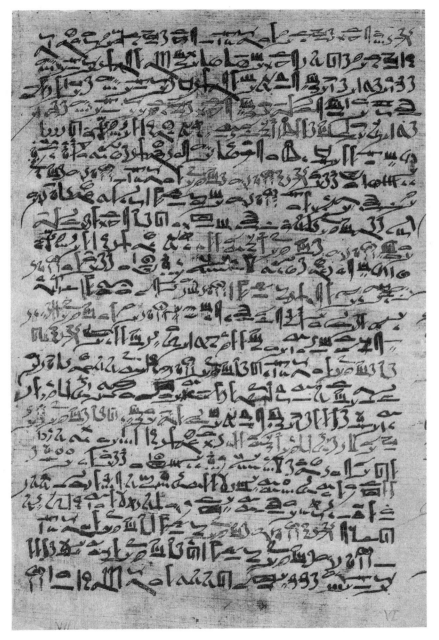

Fig. 1.2. A column from the Edwin Smith Surgical Papyrus (ca. 1600 B.C.), now in the New York Academy of Medicine.

Mesopotamian medicine displays many of the same characteristics as Egyptian healing practices. Babylonian clay tablets, like Egyptian papyri, contain case studies, systematically organized by type, many of them revealing careful observation of symptoms and intelligent prognosis. Mesopotamian healers displayed equal skill in surgery and the preparation of pharmaceutical remedies. As in Egypt, a certain amount of medical specialization developed—different categories of healers coming to have somewhat different specialties and functions. And again we find healing intimately mingled with religion and with practices that we would now view as magical. Disease was regarded as the result of invasion of the body by evil spirits (owing to fate, carelessness, sin, or sorcery). Therapy was directed toward elimination of the invading spirit through divination (including the interpretation of astrological omens), sacrifice, prayer, and magical ritual.[31]

This brief sketch of Egyptian and Mesopotamian contributions to mathematics, astronomy, and the healing arts offers us a glimpse of the beginnings of the Western scientific tradition, as well as a context within which to view the Greek achievement. There is no doubt that the Greeks were aware of the work of their Egyptian and Mesopotamian predecessors, and profited from it. In the chapters that follow, we will see how these products of Egyptian and Mesopotamian thought entered and helped shape Greek natural philosophy.

TWO

The Greeks and the Cosmos

THE WORLD OF HOMER AND HESIOD

We know nothing of Homer, reputed author of two great epic poems, the *Iliad* and the *Odyssey*. The poems, which recount heroic adventures associated with the closing days and aftermath of the Trojan War between the Greeks and Troy, are clearly the products of a long oral tradition, having roots that go far back in Greek history to the Mycenean age (before 1200); they appear also to have been influenced by non-Greek epic traditions from the Near East. They were perhaps committed to writing in the eighth century, but whether by one man (Homer) or several remains a matter of dispute. Whatever their precise origin, the *Iliad* and *Odyssey* became the foundation of Greek education and culture and remain among the best measures we have of the form and content of ancient Greek thought.[1]

Alongside Homer we must place Hesiod, who flourished around the end of the eighth century. To Hesiod, the son of a farmer, two major poetic works are attributed: *Works and Days* (which includes, among other things, a manual of farming) and the *Theogony,* which recounts the origin of the gods and the world.[2] Hesiod gave the gods a genealogy and, together with Homer, defined their character and the functions over which they presided. It was through the joint influence of Homer and Hesiod that the twelve gods of Mount Olympus were chosen from a plethora of local deities to become the gods of the Greeks.

Among the Olympians was Zeus, portrayed by Homer and Hesiod as the greatest and most powerful of the gods, lord of the sky, god of weather, wielder of lightning bolts, upholder of law and morality, and father of all. Hera, his sister and wife, presided over weddings and marriage. Poseidon, brother of Zeus, was god of both sea and earth, author of storms and earthquakes. Hades, another brother, was lord of the underworld and

Fig. 2.1. A bronze statue of Zeus in
the Museo Archeologico, Florence.
Alinari/Art Resource N.Y.

the dead. Athena, daughter of Zeus, was the goddess of warfare and the protector of cities, while Ares, son of Zeus, was the ruthless god of war.

In Homer's portrayal, the gods were intimately involved in human affairs, determining victory, defeat, misfortune, and destiny. Various instances of divine involvement appear in the *Odyssey*. Its hero, Odysseus, shipwrecked as a result of divine wrath and confined for eight years on the island of the nymph Calypso, was finally released from imprisonment at Zeus's command and set sail for Ithaca. However, Poseidon, who had not been consulted on the release of Odysseus, spotted him on his raft and decided to interfere:

> With that he gathered the clouds and troubled the waters of the deep, grasping his trident in his hands; and he roused all storms of all manner of winds, and shrouded in clouds the land and sea: and down sped night from heaven. . . . Poseidon, shaker of the earth, stirred against him a great wave, terrible and grievous, and vaulted from the crest, and therewith smote him.

And so Odysseus made his way home, sometimes aided, sometimes thwarted by the gods.[3]

In Hesiod's *Theogony* we find a brief history of the world, from primeval chaos to the orderly rule of Zeus. From chaos arose Gaia ("broad-bosomed earth") and various other offspring, including Eros (love), Erebos (a part of the underworld), and darkest Night. Erebos and Night mated to produce Day and Aither (or sky). Gaia first bore the starry heaven (Ouranos) "to cover her everywhere over and be an ever-immovable base for the gods who are blessed. And she bore the high mountains, the charming retreats of the goddess nymphs who have their abodes in the wooded glens of the mountains. And . . . she brought forth Pontos, the exhaustless sea that rages and waves."[4] Gaia (mother earth) proceeded to mate with her offspring, Ouranos (father heaven), and from that union issued Oceanus (the river that encircles the world, father of all other rivers), the twelve Titans, and a collection of monsters. Eventually Kronos, one of the Titans, castrated and otherthrew his father, Ouranos; Kronos, in turn, was deposed by his son Zeus. Zeus obtained the thunderbolt from the Cyclopes and used it to defeat the Titans and establish his own Olympian rule.

Even this brief description reveals the chasm separating the world of Homer and Hesiod from that of modern science. Theirs was a world of anthropomorphic deities interfering in human affairs and using humans as

Fig. 2.2. A shrine to the earth goddess Gaia at Delphi (4th century B.C.).

pawns in their own plots and intrigues. This was inevitably a capricious world, in which nothing could be safely predicted because of the boundless possibilities of divine intervention. Natural phenomena were personified and divinized. Sun and moon were conceived as deities, offspring of the union of Theia and Hyperion. Storms, lightning bolts, and earthquakes were not considered the inevitable outcome of impersonal, natural forces, but mighty feats, willed by the gods.

What are we to make of this? Did the ancient Greeks take the stories constituting what we now call "Greek mythology" to be true? Did they really believe in divine beings, lodged on Mount Olympus or in some other mysterious place, seducing one another and bedeviling the humans who crossed their path? Did nobody doubt that storms and earthquakes were a result of divine caprice? We have seen in the previous chapter, in the discussion of preliterate thought, how difficult these questions are.[5]

What is clear is that any attempt to measure such beliefs by modern criteria of scientific truth is a sure road to misunderstanding. We can, however, learn something from a glance at contemporary beliefs outside the scientific realm. When a political candidate, military commander, or professional athlete thanks God for victory, does he or she really believe that victory was supernaturally obtained? The answer is not entirely clear, and probably varies from case to case. What seems certain is that the public figures in question are not attempting to deal with causal questions in a philosophical or scientific manner, and it has probably never occurred to them that their assertions might be judged by philosophical or scientific criteria. By the same token, although the works of Homer and Hesiod seem to address questions of causation, we must understand that they were not intended as scientific or philosophical treatises. Homer and Hesiod—and the bards whose epic poems lie behind theirs—were recording heroic deeds in order to instruct and entertain; and if we treat them as failed philosophers, we will inevitably misunderstand their achievement.

Yet we must not dismiss these ancient sources too quickly. Homer and Hesiod, after all, are among the few sources at our disposal that reveal *anything* of archaic Greek thought; and if they do not represent primitive Greek philosophy, they were nonetheless central to Greek education and culture for centuries and cannot have been without influence on the Greek mind. It is abundantly clear that the language and the images people employ affect the reality they perceive. If the content of Homer's and Hesiod's poems was not "believed" in the same way as we believe the content of modern physics, the mythology of the Olympian gods (not to speak of the mythology of local deities) was nonetheless a central feature of Greek culture, affecting the way Greeks thought, talked, and behaved.

THE FIRST GREEK PHILOSOPHERS

Early in the sixth century, Greek philosophy made its first appearance. This was not, as some have portrayed it, the replacement of mythology by philosophy; for Greek mythology did not disappear but continued to flourish for centuries. Rather, it was the appearance of new, philosophical modes of thought alongside, or sometimes mingled with, mythology. Simply put, Homer and Hesiod were not philosophers and did not practice philosophy; Thales, Pythagoras, and Heraclitus, while living in a culture still rife with mythology, undertook a new kind of intellectual inquiry, which we are prepared to call "philosophy."

But what were the new modes of thought that we identify as philosophy?

A group of thinkers in the sixth century initiated a serious, critical inquiry into the nature of the world in which they lived—an inquiry that has stretched from their day to ours. They asked about its ingredients, its composition, and its operation. They inquired whether it is made of one thing or many. They asked about its shape and location and speculated about its origins. They sought to understand the process of change, by which things come into being and one thing seems to be transformed into another. They contemplated extraordinary natural phenomena, such as earthquakes and eclipses, and sought universal explanations applicable not only to a particular earthquake or eclipse but to earthquakes and eclipses in general. And they began to reflect on the rules of argumentation and proof.

The early philosophers did not merely pose a new set of questions; they also sought new kinds of answers. Personification of nature gradually became a less prominent feature of their discourse, and the gods disappeared from their explanations of natural phenomena. We have seen the mythological approach of Homer and Hesiod: in Hesiod's *Theogony* earth and sky are regarded as divine offspring. In Leucippus and Democritus, by contrast, the world and its various parts result from the mechanical sorting of atoms in the primeval vortex. As late as the fifth century, the historian Herodotus retained much of the old mythology, sprinkling tales of divine intervention through his *Histories.* Poseidon, by his account, used a high tide to flood a swamp the Persians were crossing. And Herodotus regarded an eclipse that coincided with the departure of the Persian army for Greece as a supernatural omen. The philosophers offered a quite different account of floods and eclipses, containing no hint of supernatural intervention. Anaximander judged eclipses to be the result of blockage of the apertures in rings of celestial fire. According to Heraclitus, the heavenly bodies are bowls filled with fire, and an eclipse occurs when the open side of a bowl turns from us. The theories of Anaximander and Heraclitus do not seem particularly sophisticated (fifty years after Heraclitus the philosophers Empedocles and Anaxagoras understood that eclipses were caused simply by cosmic shadows), but what is of critical importance is that they exclude the gods. The explanations are entirely naturalistic; eclipses do not reflect personal whim or the arbitrary fancies of the gods, but simply the nature of fiery rings or of celestial bowls and their fiery contents.[6]

The world of the philosophers, in short, was an orderly, predictable world in which things behaved according to their natures. The Greek term used to denote this ordered world was *kosmos,* from which we draw our word "cosmology." The capricious world of divine intervention was being pushed aside, making room for order and regularity; *kosmos* was being substituted for *chaos*. A distinction between the natural and the super-natural was emerging; and there was wide agreement that causes (if they are to be dealt with philosophically) are to be sought only in the nature of things. The philosophers who introduced these new ways of thinking were called by Aristotle *physikoi* or *physiologoi,* from their concern with *physis* or nature.

THE MILESIANS AND THE QUESTION OF ULTIMATE REALITY

These philosophical developments seem to have emerged first in Ionia, on the west coast of Asia Minor (present-day Turkey, just across the Aegean Sea from the Greek mainland; see map 1). There Greek colonists had established thriving cities, such as Ephesus, Míletus, Pergamum, and Smyrna, whose prosperity was built on trade and the exploitation of local natural resources. Ionia may, like many frontier societies, have encouraged hard work and self-sufficiency; in return, it offered prosperity and opportunity. It also brought Greeks into contact with the art, religion, and learning of the Near East, with which Ionia had cultural, commercial, diplomatic, and military contact. Important though these influences undoubtedly were, the critical factor was surely the availability of fully alphabetic writing and its wide dissemination among the Greek population. The result was a burst of creativity in lyric poetry and philosophy.

The earliest philosophers of whom we have any knowledge were from the city of Miletus on the coast of southern Ionia. The names Thales, Anaximander, and Anaximenes have come down to us from the sixth century, Leucippus from the fifth. The available fragments portray the earliest Milesian philosopher, Thales, as a geometer, astronomer, and engineer. He is alleged to have successfully predicted a solar eclipse in 585; however, the sources of the legend do not appear particularly reliable, and it is unlikely that Greek astronomical knowledge had reached the point in Thales' lifetime where such a prediction was possible. Other fragments assign him the theory that the earth (a flat disk) floats on water, a notion that may be a truer measure of his astronomical and cosmological sophistication.[7]

Our knowledge of all of the Milesians is plagued by the problem of questionable, fragmentary sources; and we must approach all claims about the early Greek philosophers with healthy skepticism. What seems undeniable, however, is their interest in the problem of the fundamental reality, the basic stuff of which the universe was made or out of which it emerged. Aristotle, writing in the fourth century B.C. (with his own axes to

Map 1. The Greek World about 450 B.C.

grind, and in possession of only fragmentary and indirect evidence himself), gave us the following account:

> For the original source of all existing things, that from which
> a thing first comes-into-being and into which it is finally destroyed, the substance persisting but changing in its qualities,
> this [the first philosophers] declare is the element and first
> principle of existing things, and for this reason they consider
> that there is no absolute coming-to-be or passing away, on the
> ground that such a nature is always preserved.[8]

Thales, according to Aristotle, considered water to be this most fundamental reality, though Aristotle could do no more than speculate on the reasoning that lay behind Thales' choice.

Other Milesians of the sixth century, presumably Thales' students or disciples (we have no exact knowledge of their lives), seem to have given different answers to the same question. Anaximander (fl. 550), according to a variety of late reports, believed the origin of things was to be found in the *apeiron,* the unlimited or boundless—"a huge, inexhaustible mass, stretching away endlessly in every direction," according to one of his modern interpreters.[9] From the *apeiron* emerged a seed, which gave rise to the cosmos. Finally, Anaximenes (fl. 545) appears to have argued that the underlying stuff was air, which can be rarefied or condensed to produce the variety of substances found in the world as we know it. It is worth noting that the Milesians were materialists and monists: that is, they judged the primary substance to be some kind of material stuff, and to be one.

All of this may seem primitive. And in one sense it is; it cannot be equated with, nor does it anticipate, any modern theory. But comparing the past with the present is a sure recipe for distorting the achievements of the past. When the Milesians are compared with their immediate predecessors, their importance becomes immediately apparent. In the first place, the Milesians asked a new sort of question: what is the origin of things, or what is the simple underlying reality that can take on a variety of forms to produce the diversity of substances that we perceive? This is a search for unity behind diversity and order behind change. Second, the answers offered by the Milesians contain none of the personification or deification of nature that we saw in Homer and Hesiod. The Milesians left the gods out. What they may have thought about the Olympian gods we do not (in most cases) know; but they did not invoke the gods to explain the origin and nature of things. Third, the Milesians seem to have been aware of the need not simply to state their theories, but also to defend them against critics or competitors. We thus see the beginnings of a tradition of critical assessment.[10]

Milesian speculations about the underlying stuff were only the beginning of a quest that has continued to our own day. In antiquity, the Milesians were succeeded by various schools of thought. Fifty years later Heraclitus (fl. 500) of Ephesus (an Ionian city not far from Miletus) associated the origin of things with fire: "this world-order did none of [the] gods or men make, but it always was and is and shall be: an everliving fire, kindling in measures and going out in measures."[11] In the second half of the fifth century, the materialism of the sixth century was adopted and ex-

Fig. 2.3. The ruins of ancient Ephesus. SEF/Art Resource N.Y.

tended by the atomists Leucippus of Miletus (fl. 440) and Democritus of Abdera (fl. 410). Leucippus and Democritus argued that the world consisted of an infinity of tiny atoms moving randomly in an infinite void. The atoms, solid corpuscles too small to be seen, come in an infinitude of shapes; by their motions, collisions, and transient configurations, they account for the great diversity of substances and the complex phenomena that we experience. Leucippus and Democritus even attempted to explain the formation of worlds out of vortices or whirlpools of atoms.[12]

The atomists offered ingenious accounts of many other natural phenomena, but we must not allow ourselves to be diverted from the main point. What is important about the atomists is their vision of reality as a lifeless

piece of machinery, in which everything that occurs is the necessary outcome of inert, material atoms moving according to their nature. No mind
and no divinity intrude into this world. Life itself is reduced to the motions
of inert corpuscles. There is no room for purpose or freedom; iron necessity alone rules. This mechanistic worldview would fall out of favor with
Plato and Aristotle and their followers; but it returned with a vengeance
(and with a few novel twists) in the seventeenth century and has been a
powerful force in scientific discussions ever since.

Not all who investigated the underlying stuff were monists or materialists. Nor were the gods altogether absent from their explanations. Empedocles of Acragas (fl. 450), a rough contemporary of Leucippus in the
second half of the fifth century, identified four elements or "roots" (as he
called them) of all material things: fire, air, earth, and water (introduced in
mythological garb as Zeus, Hera, Aidoneus, and Nestis). From these four
roots, Empedocles wrote, "sprang all things that were and are and shall be,
trees and men and women, beasts and birds and water-bred fishes, and
the long-lived gods too, most mighty in their prerogatives. For there are
these things alone, and running through one another they assume many
a shape."[13] But material ingredients alone cannot explain motion and
change. Empedocles therefore introduced two additional, immaterial principles: love and strife, which induce the four roots to congregate and
separate.

Empedocles was not the only ancient philosopher to include immaterial
principles among the most fundamental things. The Pythagoreans of the
sixth and fifth centuries (concentrated especially in the Greek colonies of
southern Italy and known to us not as individuals but as a "school" of
thought) seem to have argued, if we understand their doctrine, that the
ultimate reality is numerical rather than material—not matter, but number. Aristotle reports that in the course of their mathematical studies the
Pythagoreans were struck by the power of numbers to account for phenomena such as the musical scale. According to Aristotle, "since . . . all
other things seemed in their whole nature to be modelled after numbers,
and numbers seemed to be the first things in the whole of nature, they
supposed the elements of numbers to be the elements of all things, and
the whole heaven to be a musical scale and a number."[14] Now this is an
obscure passage, and our uncertainty is compounded by the probability
that Aristotle did not fully understand the Pythagorean teaching or was not
altogether fair to it. Did the Pythagoreans literally believe that material
things were constructed out of numbers? Or did they mean only to claim
that material things have fundamental numerical properties and that such

properties offer insight into the nature of things? We will never know for certain. A sensible reading of the Pythagorean position is that in some sense numbers come first, and everything else is their offspring; number is in that sense the fundamental reality, and material things derive their existence, or at least their properties, from number. If we wish to be more cautious, we can affirm at the very least that the Pythagoreans regarded number as a fundamental aspect of reality and mathematics as a basic tool for investigating this reality.[15]

THE PROBLEM OF CHANGE

If the most prominent philosophical problem of the sixth century was this question of the origins and fundamental ingredients of the world, a related issue came to dominate the philosophical enterprise in the fifth century. When we have truly discovered the fundamental ingredients of the world, can there be any doubt that we will find them to be unchangeable? It seems not: would something thought to be the ultimate reality be judged truly ultimate if it changed form or came into and passed out of existence? Would we not insist on explaining change in this entity by reference to something even more ultimate? At the end of the explanatory road, there must be something fixed and unchangeable. If we agree, then, that the ultimate reality must be unchangeable, is it possible to account for, or even to accept, the reality of change? Is stability at the level of ultimate reality compatible with genuine change on some other level? How can the world be both stable and changeable?

One of the earliest philosophers to address this question was Heraclitus, who offered a ringing declaration of the reality of change. Heraclitus is reputed to have claimed that nobody can step twice into the same river (because the second time it is no longer exactly the same river), and this aphorism made him the symbol, even in antiquity, of the opinion that everything is in a state of flux. Heraclitus also argued that a condition of overall equilibrium or stability may conceal underlying change in the form of counterbalancing forces or the struggle of opposites. For example, there is a perpetual struggle between the substances earth, water, and fire, each endeavoring to consume the others; however, dynamic equilibrium is achieved through overall balance or reciprocity.[16]

What Heraclitus affirmed, Parmenides (fl. 480, from the Greek city-state at Elea in southern Italy) denied. Parmenides wrote a long philosophical poem (philosophy had not yet settled on prose as its exclusive form of discourse), large sections of which have survived. In it, Parmenides adopted the radical position that change—all change—is a logical impossibility. Parmenides began by denying, on various logical grounds, the possibility that a thing should pass from non-being to being: for example, if a thing were to come into being, why at one moment rather than another, and by what means? His conclusion was that out of nothing comes nothing. "For never shall this be proved," he wrote, "that things that are not are."[17] Parmenides proceeded, on analogous grounds, to deny all other forms of change. He also denied the existence of time and plurality; what exists is one and now.

Parmenides' pupil Zeno (fl. 450) extended and defended the Parmenidean doctrine with a set of proofs against the possibility of one kind of change—motion, or change of place. One of these proofs, the "stadium paradox," will illustrate Zeno's approach. It is impossible, Zeno argued, ever to traverse a stadium, because before you cover the whole you must cover the half; and before you cover the half, you must cover the quarter; before the quarter, the eighth; and so on to infinity. To traverse a stadium is therefore to traverse an infinite sequence of halves, and it is impossible to traverse, or even "to come into contact with" (as Aristotle put it in his discussion of the paradox), an infinity of intervals in a finite time. The same argument may be applied to any spatial interval whatsoever—from which it follows that all motion is impossible.[18]

Now all of this may seem preposterous. With a little effort, Parmenides and Zeno could have opened their eyes and observed changes all around them. Did they not get up in the morning, enjoy a good breakfast, and make their way to the agora (the public square) for a hard day's philosophizing? And didn't they recognize that doing so required them to move? No doubt. Parmenides and Zeno knew perfectly well what experience taught, but the question was whether experience could be trusted. What does one do if experience suggests the reality of change, while careful argumentation (with due attention to the rules of logic) unambiguously teaches its impossibility? For Parmenides and Zeno, the answer was clear: the rational process must prevail. Parmenides distinguished between "the way of seeming," associated with observation, and "the way of truth," trod by reason. In his poem he warned against letting "custom, born of much experience, force thee to let wander along this road thy aimless eye, thy echoing ear or thy tongue; but do thou judge by reason the strife-encompassed proof that I have spoken."[19] So, yes, Parmenides and Zeno acknowledged that experience teaches the reality of change. But they knew on rational grounds that this was an illusion—a pleasant and powerful illusion, perhaps, but an illusion nonetheless.

Parmenides' denial of the possibility of change was enormously influential, offering a challenge that generations of philosophers felt compelled to address. Empedocles answered with his theory of four material "roots" or elements, plus love and strife. The elements do not come into being or pass away, and so the fundamental Parmenidean requirement is met; but they do congregate and separate and mix in various proportions, and thus change is genuine. The atomists Leucippus and Democritus granted that the individual atom is absolutely immutable, so that at the atomic level there is no generation, corruption, or alteration of any kind. However, the atoms are perpetually moving, colliding, and congregating; and through the motions and configurations of the atoms the endless variety in the world of sense experience is produced. According to the atomists, therefore, fundamental stability underlies superficial change; both are present, and both are real.[20]

THE PROBLEM OF KNOWLEDGE

Poking through these discussions of the underlying reality and the problem of change and stability has been a third basic issue, which early Greek philosophers also addressed—namely, the problem of knowledge (more technically known as epistemology). It is implicit in the quest for the fundamental reality underlying the variety of substances revealed by the senses: if the senses do not reveal the unity of things, then we must find other guides to the truth. The problem of knowledge is explicit in fifth-century discussions of change and stability. Parmenides' radical stance on the question of change had clear-cut epistemological implications: if the senses reveal change, their unreliability is thereby demonstrated; truth is to be gained only by the exercise of reason. The atomists, too, tended to denigrate sense experience. After all, the senses revealed the "secondary" qualities—colors, tastes, odors, and the tactile qualities—whereas reason taught that only atoms and the void truly exist. In a surviving fragment, Democritus identifies "two forms of knowledge, one genuine, one obscure. To the obscure belong all the following: sight, hearing, smell, taste, touch."[21] The fragment breaks off before the idea is completed, but we may assume that in Democritus's judgment genuine knowledge is rational knowledge.

If the early philosophers were inclined to favor reason over sense, this tendency was neither universal nor without qualification. Empedocles defended the senses against the attack of Parmenides. The senses may not be perfect, he argued, but they are useful guides if employed with discrimination. "But come, consider with all thy powers how each thing is manifest," he wrote, "neither holding sight in greater trust as compared with hearing, nor loud-sounding hearing above the clear evidence of thy tongue, nor withholding thy trust from any of the other limbs, wheresoever there is a path for understanding." And Anaxagoras (fl. 450) of Clazomenae (another Ionian coastal city) argued in a brief fragment that the senses offer "a glimpse of the obscure."[22]

One of the benefits gained from Greek epistemological concerns (from Greek rationalism in particular) was that they directed attention to the rules of reasoning, argumentation, and theory-assessment. Formal logic would be the creation of Aristotle; but his sixth- and fifth-century predecessors became increasingly aware of the need to test the soundness of an argument and to assess the grounds on which a theory rested. The sophistication with which Parmenides and Zeno could argue—their sensitivity, for example, to the rules of inference and the criteria of proof—demonstrates how far Greek philosophy had come in a century and a half.

PLATO'S WORLD OF FORMS

The death of Socrates in 399 B.C., coming as it did around the turn of the century (not on their calendar, of course, but on ours), has made it a convenient point of demarcation in the history of Greek philosophy. Thus Socrates' predecessors of the sixth and fifth centuries (the philosophers who have occupied us up to now in this chapter) are commonly called the "pre-Socratic philosophers." But Socrates' prominence is more than an accident of the calendar, for Socrates represents a shift in emphasis within Greek philosophy, away from the cosmological concerns of the sixth and fifth centuries toward political and ethical matters. Nonetheless, the shift was not so dramatic as to preclude continuing attention to the major problems of pre-Socratic philosophy. We find both the new and the old in the work of Socrates' younger friend and disciple, Plato.

Plato (427-348/47) was born into a distinguished Athenian family, active in affairs of state; he was undoubtedly a close observer of the political events that led up to Socrates' execution. After Socrates' death, Plato left Athens and visited Italy and Sicily, where he seems to have come into contact with Pythagorean philosophers. In 388 Plato returned to Athens and founded a school of his own, the Academy, where young men could pursue advanced studies (see fig. 4.2). Plato's literary output appears to have consisted almost entirely of dialogues, the majority of which have survived. We will find it necessary to be highly selective in our examination of Plato's philosophy; let us begin with his quest for the underlying reality.[23]

In a passage in one of his dialogues, the *Republic,* Plato reflected on the

Fig. 2.4. Plato (1st century A.D. copy), Museo Vaticano, Vatican City. Alinari/Art Resource N.Y.

relationship between the actual tables constructed by a carpenter and the idea or definition of a table in the carpenter's mind. The carpenter replicates the mental idea as closely as possible in each table he makes, but always imperfectly. No two manufactured tables are alike down to the smallest detail, and limitations in the material (a knot here, a warped board there) insure that none will fully measure up to the ideal.

Now, Plato argued, there is a divine craftsman who bears the same relationship to the cosmos as the carpenter bears to his tables. The divine craftsman (the Demiurge) constructed the cosmos according to an idea or plan, so that the cosmos and everything in it are replicas (and always imperfect ones because of limitations inherent in the materials) of eternal ideas or forms. In short, there are two realms: a realm of forms or ideas, containing the perfect idea of every single thing; and the material realm in which these forms or ideas are imperfectly replicated.

Plato's notion of two distinct realms will seem strange to many people, and we must therefore stress several points of importance. The forms are incorporeal, intangible, and insensible; they have always existed, sharing the property of eternality with the Demiurge; and they are absolutely changeless. They include the form, the perfect idea, of everything in the material world. One does not speak of their location, since they are incorporeal and therefore not spatial. Although incorporeal and imperceptible by the senses, they objectively exist; indeed, true reality (reality in its fullness) is located only in the world of forms. The sensible, corporeal world, by contrast, is imperfect and transitory. It is less real in the sense that the corporeal object is a replica of, and therefore dependent for its existence upon, the form. Corporeal objects exist secondarily, while the forms exist primarily.

Plato illustrated this conception of reality in his famous "allegory of the cave," found in Book VII of the *Republic*. Men are imprisoned within a deep cave, chained so as to be incapable of moving their heads. Behind them is a wall, and beyond that a fire. People walk back and forth beneath the wall, holding above it various objects, including statues of humans and animals, which cast shadows on the wall of the cave visible to the prisoners. The prisoners see only the shadows cast by these statues and other objects; and, having lived in the cave from childhood, they no longer recall any other reality. They do not suspect that these shadows are but imperfect images of objects that they cannot see; and consequently they mistake the shadows for the real.

So it is with all of us, says Plato. We are souls imprisoned in bodies. The shadows of the allegory represent the world of sense experience. The soul, peering out from its prison, is able to perceive only these flickering shadows, and the ignorant claim that this is all there is to reality. However, there do exist the statues and other objects of which the shadows are feeble representations and also the humans and animals of which the statues are imperfect replicas. To gain access to these higher realities, we must escape the bondage of sense experience and climb out of the cave until we find ourselves able, finally, to gaze on the eternal realities, thereby entering the realm of true knowledge.[24]

What are the implications of these views for the concerns of the pre-Socratic philosophers? First, Plato equated his forms with the underlying reality, while assigning derivative or secondary existence to the corporeal world of sensible things. Second, Plato has made room for both change and stability by assigning each to a different level of reality: the corporeal realm is the scene of imperfection and change, while the realm of forms is characterized by eternal, changeless perfection. Both change and stability are therefore genuine; each characterizes something; but stability belongs to the forms and thus shares their fuller reality.

Third, as we have seen, Plato addressed epistemological issues, placing observation and true knowledge (or understanding) in opposition. Far from leading upward to knowledge or understanding, the senses are chains that tie us down; the route to knowledge is through philosophical reflection. This is explicit in the *Phaedo,* where Plato maintains the uselessness of the senses for the acquisition of truth and points out that when the soul attempts to employ them it is inevitably deceived.

Now the short account of Plato's epistemology frequently ends here; but there are important qualifications that it would be a serious mistake to omit. Plato did not, in fact, dismiss the senses altogether, as Parmenides had done and as the passage from the *Phaedo* might suggest Plato did. Sense experience, in Plato's view, served various useful functions. First, sense experience may provide wholesome recreation. Second, observation of certain sensible objects (especially those with geometrical aspects) may serve to direct the soul toward nobler objects in the realm of forms; Plato used this argument as justification for the pursuit of astronomy. Third, Plato argued (in his theory of reminiscence) that sense experience may actually stir the memory and remind the soul of forms that it knew in a prior existence, thus stimulating a process of reflection that will lead to actual knowledge of the forms. Finally, although Plato firmly believed that knowledge of the eternal forms (the highest, and perhaps the only true, form of knowledge) is obtainable only through the exercise of reason, the changeable realm of matter is also an acceptable object of study. Such studies serve the purpose of supplying examples of the operation of reason in the cosmos. If this is what interests us (as it sometimes did Plato), the best method of exploring it is surely through observation. The legitimacy and utility of sense experience are clearly implied in the *Republic,* where Plato acknowledged that a prisoner emerging from the cave first employs his sense of sight to apprehend living creatures, the stars, and finally the most noble of visible (material) things, the sun. If he aspires to apprehend "the essential reality," then he must proceed "through the discourse of reason unaided by any of the senses." Both reason and sense are thus instruments worth having; which one we employ on a particular occasion will depend on the object of study.[25]

There is another way of expressing all of this, which may shed light on Plato's achievement. When Plato assigned reality to the forms, he was, in fact, identifying reality with the properties that classes of things have in common. The bearer of true reality is not (for example) this dog with the droopy left ear or that one with the menacing bark, but the idealized form of a dog shared (imperfectly, to be sure) by every individual dog—that by virtue of which we are able to classify all of them as dogs. Therefore, to gain true knowledge, we must set aside all characteristics peculiar to things as individuals and seek the shared characteristics that define them into classes. Now stated in this modest fashion, Plato's view has a distinctly modern ring: idealization is a prominent feature of a great deal of modern science; we develop models or laws that overlook the incidental in favor of the essential. (Galileo's inertial principle, for example, was an attempt to describe motion under ideal circumstances, all resistance or interference being excluded.) However, Plato went beyond this, maintaining not merely that true reality is to be found in the common properties of classes of things, but also that this common property (the idea or form) has objective, independent, and indeed prior existence.

PLATO'S COSMOLOGY

The doctrines that we have been considering—Plato's response to the pre-Socratics, found in his *Republic, Phaedo,* and various other dialogues—represent only a small portion of his total philosophy. Plato also wrote a dialogue, the *Timaeus,* which reveals his interest in the world of nature. Here we find his views on astronomy, cosmology, light and color, the elements, and human physiology. Since the *Timaeus* gave the early Middle Ages (before the twelfth century) its only coherent natural philosophy, this work represents one of the principal channels of Platonic influence and therefore demands our attention.

Plato referred to the contents of the *Timaeus* as a "likely story," and this has misled some readers to view it as a myth in which Plato himself placed no stock. In fact, Plato stated quite clearly that this was the best account possible, that anything better than a likely account was precluded by the subject matter. Certainty is attainable only when we give an account of the eternal and unchanging forms; when we describe the imperfect and changeable, our description will inevitably share in the imperfection and changeability of its subject—and will therefore be no better than "likely."

What do we find in the *Timaeus?* One of its most striking characteristics is Plato's vehement opposition to certain features of pre-Socratic thought. The *physikoi* had deprived the world of divinity; in the process, they had also deprived it of plan and purpose. According to these philosophers, things behave according to their inherent natures, and this alone accounts for the order and regularity of the cosmos. Order, then, is intrinsic, rather than extrinsic; it is not imposed by an outside agent but arises from within.

Now Plato found such an opinion not only foolish but dangerous. He had no intention of restoring the gods of Mount Olympus, who interfered in the day-to-day operation of the universe, but he was convinced that the order and rationality of the cosmos could only be explained as the imposition of an outside mind. If the *physikoi* found the source of order in *physis* (nature), he would locate it in *psychē* (mind).[26]

Plato depicted the cosmos as the handiwork of a divine craftsman, the Demiurge. The Demiurge, according to Plato, is a benevolent craftsman, a rational god (indeed, the very personification of reason), who struggles against the limitations inherent in the materials with which he must work in order to produce a cosmos as good, beautiful, and intellectually satisfying as possible. The Demiurge takes a primitive chaos, filled with the unformed material out of which the cosmos will be constructed, and imposes order according to a rational plan. This is not creation from nothing, as in the Judeo-Christian account of creation, for the raw materials are already present and contain properties over which the Demiurge has no control; nor is the Demiurge omnipotent, for he is constrained and limited by the materials that he confronts. Nevertheless, Plato clearly intended to portray the Demiurge as a supernatural being, distinct from, and outside of, the cosmos that he constructed. Whether Plato meant his readers to take the Demiurge literally is another matter, much debated, and perhaps incapable of ever being resolved. What is not open to dispute is Plato's wish to declare that the cosmos is the product of reason and planning, that the order in the cosmos is rational order, imposed on recalcitrant materials from outside.

The Demiurge is not only a rational craftsman but also a mathematician, for he constructs the cosmos on geometrical principles. Plato took over the four roots or elements of Empedocles: earth, water, air, and fire. But under Pythagorean influence, he reduced them to something more fundamental—triangles. He thus formulated a "geometrical atomism." As two-dimensional figures, triangles are of course incorporeal; however, if suitably combined, they can be made into three-dimensional corpuscles, each different shape corresponding to one of the elements. It was already known in Plato's day that there are five, and only five, regular geometrical solids (symmetrical solid figures formed of plane surfaces, all identical); these are the tetrahedron (four equilateral triangles), the cube (six squares), the octahedron (eight equilateral triangles), the dodecahedron (twelve pentagons), and the icosahedron (twenty equilateral triangles). (See fig. 2.5.) Plato associated each of the elements with one of these figures—fire with the tetrahedron (the smallest, sharpest, and most mobile of the regular solids), air with the octahedron, water with the icosahedron, and earth with the most stable of the regular solids, the cube. Finally, Plato found a function for the dodecahedron (the regular solid closest to the sphere) by identifying it with the cosmos as a whole.[27]

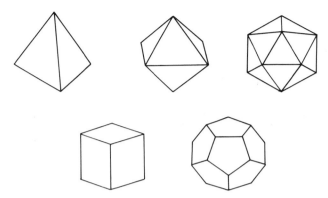

Fig. 2.5. The five Platonic solids: tetrahedron, octahedron, icosahedron, cube, and dodeca-hedron. Courtesy of J. V. Field.

Three features of this scheme deserve discussion. First, it accounts for change and diversity in the same way as does Empedocles' theory: the elements can mix in various proportions to produce variety in the material world. Second, it allows for transmutation of the elements from one to another, thus further accounting for change. For example, a single corpuscle of water (the icosahedron) can be dissolved into its twenty constituent equilateral triangles, which can then recombine into, say, two corpuscles of air (the octahedron) and one of fire (the tetrahedron). Only earth, which is composed of squares (and the square divided diagonally does not yield equilateral triangles), is excluded from this process of transmutation. Third, Plato's geometrical corpuscles represent a significant step toward the mathematization of nature. Indeed, it is important for us to see just how large a step it is. Plato's elements are not material substance packaged as the regular solids; in such a scheme matter would still be acknowledged as the fundamental stuff. For Plato, the shape is all there is; corpuscles are entirely reducible (without residue) to the regular solids, which are reducible to plane geometrical figures. Water, air, and fire are not *triangular;* they are simply *triangles.* The Pythagorean program of reducing everything to mathematical first principles has been fulfilled.

Plato proceeded to describe many features of the cosmos; let us glance at a few of them. He demonstrated a rather sophisticated command of cosmology and astronomy. He proposed a spherical earth, surrounded by the spherical envelope of the heavens. He defined various circles on the celestial sphere, marking the paths of the sun, moon, and other planets. He understood that the sun moves around the celestial sphere once a year on a circle (which we call the ecliptic) tilted in relation to the celestial equator (see fig. 2.6). He knew that the moon makes a monthly circuit of approximately the same path. He knew that Mercury, Venus, Mars, Jupiter, and Saturn do the same, each at its own pace and with occasional reversals, and that Mercury and Venus never stray far from the sun. He even knew that the overall motion of the planetary bodies (if we combine their slow motion around the ecliptic with the daily rotation of the celestial sphere) is spiral. And what is perhaps most important of all, Plato seems to have understood that the irregularities of planetary motion can be explained by the compounding of uniform circular motions.[28]

When Plato descended from the cosmos to the human frame he offered an account of respiration, digestion, emotion, and sensation. He had a theory of sight, for example, which supposed that visual fire issues from the eye, interacting with external light to create a visual pathway that could transmit motions from the visible object to the observer's soul. The *Timaeus* even offered a theory of disease and outlined a regimen that was to insure health.

It is an admirable cosmos that Plato has portrayed. What are its most prominent features? From triangles and regular solids the Demiurge fashioned a final product of the utmost rationality and beauty; and that means, according to Plato, that the cosmos must be a living creature. The Demiurge, we read in the *Timaeus,* "wishing to make the world most nearly like that intelligible thing which is best and in every way complete, fashioned it as a single visible living creature." But if the world is a living creature, it must possess a soul. And indeed it does; in the center of the cosmos the Demiurge "set a soul and caused it to extend throughout the whole and further wrapped its body round with soul on the outside; and so he established one world alone, round and revolving in a circle, solitary but able by reason of its excellence to bear itself company, needing no other acquaintance or friend but sufficient to itself." The world soul is ultimately responsible for all motions in the cosmos, just as the human soul is responsible for the motions of the human body. We see here the origins of the strong animistic strain that was to remain an important feature of the Platonic tradition. Repelled by the lifeless necessity of the atomistic world, Plato has described an animated cosmos, permeated by rationality, replete with purpose and design.[29]

Nor is deity absent. There is the Demiurge, of course; but in addition Plato assigned divinity to the world soul and considered the planets and the fixed stars to be a host of celestial gods. However, unlike the gods of traditional Greek religion, Plato's deities never interrupt the course of na-

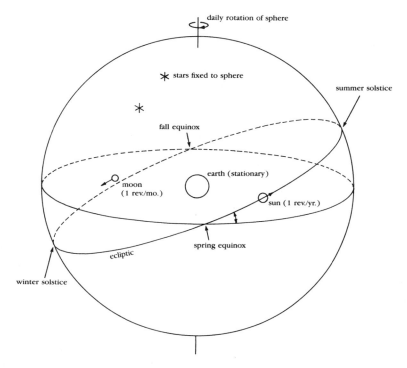

Fig. 2.6. The celestial sphere according to Plato.

ture. Quite the contrary, it is the very steadfastness of the gods which, in Plato's view, guarantees the regularity of nature; the sun, moon, and other planets *must* move with some combination of uniform circular motions precisely because such motion is most perfect and rational, and consequently such motion is the only kind conceivable for a divine being. Thus Plato's reintroduction of divinity does not represent a return to the unpredictability of the Homeric world. Quite the contrary, the function of divinity for Plato was to undergird and account for the order and rationality of the cosmos. Plato restored the gods in order to account for precisely those features of the cosmos which, in the view of the *physikoi,* required the banishing of the gods.[30]

THE ACHIEVEMENT OF EARLY GREEK PHILOSOPHY

If we survey early Greek philosophy with a modern scientific eye, certain pieces of it look familiar. The pre-Socratic inquiry into the shape and arrangement of the cosmos, its origin, or its fundamental ingredients reminds us of questions still investigated in modern astrophysics, cosmology, and particle physics. However, other pieces of early philosophy look considerably more foreign. Working scientists today do not inquire whether change is logically possible or where true reality is to be found; and it would be a considerable feat to turn up, say, a physicist or chemist who worries about how to balance the respective claims of reason and observation. These matters are no longer talked about by scientists. Does it follow that the early philosophers who devoted their lives to such questions were "unscientific," perhaps even misguided or dim-witted?

This question needs to be handled with some delicacy. Surely the fact that the *physikoi* were concerned about some matters no longer of interest is no indictment of their enterprise; in the course of any intellectual endeavor some problems get resolved, while others go out of fashion. But the objection may go deeper than that: are there issues that are intrinsically inappropriate or illegitimate, questions that were futile from the beginning? And did Plato and the *physikoi* waste their time on any of these? Perhaps we can answer in this way. Themes such as the identity of the ultimate reality, the distinction between natural and supernatural, the source of order in the universe, the nature of change, and the foundations of knowledge are quite different from the explanation of small-scale observational data (say, the descent of a heavy body, a chemical reaction, or a physiological process) that have occupied scientists for the past few centuries; but to be different is not to be insignificant. At least until Isaac Newton, these larger themes demanded as much attention from the student of nature as did the problems that now fill up a university course in science. Such questions were interesting and essential precisely because they were part of the effort to create a conceptual framework and a vocabulary for investigating the world. They were foundational questions; and it is often the fate of foundational questions to seem pointless to later generations who take the foundations for granted. Today, for example, we may find the distinction between the natural and supernatural obvious; but until the distinction was carefully drawn, the investigation of nature could not properly begin.

Thus the early philosophers began at the only possible place: the beginning. They created a conception of nature that has served as the foundation of scientific belief and investigation in the intervening centuries—the conception of nature presupposed, more or less, by modern science. In the meantime many of the questions they asked have been resolved—often with rough-and-ready solutions, rather than definitive ones, but resolved sufficiently to slip from the forefront of scientific attention. As they have sunk from view, their place has been taken by a collection of much narrower investigations. If we would understand the scientific enterprise in all of its richness and complexity, we must see that its two parts—the foundation and the superstructure—are complementary and reciprocal. Modern laboratory investigation occurs within a broad conceptual framework and cannot even begin without expectations about nature or the underlying reality; in turn, the conclusions of laboratory research reflect back on these most fundamental notions, forcing refinement and (occasionally) revision. The historian's task is to appreciate the enterprise in all of its diversity. If the garden of the *physikoi* is situated at the beginning of the road to modern science, then the historian of science may profitably dally in its shady corners before embarking on his journey.

THREE

Aristotle's Philosophy of Nature

LIFE AND WORKS

Aristotle was born in 384 B.C. in the northern Greek town of Stagira, into a privileged family. His father was personal physician to the Macedonian king, Amyntas II (grandfather of Alexander the Great). Aristotle had the advantage of an exceptional education: at age seventeen, he was sent to Athens to study with Plato. He remained in Athens as a member of Plato's Academy for twenty years, until Plato died about 347. Aristotle then spent several years in travel and study, crossing the Aegean Sea to Asia Minor (modern Turkey) and its coastal islands. During this period he undertook biological studies, and he encountered Theophrastus, who was to become his pupil and lifetime colleague, before returning to Macedonia to become the tutor of the young Alexander (later "the Great"). In 335, when Athens fell under Macedonian rule, Aristotle returned to the city and began to teach in the Lyceum, a public garden frequented by teachers. He remained there, establishing an informal school, until shortly before his death in 322.[1]

In the course of his long career as student and teacher, Aristotle systematically and comprehensively addressed the major philosophical issues of his day. He is credited with more than 150 treatises, approximately thirty of which have come down to us. The surviving works seem to consist mainly of lecture notes or unfinished treatises not intended for wide circulation; whatever their exact origin, they were obviously directed to advanced students or other philosophers. In modern translation they occupy well over a foot of bookshelf, and they contain a philosophical system overwhelming in power and scope. It is out of the question for us to survey the whole of Aristotle's philosophy, and we must be content with examining the funda-

Fig. 3.1. Aristotle, Museo Nazionale, Rome. Alinari/Art Resource N.Y.

mentals of his philosophy of nature—beginning with his response to positions taken by the pre-Socratics and Plato.[2]

METAPHYSICS AND EPISTEMOLOGY

Through his long association with Plato, Aristotle had, of course, become thoroughly versed in Plato's theory of forms. Plato had drastically diminished (without totally rejecting) the reality of the material world observed by the senses. Reality in its perfect fullness, Plato argued, is possessed only by the eternal forms, which are dependent on nothing else for their existence. The objects that make up the sensible world, by contrast, derive their characteristics and their very being from the forms; it follows that sensible objects exist only derivatively or dependently.

Aristotle refused to accept this dependent status that Plato assigned to sensible objects. They must have autonomous existence, for in his view they were what make up the real world. Moreover, the traits that give an individual object its character do not, Aristotle argued, have a prior and separate existence in a world of forms, but belong to the object itself. There is no perfect form of a dog, for example, existing independently and replicated imperfectly in individual dogs, imparting to them their attributes. For Aristotle, there were just individual dogs. These dogs certainly shared a set of attributes—for otherwise we would not be entitled to call them "dogs"—but these attributes existed in, and belonged to, individual dogs.

Perhaps this way of viewing the world has a familiar ring. Making individual sensible objects the primary realities ("substances," Aristotle called them) will seem like good common sense to most readers of this book, and probably struck Aristotle's contemporaries the same way. But if it makes good common sense, can it also be good philosophy? That is, can it deal successfully, or at least plausibly, with the difficult philosophical issues raised by the pre-Socratics and Plato—the nature of the fundamental reality, epistemological concerns, and the problem of change and stability? Let us take up these problems one by one.[3]

The decision to locate reality in sensible, corporeal objects does not yet tell us very much about reality—only that we should look for it in the sensible world. Already in Aristotle's day, any philosopher would demand to know more: one thing he would demand to know was whether corporeal objects are irreducible or must be considered composites of more fundamental constituents. Aristotle addressed this question by drawing a distinction between properties and their subjects (warmth and the warm object, for example). He maintained (as most of us would) that a property has to be the property *of* something; we call that something its "subject." To be a property is to belong to a subject; properties cannot exist independently.

Individual corporeal objects, then, have both properties (color, weight, texture, and the like) and something other than properties to serve as their subject. These two roles are played, respectively, by "form" and "matter" (technical terms that did not mean for Aristotle exactly what they mean for us). Corporeal objects are "composites" of form and matter—form consisting of the properties that make the thing what it is, matter serving as the subject or substratum for form. A white rock, for example, is white, hard, heavy, and so forth, by virtue of its form; but matter must also be present, to serve as subject for the form, and this matter brings no properties of its own to its union with form.[4] (Aristotle's doctrine will be further discussed in chapter 12, below, in connection with medieval attempts to clarify and extend it.)

We can never, in actuality, separate form and matter; they are presented to us only as a unified whole. If they were separable, we should be able to put the properties (no longer the properties *of* anything) in one pile, the matter (absolutely propertyless) in another—an obvious impossibility. But if form and matter can never be separated, is it not meaningless to speak of them as the *real* constituents of things? Isn't this a purely logical distinction, existing in our minds, but not in the external world? Surely not for Aristotle, and perhaps not for us; most of us would think twice before denying the reality of cold or red, although we can never collect a bucket of either one. In short, Aristotle once again surprises us by using common-sense notions to build a persuasive philosophical edifice.

Aristotle's claim that the primary realities are concrete individuals surely has epistemological implications, since true knowledge must be of the truly real. By this criterion, Plato's attention was naturally directed toward the eternal forms, knowable through reason or philosophical reflection. Aristotle's metaphysics of concrete individuals, by contrast, directed his quest for knowledge toward the world of individuals, of nature, and of change—a world encountered through the senses.

Aristotle's epistemology is complex and sophisticated. It must suffice here to indicate that the process of acquiring knowledge begins with sense experience; from repeated sense experience follows memory; and from memory, by a process of "intuition" or insight, the experienced investigator is able to discern the universal features of things. By the repeated observation of dogs, for example, the experienced dog breeder comes to know what a dog really is; that is, he comes to understand the form or definition of a dog, the crucial traits without which an animal cannot be a dog. Note that Aristotle, no less than Plato, was determined to grasp the universal; but, unlike his teacher, Aristotle argued that one must do so by starting with the individual. Once we possess the universal definition, we can put it to use as the premise of deductive demonstrations.[5]

Knowledge is thus gained by a process that begins with experience (a term broad enough, in some contexts, to include common opinion or the reports of distant observers). In that sense knowledge is empirical; nothing can be known apart from such experience. But what we learn by this "inductive" process does not acquire the status of true knowledge until put into deductive form; the end product is a deductive demonstration (nicely illustrated in a Euclidean proof) beginning from universal definitions as premises. Although Aristotle discussed both the inductive and deductive phases (the latter far more than the former) in the acquisition of knowledge, he stopped considerably short of later methodologists, especially in the analysis of induction.

This is the theory of knowledge outlined by Aristotle in the abstract. Is it also the method actually employed in Aristotle's own scientific investigations? Probably not—with perhaps an occasional exception. Like the modern scientist, Aristotle did not proceed by following a methodological recipe book, but rather by rough and ready methods, familiar procedures that had proved themselves in practice. Somebody has defined science as "doing your damnedest, no holds barred"; when it came (for example) to his extensive biological researches, this is doubtless exactly what Aristotle did. It is not a surprise, and certainly no character defect, that Aristotle should, in the course of thinking about the nature and the foundations of knowledge, formulate a theoretical scheme (an epistemology) not perfectly consistent with his own scientific practice.[6]

NATURE AND CHANGE

The problem of change had become a celebrated philosophical issue in the fifth century B.C. In the fourth century, Plato had dealt with it by restricting change to the imperfect material replica of the changeless world of forms. For Aristotle, a distinguished naturalist who was philosophically committed to the full reality of the changeable individuals that make up the sensible world, the problem of change was a most pressing one.[7]

Aristotle's starting point was the commonsense assumption that change is genuine. But this does not, by itself, get us very far; it remains to be demonstrated that the idea of change can withstand philosophical scrutiny, and to be shown how change can be explained. Aristotle had various weapons in his arsenal by which to achieve these ends. The first was his doctrine of form and matter. If every object is constituted of form and matter, then Aristotle could make room for both change and stability by arguing that when an object undergoes change, its form changes (by a process of replacement, the new form replacing the old one) while its matter remains. Aristotle went on to argue that change in form takes place between a pair of contraries, one of which is the form to be achieved, the other its privation or absence. When the dry becomes wet or the cold becomes hot, this is change from privation (dry or cold) to the intended form (wet or hot). Change, for Aristotle, is thus never open-ended, but confined to the narrow corridor connecting pairs of contrary qualities; order is thus discernible even in the midst of change.

A determined Parmenidean might protest that to this point the analysis does nothing to escape Parmenides' objection to all change on the ground that inevitably it calls for the emergence of something out of nothing. Aristotle's reply is found in his doctrine of potentiality and actuality. Aristotle would undoubtedly have granted that *if* the only two possibilities are being and nonbeing—that is, if things either exist or do not exist—then the transition from nonhot to hot would indeed involve passage from nonbeing to being (the nonbeing of hot to the being of hot) and would thus be vulnerable to Parmenides' objection. But Aristotle believed that the objection could be successfully circumvented by supposing that there are three categories associated with being: (1) nonbeing, (2) potential being, and

(3) actual being. If such is the state of things, then change can occur between potential being and actual being without nonbeing ever entering the picture. A seed, for example, is potentially, but not actually, a tree. In becoming a tree, it becomes actually what it already was potentially. The change thus involves passage from potentiality to actuality—not from nonbeing to being, but from one kind of being to another kind of being. This doctrine is perhaps best illustrated from the biological realm, but it has general applicability. A heavy body held above the earth falls in order to fulfill its potential (of being situated with other heavy things about the center of the universe); and a block of marble has the potential to assume whatever shape the sculptor chooses to give it.

If these arguments allow us to escape the logical dilemmas associated with the idea of change, and therefore to believe in the possibility of change, they do not tell us anything about the cause of change. Why should a seed move from being a potential tree to being an actual tree, or an object change from black to white, rather than remaining in its original state? This brings us to Aristotle's ideas about nature and causation.

The world we inhabit is an orderly one, in which things generally behave in predictable ways, Aristotle argued, because every natural object has a "nature"—an attribute (associated primarily with form) that makes the object behave in its customary fashion, provided no insurmountable obstacle intervenes. For Aristotle, a brilliant zoologist, the growth and development of biological organisms were easily explained by the activity of such an inner driving force. An acorn becomes an oak tree, because that is its nature. But the theory was applicable beyond biological growth and, indeed, beyond the biological realm altogether. Dogs bark, rocks fall, and marble yields to the hammer and chisel of the sculptor because of their respective natures. Ultimately, Aristotle argued, all change and motion in the universe can be traced back to the natures of things. For the natural philosopher, who by definition is interested in change and things capable of undergoing change, these natures are the central object of study. To this general statement of Aristotle's theory of "nature," we need only add two reminders: that it does not apply to objects produced artificially, for such objects possess no inner source of change, but are merely the recipients of external influence; and that the nature of a complex organism does not result from a summation or mixture of the natures of the constituent materials, but is a unique nature characteristic of that organism as a unified whole.[8]

With this theory of nature in mind, we can understand a feature of Aristotle's scientific practice that has puzzled and distressed modern commentators and critics—namely, the absence from his work of anything resembling controlled experimentation. Unfortunately, such criticism over-looks Aristotle's aims—aims that drastically limited his methodological options. If, as Aristotle believed, the nature of a thing is to be discovered through the behavior of that thing in its natural, unfettered state, then artificial constraints will merely interfere. If, despite interference, the object behaves in its customary fashion, we have troubled ourselves for no purpose. If we set up conditions that prevent the nature of an object from revealing itself, all we have learned is that it can be interfered with to the point of remaining concealed. Experiment reveals nothing about natures that we cannot learn better in some other way. Aristotle's scientific practice is not to be explained, therefore, as a result of stupidity or deficiency on his part—failure to perceive an obvious procedural improvement—but as a method compatible with the world as he perceived it and well suited to the questions that interested him. Experimental science emerged not when, at long last, the human race produced somebody clever enough to perceive that artificial conditions would assist in the exploration of nature, but when natural philosophers began asking questions to which such a procedure promised to offer answers.[9]

To complete our analysis of Aristotle's theory of change, we must briefly consider the celebrated four Aristotelian causes. To understand a change or the production of an artifact is to know its causes (perhaps best translated "explanatory conditions and factors"). There are four of these: the form received by a thing; the matter underlying that form, which persists through the change; the agency that brings about the change; and the purpose served by the change. These are called, respectively, formal cause, material cause, efficient cause, and final cause. To take an extremely simple example—the production of a statue—the formal cause is the shape given the marble, the material cause is the marble that receives this shape, the efficient cause is the sculptor, and the final cause is the purpose for which the statue is produced (perhaps the beautification of Athens or the celebration of one of its heroes). There are cases in which identifying one or another of the causes is difficult, or in which one or more causes merge, but Aristotle was convinced that his four causes provided an analytical scheme of general applicability.

We have said enough about the form-matter distinction to make clear what was meant by formal and material causes, and efficient cause is close enough to modern notions of causation to require no further comment; but final cause requires a few words of explanation. In the first place, the expression "final cause" is an English cognate derived from the Latin word *finis,* meaning "goal," "purpose," or "end," and has nothing to do with the fact that it often appears last in the list of Aristotelian causes. Aristotle argued, quite rightly, that many things cannot be understood without knowledge of purpose or function. To explain the arrangement of teeth in the mouth, for example, we must understand their functions (sharp teeth in front for tearing, molars in back for grinding). Or to take an example from the inorganic realm, it is not possible to grasp why a saw is made as it is without knowing the function the saw is meant to serve. Aristotle went so far as to give final cause priority over material cause, noting that the purpose of the saw determines the material (iron) of which it must be made, whereas the fact that we possess a piece of iron does nothing to determine that we will make it into a saw.[10]

Perhaps the most important point to be made about final cause is its clear illustration of the role of purpose (the more technical term is "teleology") in Aristotle's universe. The world of Aristotle is not the inert, mechanistic world of the atomists, in which the individual atom pursues its own course mindless of all others. Aristotle's world is not a world of chance and coincidence, but an orderly, organized world, a world of purpose, in which things develop toward ends determined by their natures. It would be unfair and pointless to judge Aristotle's success by the degree to which he anticipated modern science (as though his goal was to answer our questions, rather than his own); it is nonetheless worth noting that the emphasis on functional explanation to which Aristotle's teleology leads would prove to be of profound significance for all of the sciences and remains to this day a dominant mode of explanation within the biological sciences.

COSMOLOGY

Aristotle not only devised methods and principles by which to investigate and understand the world: form and matter, nature, potentiality and actuality, and the four causes. In the process, he also developed detailed and influential theories regarding an enormous range of natural phenomena, from the heavens above to the earth and its inhabitants below.[11]

Let us start with the question of origins. Aristotle adamantly denied the possibility of a beginning, insisting that the universe must be eternal. The alternative—that the universe came into being at some point in time—he regarded as unthinkable, violating (among other things) Parmenidean strictures about something coming from nothing. Aristotle's position on this question would prove troublesome for his medieval commentators.

Aristotle considered this eternal universe to be a great sphere, divided into an upper and a lower region by the spherical shell in which the moon is situated. Above the moon is the celestial region; below is the terrestrial region; the moon, spatially intermediate, is also of intermediate nature.

The terrestrial or sublunar region is characterized by birth, death, and transient change of all kinds; the celestial or supralunar region, by contrast, is a region of eternally unchanging cycles. That this scheme had its origin in observation would seem clear enough; in his *On the Heavens,* Aristotle noted that "in the whole range of time past, so far as our inherited records reach, no change appears to have taken place either in the whole scheme of the outermost heaven or in any of its proper parts."[12] If in the heavens we observe eternally unvarying circular motion, he continued, we can infer that the heavens are not made of the terrestrial elements, the nature of which (observation reveals) is to rise or fall in transient rectilinear motions. The heavens must consist of an incorruptible fifth element (there are four terrestrial elements): the quintessence (literally, the fifth essence) or aether. The celestial region is completely filled with aether (no void space) and divided, as we shall see, into concentric spherical shells bearing the planets. It had, for Aristotle, a superior, quasi-divine status.[13]

The sublunar region is the scene of generation, corruption, and impermanence. Aristotle, like his predecessors, inquired into the basic element or elements to which the multitude of substances found in the terrestrial region can be reduced. He accepted the four elements originally proposed by Empedocles and subsequently adopted by Plato—earth, water, air, and fire. He agreed with Plato that these elements are in fact reducible to something even more fundamental; but he did not share Plato's mathematical inclination and therefore refused to accept Plato's regular solids and their constituent triangles. Instead, he expressed his own commitment to the reality of the world of sense experience by choosing *sensible qualities* as the ultimate building blocks. Two pairs of qualities are crucial: hot-cold and wet-dry. These combine in four pairs, each of which gives rise to one of the elements (see fig. 3.2):

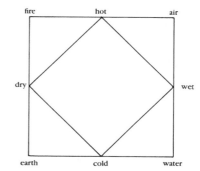

cold and dry = earth
cold and wet = water
hot and wet = air
hot and dry = fire

Fig. 3.2. Square of opposition of the Aristotelian elements and qualities. For a medieval (9th-century) version of this diagram, see John E. Murdoch, *Album of Science: Antiquity and the Middle Ages,* p. 352.

Notice the use made once again of contraries. There is nothing to forbid any of the four qualities being replaced by its contrary (as the result of outside influence). If water is heated, so that the cold of water yields to hot, the water is transformed into air. Such a process easily explains changes of state (from solid to liquid to gas, and conversely), but also more general transmutation of one substance into another. On such a theory as this, alchemists could easily build.[14]

The various substances that make up the cosmos totally fill it, leaving absolutely no empty space. To appreciate Aristotle's view, we must lay aside our almost automatic inclination to think atomistically; we must conceive material things not as aggregates of tiny particles but as continuous wholes. If it is obvious that, say, a loaf of bread is composed of crumbs separated by small spaces, there is no reason not to suppose that those spaces are filled by some finer substance, such as air or water. And there is certainly no simple way of demonstrating, nor indeed any obvious reason for believing, that water and air are anything but continuous. Similar reasoning applied to the whole of the universe led Aristotle to the conclusion that the universe is full, a *plenum,* containing no void space.

Aristotle defended this conclusion with a variety of arguments, such as the following. There must always be a ratio between any two motions (measured by the times required to traverse a given space); if this difference in time results from the difference in density between two media, the ratio of times will equal the ratio of densities. However, if one of the media were a void space, its density (zero) would have no ratio to the density of the other medium, and therefore the one time would have no ratio to the other, thereby violating the assumption with which the argument begins. Today we might make the same point by arguing that if resistance is what checks the speed of a moving body, then in the absence of resistance the body would move with infinite speed—a nonsensical notion. Critics have frequently noted that this argument can just as well be taken to prove that the absence of resistance does not entail infinite speed as to prove that void does not exist. The point is, of course, well taken. However, we need to understand that Aristotle's denial of the void did not rest on this single piece of reasoning. In fact, this was but one small part of a lengthy campaign against the atomists, in which Aristotle battled the notion of void space (or void place) with a variety of arguments, some more and some less persuasive.[15]

In addition to being hot or cold and wet or dry, each of the elements is also heavy or light. Earth and water are heavy, but earth is the heavier of the two. Air and fire are light, fire being the lighter of the two. In assigning

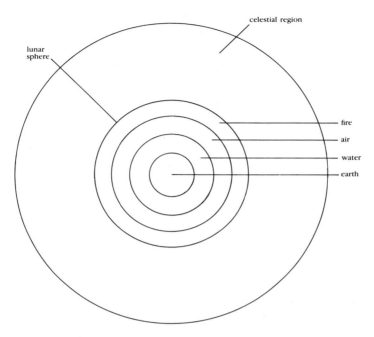

Fig. 3.3. The Aristotelian cosmos.

levity to two of the elements, Aristotle did not mean (as we might, if we were making the claim) simply that they are less heavy, but that they are light in an absolute sense; levity is not a weaker version of gravity, but its contrary. Because earth and water are heavy, it is their nature to descend toward the center of the universe; because air and fire are light, it is their nature to ascend toward the periphery (that is, the periphery of the terrestrial region, the spherical shell that contains the moon). If unhindered, therefore, earth and water would descend toward the center; because of its greater heaviness, earth would collect at the center, with water in a concentric spherical shell outside it. Air and fire ascend, but fire, owing to its greater levity, occupies the outermost region, with air as a concentric sphere just inside it. In the ideal case (in which there are no mixed bodies and nothing prevents the natures of the four elements from fulfilling themselves), the elements would form a set of concentric spheres: fire on the outside, followed by air and water, and finally earth at the center (see fig. 3.3). But in reality, the world is composed largely of mixed bodies, one always interfering with another, and the ideal is never attained. Nonetheless, the ideal arrangement defines the natural place of each of the elements; the natural place of earth is at the center of the universe, of fire just inside the sphere of the moon, and so forth.[16]

It must be emphasized that the arrangement of the elements is spherical. Earth collects at the center to form *the earth,* and it too is spherical. Aristotle defended this belief with a variety of arguments. Arguing from his natural philosophy, he pointed out that since the natural tendency of earth is to move toward the center of the universe, it must arrange itself symmetrically about that point. But he also called attention to observational evidence, including the circular shadow cast by the earth during a lunar eclipse and the fact that north-south motion by an observer on the surface of the earth alters the apparent position of the stars. Aristotle even reported an estimate by mathematicians of the earth's circumference (400,000 stades = about 45,000 miles, roughly 1.8 times the modern value). The sphericity of the earth, thus defended by Aristotle, would never be forgotten or seriously questioned. The widespread myth that medieval people believed in a flat earth is of modern origin.[17]

Finally, we must note one of the implications of this cosmology—namely, that space, instead of being a neutral, homogeneous backdrop (analogous to our modern notion of geometrical space) against which events occur, has properties. Or to express the point more precisely, ours is a world of *space,* whereas Aristotle's was a world of *place.* Heavy bodies move toward their place at the center of the universe not because of a tendency to unite with other heavy bodies located there, but simply because it is their nature to seek that central point; if by some miracle the center happened to be vacant (a physical impossibility in an Aristotelian universe, but an interesting imaginary state of affairs), it would remain the destination of heavy bodies.[18]

MOTION, TERRESTRIAL AND CELESTIAL

The best way of approaching Aristotle's theory of motion is through its two most basic principles. The first is that motion is never spontaneous: there is no motion without a mover. The second is the distinction between two types of motion: motion toward the natural place of the moving body is "natural motion"; motion in any other direction is "forced or violent motion."

The mover in the case of natural motion is the nature of the body, which is responsible for its tendency to move toward its natural place as defined by the ideal spherical arrangement of the elements. Mixed bodies have a directional tendency that depends on the proportion of the various elements in their composition. When a body undergoing natural motion reaches its natural place, its motion ceases. The mover in the case of forced motion is an external force, which compels the body to violate its natural tendency and move in some direction other than toward its natural place. Such motion ceases when the external force is withdrawn.[19]

So far, this seems sensible. One obvious difficulty, however, is to explain why a projectile hurled horizontally, and therefore undergoing forced motion, does not come to an immediate halt when it loses contact with whatever propelled it. Aristotle's answer was that the medium takes over as mover. When we project an object, we also act on the surrounding medium (air, for instance), imparting to it the power to move objects; this power is communicated from part to part, in such a way that the projectile is always in contact with a portion of the medium capable of keeping it in motion. If this seems implausible, consider the greater implausibility (from Aristotle's standpoint) of the alternative—that a projectile, which is inclined by nature to move toward the center of the universe, moves horizontally or upward despite the fact that there is no longer anything causing it to do so.

Force is not the only determinant of motion. In all real cases of motion in the terrestrial realm, there will also be a resistance or opposing force. And it seemed clear to Aristotle that the quickness of motion must depend on these two determining factors—the motive force and the resistance. The question arose: what is the relationship between force, resistance, and speed or quickness? Although it probably did not occur to Aristotle that there might be a quantitative law of universal applicability, he was not without interest in the question and did make several forays into quantitative territory. In reference to natural motion in his *On the Heavens* and again in his *Physics*, Aristotle claimed that when two bodies of differing weight descend, the times required to cover a given distance will be inversely proportional to the weights (a body twice as heavy will require half the time). In the same chapter of the *Physics*, Aristotle introduced resistance into the analysis of natural motion, arguing that if bodies of equal weight move through media of different densities, the times required to traverse a given distance are proportional to the densities of the respective media; that is, the greater the resistance the slower the body moves. Finally, Aristotle also dealt with forced motion in his *Physics*, claiming that if a given force moves a given weight (against its nature) for a given distance in a given time, the same force will move half that weight twice the distance in that same time (or the same distance in half that time); alternatively, half the force will move half the weight the same distance in the same time.[20]

From such statements, some of Aristotle's successors have made a determined effort to extract a general law. This law is customarily stated as:

$$v \propto F/R$$

That is, velocity (v) is proportional to the motive force (F) and inversely proportional to the resistance (R). For the special case of the natural descent of a heavy body, the motive force is the weight (W) of the body; the relationship then becomes:

$$v \propto W/R$$

Such relationships probably do no great violence to Aristotle's intent for most cases of motion; however, giving them mathematical form, as we have done, suggests that they hold for all values of v, F, and R—a claim Aristotle would certainly have denied. He stated explicitly, for example, that a resistance equal to the motive force will prevent motion altogether, whereas the formula above offers no such result. Moreover, the appearance of velocity in these relationships seriously misrepresents Aristotle's conceptual framework, which contained no concept of velocity as a quantifiable measure of motion, but described motion only in terms of distances and times. Velocity as a technical scientific term to which numerical values might be assigned was a contribution of the Middle Ages.

Aristotle has been severely criticized for this theory of motion, the assumption being that any sensible person should have recognized that it was fatally flawed. Is such criticism justified? In the first place, few historians consider the assignment of credit and blame to be their primary mission; understanding the past seems a far more useful goal. Second, some of the criticisms apply only to the theory foisted onto Aristotle by followers and critics, not to his own. Third, the theory in its genuinely Aristotelian version makes quite good sense; for example, various surveys have shown the majority of modern, university-educated people to be prepared to assent to many of the basics of Aristotle's theory of motion. Fourth, the relatively modest level of quantitative content in Aristotle's theory is easily explained as the outcome of his larger philosophy of nature. His primary goal was to understand essential natures, not to explore quantitative relationships between such incidental factors as the space-time (or place-time) coordinates applicable to a moving body; even an exhaustive investigation of the latter gives us no useful information about the former. (One of the important features of modern mechanics is precisely its determination to treat all bodies identically, refusing to acknowledge differences in essential nature: whatever the body is made of, the same laws apply and the same behaviors result.) We may criticize Aristotle, if we like, for not being interested in whatever interests us, but we do not thereby learn anything significant about Aristotle.

Motion in the celestial sphere is an altogether different sort of phenomenon. The heavens are composed of the quintessence, an incorruptible substance, possessing no contraries and therefore incapable of qualitative change. It might seem fitting for such a region to be absolutely motionless, but this hypothesis is defeated by the most casual observation of the heavens. Aristotle therefore assigned to the heavens the most perfect of motions— continuous uniform circular motion. Besides being the most perfect of motions, uniform circular motion appears to have the capability of explaining the observed celestial cycles.

By Aristotle's day, these cycles had been an object of study for centuries. It was understood that the "fixed" stars move with perfect uniformity, as though fixed to a uniformly rotating sphere, with a period of rotation of approximately one day. But there were seven stars, the wandering stars or planets, that displayed a more intricate motion; these seven were the sun, moon, Mercury, Venus, Mars, Jupiter, and Saturn. The sun moves slowly (about 1°/day), west to east, with small variations in speed, through the sphere of fixed stars along a path called the ecliptic, which passes through the center of the zodiac (see fig. 2.6). The moon follows approximately the same course, but more rapidly (about 12°/day). The remaining planets also move along the ecliptic, with variable speed and with an occasional change of direction.

Are such complex motions compatible with the requirement of uniform circular motion in the heavens? Eudoxus, a generation before Aristotle, had already shown that they are. I will return to this subject in chapter 5, below; for the moment, it will be sufficient to point out that Eudoxus treated each complex planetary motion as a composite of a series of simple uniform circular movements. He did this by assigning to each planet a series of concentric spheres, and to each sphere one component of the complex planetary motion. Aristotle took over this Eudoxan scheme, with various modifications. When he was finished, he had produced an intricate piece of celestial machinery, consisting of fifty-five planetary spheres plus the sphere of the fixed stars.

What is the cause of movement in the heavens? Aristotle's natural philosophy would not allow such a question to go unasked. The celestial spheres are composed, of course, of the quintessence; their motion, being eternal, must be natural rather than forced. The cause of this eternal motion must itself be unmoved, for if we do not postulate an unmoved mover, we quickly find ourselves trapped in an infinite regress: a moving mover must have acquired its motion from yet another moving mover, and so on. Aristotle identified the unmoved mover for the planetary spheres as the "Prime Mover," a living deity representing the highest good, wholly actualized, totally absorbed in self-contemplation, nonspatial, separated from the spheres it moves, and not at all like the traditional anthropomorphic Greek gods. How, then, does the Prime Mover or Unmoved Mover cause motion in the heavens? Not as efficient cause, for that would require contact between the mover and the moved, but as final cause. That is, the Prime Mover is the object of desire for the celestial spheres, which endeavor to imitate its changeless perfection by assuming eternal, uniform circular motions. Now any reader who has followed this much of Aristotle's discussion would be justified in assuming that there is a single Unmoved Mover for the entire cosmos; it comes as something of a surprise, therefore, when Aristotle announces that, in fact, each of the celestial spheres has its own Unmoved Mover, the object of its affection and the final cause of its motion.[21]

ARISTOTLE AS A BIOLOGIST

There is no way of determining when or how Aristotle became interested in the biological sciences. That his father was a physician is a factor that we must surely take into account. Aristotle's biological studies no doubt occurred over an extended period, but several years on the island of Lesbos (off the coast of Asia Minor) offered him an exceptional opportunity for the observation of marine life. He was probably assisted in the gathering of biological data by his students, and he certainly relied on the reports of other observers, including physicians, fishermen, and farmers. The product of this research effort was a series of large zoological treatises and short works on human physiology and psychology that occupy well over 400 pages in a modern translation; these works laid the foundations of systematic zoology and have profoundly shaped thought on human biology for some two thousand years.[22]

In Aristotle's day human anatomy and physiology had long attracted attention for their medical import and presumably required no further justification, but Aristotle felt obliged to defend zoological research. In *On the Parts of Animals* he admitted that animals are ignoble by comparison with the heavens and acknowledged that zoological studies are distasteful to many. However, he considered this distaste to be childish, and he argued that in zoological studies the quantity and richness of the available data compensate for the ignobility of the object of study. He argued, moreover, that zoological studies contribute to knowledge of the human frame owing to the close resemblance between animal and human nature; he noted the pleasure of discovering causes in the zoological realm; and he pointed out that order and purpose are displayed with particular clarity in the animal kingdom, providing us with a golden opportunity to refute the notion that the "works of nature" are products of chance alone.[23]

Aristotle saw that biology has both a descriptive and an explanatory side. He considered the explanation of biological phenomena the ultimate goal, but acknowledged the gathering of biological data as the first order of business. His *History of Animals,* which was intended to meet this first need, is a vast storehouse of biological information. Aristotle began with the human body, as a standard by comparison with which other animals could be understood. He subdivided the human body into head, neck, thorax, arms, and legs; and he proceeded to a discussion of both internal and external features, including brain, digestive system, sexual organs, lungs, heart, and blood vessels.

However, Aristotle made his greatest contribution not in the area of human anatomy but in descriptive zoology. More than 500 species of animals are mentioned in his *History of Animals;* the structure and behavior of many are described in considerable detail, often on the basis of skillful dissection. Although he devoted considerable attention to the theoretical problems of classification, in practice Aristotle adopted "natural" or popular groupings based on multiple attributes. He divided animals into two major categories—"blooded" (that is, red-blooded) and "bloodless." The former category he subdivided into viviparous quadrupeds (four-footed mammals that bring forth living young), oviparous (or egg-laying) quadrupeds, marine mammals, birds, and fish; the latter into mollusks (such as the octopus and cuttlefish), crustacea (including crabs and crayfish), testacea (including the snail and oyster), and insects. These major categories Aristotle arranged hierarchically in a scale of being according to degree of vital heat.[24]

Although he ranged over the whole of the animal kingdom, Aristotle was no doubt most at home when it came to marine life, of which he exhibited intimate firsthand knowledge. It has often been noted, for example, that he described the placenta of the dogfish (*Mustelus laevis*) in terms that were not confirmed until the nineteenth century. But Aristotle displayed impressive skill in other parts of the animal kingdom as well. His description of the incubation of birds' eggs is an excellent example of meticulous observation:

> Generation from the egg proceeds in an identical manner with all birds, but the full periods from conception to birth differ. . . . With the common hen after three days and three nights there is the first indication of the embryo. . . . Meanwhile the yolk comes into being, rising towards the sharp end, where the primal element of the egg is situated, and where the egg gets hatched; and the heart appears, like a speck of blood, in the white of the egg. This point beats and moves as though endowed with life, and from it . . . two vein-ducts with blood in them trend in a convoluted course . . . ; and a membrane carrying bloody fibres now envelops the white, leading off from the vein-ducts. A little afterwards the body is differentiated, at first very small and white. The head is clearly distinguished, and in it the eyes, swollen out to a great extent. . . .[25]

Natural history, which enumerates and describes the population of the universe, is no doubt an appealing occupation and may be regarded by some as an end in itself. But for Aristotle it was a means to a higher end—the source of factual data that would lead to physiological understanding and causal explanations. And for him, true knowledge was always causal knowledge.

Aristotle applied to the understanding of physiology the same principles that functioned in other realms of his natural philosophy. (Whether they were first developed in the biological realm and then applied to metaphysics, physics, and cosmology, or vice versa, is a matter of dispute among scholars.)[26] Thus form and matter, actuality and potentiality, the four causes, and especially the element of purpose or function associated with final cause are central to his biology. The ingredients of a proper biological explanation are nicely summarized in Aristotle's *On the Generation of Animals:* "Everything that comes into being or is made must [1] be made out of something, [2] be made by the agency of something, and [3] must become something."[27] That out of which an organism is made is, of course, its material cause; the agency by which it is made is its formal or efficient cause (which often merge in Aristotle's biology); and that which it becomes, the goal of its development, is its final cause.

Each organism, then, is constituted of matter and form: the matter consists of the various organs that make up the body; the form is the organizing principle that molds these organs into a unified organic whole. Aristotle identified form with soul and assigned it responsibility for the vital characteristics of living things—nutrition, reproduction, growth, sensation, movement, and so forth. Indeed, Aristotle arranged living things in a hierarchy on the basis of their participation in several kinds of soul, each of which performs certain functions. Plants possess a nutritive soul, which enables them to obtain nourishment, grow, and reproduce. Animals possess, in addition, a sensitive soul, which accounts for sensation and (indirectly) for movement. Finally, humans add to these a rational soul, which supplies the higher capacities of reason. If, as Aristotle maintained, soul is but the form of the organism, then it is clear that soul (including the human soul) is not immortal; at death the organism disintegrates, and its form evaporates into nonbeing.[28]

How is soul, the form of the living organism, communicated from parent to offspring? This brings us to one of the central questions of Aristotle's physiology—the problem of organic generation. First, Aristotle argued that the existence of two genders—male and female—reflects the distinction between formal or efficient cause (here merged) and the matter on which this cause works. In humans and higher animals the female supplies the matter as menstrual blood. Male semen bears the form and impresses this on the menstrual blood to produce a new organism. The young in higher animals, which have a large measure of vital heat, are brought forth live, as fully developed members of the species; in animals somewhat deficient in vital heat, the offspring are eggs hatched internally; as we descend the scale of perfection, we come to animals that produce eggs hatched externally, the eggs being more or less perfect depending on the exact degree of heat; at the bottom of the scale, bloodless animals produce a grub or maggot:

> We must observe how rightly Nature orders generation in
> regular gradation. The more perfect and hotter animals
> produce their young perfect in respect of quality . . . , and
> these generate living animals within themselves from the first.
> The second class do not generate perfect animals within
> themselves from the first (for they are only viviparous after
> first laying eggs). . . . The third class do not produce a perfect
> animal, but an egg, and this egg is perfect. Those whose na-
> ture is still colder than these produce an egg, but an im-
> perfect one, which is perfected outside the body. . . . The
> fifth and coldest class does not even lay an egg from itself; but

so far as the young ever attain to this condition at all, it is
outside the body of the parent. . . . For insects produce a grub
first; the grub after developing becomes egg-like. . . .[29]

The idea of perfection, so prominent in Aristotle's theory of generation,
brings us to the third and last element of biological explanation—final
cause or, as Aristotle put it in a passage quoted above, that which a biological organism is in the process of becoming. The biologist, in Aristotle's
view, always needs to know the complete, mature form or nature of an
organism. Only such knowledge will enable him to understand the structure of the organism and the existence and interrelations of its parts. For
example, Aristotle explained the presence of lungs in land animals by reference to the needs of the organism as a whole. Blooded animals, he
argued, require an external cooling agent because of their warmth. In
fish, this agent is water, and consequently fish have gills instead of lungs.
Animals that breathe, however, are cooled by air and consequently come
equipped with lungs.[30] Knowledge of the mature form is also part of the
explanation of the organism's development, for there is an upward movement in the organic realm, as organisms strive to actualize the potentialities that exist within them. We cannot understand the changes that
occur within an acorn, for example, if we do not understand the oak tree
that is its final destination. Finally, purpose and function enter Aristotle's
biology not merely as an explanation of the form or development of the
individual or species, but on a universal or cosmic level, to explain the
interdependence and interrelationships of species in the order of nature.

There is, of course, much more to Aristotle's biological system. He explained nutrition, growth, locomotion, and sensation. He considered the
functions of the principal organs, including brain, heart, lungs, liver, and
reproductive organs. It is important to note that he made the heart the central organ of the body, the seat of emotion and sensation as well as of vital
heat. He developed the notion of hierarchy in the biological realm: form,
he believed, is superior to matter, living to nonliving, male to female,
blooded to bloodless, the mature to the immature. Indeed, he arranged
living things in a single, hierarchical scale of being, beginning with the
Prime Mover at the top and descending through the human race to viviparous, oviparous, and vermiparous animals, and finally to plants.

Let us conclude this discussion with a brief analysis of method in Aristotle's biological works. If there is any branch of the scientific enterprise
that demands observation, surely it is biology (and especially natural history). It is inconceivable that Aristotle would have attempted to describe

the structure and habits of animals on any other basis. The observation in question was frequently his own, and we find in his works plentiful evidence of empirical method, including dissection. However, no naturalist working alone could amass the quantity of data contained in Aristotle's biological works, and it is apparent that he relied on the reports of travelers, farmers, and fishermen, the help of assistants, and the writings of his predecessors. Aristotle was generally critical of his sources, and displayed a healthy skepticism even about his own observations. However, he was not always skeptical enough, and there are many examples of descriptive error in his biological works. When it came to biological theory, Aristotle (like any theorist) was obliged to make inferences from the observational data; if his inferences were not always the ones we would make, they nonetheless display the insight of one of the most brilliant biologists ever to live. They also, of course, display the powerful influence of Aristotle's larger philosophical system, which continually influenced the questions he asked, the details he noticed, and the theoretical interpretation he placed upon them.[31]

ARISTOTLE'S ACHIEVEMENT

The proper measure of a philosophical system is not the degree to which it anticipated modern thought, but its degree of success in treating the philosophical problems of its own day. If a comparison is to be made, it must be between Aristotle and his predecessors, not Aristotle and the present. Judged by such criteria, Aristotle's philosophy is an astonishing achievement. In natural philosophy, he offered a subtle and sophisticated treatment of the major problems posed by the pre-Socratics and Plato: the nature of the fundamental stuff, the proper means of knowing it, the problems of change and causation, the basic structure of the cosmos, and the nature of deity and its relationship to material things.

But Aristotle also went far beyond any predecessor in the analysis of specific natural phenomena. It is no exaggeration to claim that, almost single-handedly, he created entirely new disciplines. His *Physics* contains a detailed discussion of terrestrial dynamics. He devoted the better part of his *Meteorology* to phenomena of the upper atmosphere, including comets, shooting stars, rain and the rainbow, thunder, and lightning. His *On the Heavens* developed the work of certain predecessors into an influential account of planetary astronomy. He touched upon geological phenomena, including earthquakes and mineralogy. He undertook a thorough analysis of sensation and the sense organs, particularly vision and the eye, developing a theory of light and vision that would remain influential until the seventeenth century. He concerned himself with what we might regard as the basic chemical processes—mixtures and combinations of substances. He wrote a book on the soul and its faculties. And, as we have seen, he contributed monumentally to developments in the biological sciences.

We will consider Aristotle's influence in subsequent chapters. Let us conclude here simply by stating that his powerful influence in late antiquity and his dominance from the thirteenth century through the Renaissance resulted not from intellectual subservience on the part of scholars during those periods or from interference on the part of the church, but from the overwhelming explanatory power of his philosophical and scientific system. Aristotle prevailed through persuasion, not coercion.

FOUR

Hellenistic Natural Philosophy

Aristotle's death in 322 B.C. nearly coincided with the end of the military campaigns of Alexander the Great (334–323 B.C.), which established a far-flung Greek empire and sounded the death knell for the autonomous Greek city-states. Alexander dramatically enlarged Greek territory, carrying Greek language and culture as far east as Bactria (now part of Afghanistan) and the Indus River and as far south as Egypt (see map 2). However, Alexander and his successors also borrowed from the conquered peoples, creating a synthesis of Greek and foreign elements designated by the adjective "Hellenistic"—meaning "Greekish." Although Greek elements were overwhelmingly dominant, the historians who coined this term wished to distinguish the Hellenistic period from what they regarded as the unadulterated Greek culture of the older, "Hellenic" times. The phrase "Hellenistic natural philosophy," then, denotes thought about nature among scholars and educated people throughout this Greek empire. In the short run, the center of gravity remained in the traditional Greek territories; in the long run leadership shifted southward to Alexandria in Egypt and westward to Rome.

SCHOOLS AND EDUCATION

Before we examine the content of Hellenistic natural philosophy, we need to examine its social underpinnings—the social and institutional mechanisms by which learning in general and natural philosophy in particular were transmitted. Knowledge can, of course, be transmitted individually, from parent to child, friend to friend, or master to apprentice. But as that knowledge increases in complexity and sophistication, pressure for a more formalized, collective educational system is likely to grow. Did this occur in ancient Greece? If so, what was the nature of the educational system that resulted?[1]

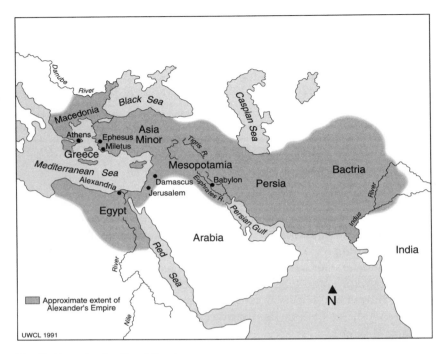

Map 2. Alexander the Great's Empire

No formal education was required in any ancient society, but several years of instruction at the elementary level became an ideal among the early Greek aristocracy. Because it was intended for pre-adolescent children (*paides*), this education was referred to as *paideia*. Traditionally, *paideia* consisted of two parts: *gymnastikē* for the body and *mousikē* for the mind or spirit. *Gymnastikē* included physical culture and athletics. *Mousikē* covered all of the arts presided over by the muses, especially music and poetry. However, social needs eventually outran this bipartite system, and by the beginning of the fifth century B.C. there were also schools for reading and writing.

Instruction in *gymnastikē* took place most frequently in a sports ground or wrestling school, or possibly in a public gymnasium. *Mousikē* and literary education could be carried on almost anywhere, including a public building or the teacher's home. It must be understood that there was nothing resembling modern, compulsory, mass education. Teachers entered the educational enterprise privately, on their own initiative; and the aristocracy took advantage of teachers' services according to individual need and inclination.

A major change in this educational pattern occurred in the fifth century B.C., with the coming of the sophists. Education to this point had been strictly elementary, largely athletic and artistic in its orientation. About the middle of the fifth century, itinerant teachers known as sophists made their appearance in Athens, offering something new. First, they offered education at a more advanced level. Second, their goal was the training of citizens and statesmen, and this called for a shift in the content of education toward intellectual, and especially political, matters. The sophists offered what we would regard as group tutorials, with no fixed curriculum or universal pattern, and certainly no shared philosophical system, for a duration negotiated by the parties concerned. (A figure of three or four years is often suggested by historians, but it has recently been argued that in some cases the duration of instruction may have been "as little as a week or an hour.")[2] In order to attract business, sophist teachers needed to be visible, and it therefore became customary for them to teach in a public place, such as the agora (public marketplace) or a large public gymnasium (of which Athens at the time had three). When business dried up or the teacher wore out his welcome, he moved on.

Against this background, we can begin to understand the teaching of Socrates and Plato. Socrates and Plato doubtless differed from the sophists in various respects—they were not itinerant but remained in Athens, and they departed from sophistic methods of instruction—but these distinctions were probably lost on contemporary Athenians, who must have viewed both men as typical representatives of the sophistic movement. When Plato returned to Athens in 388 after his travels in Italy, he established a school in the Academy, a monumental public gymnasium just outside the city walls, which had long been used for educational purposes. If there was anything unusual about this venture it was that Plato's school acquired permanency, so that it endured long beyond his death.[3]

Plato's school was a philosophical community, consisting of scholars who had reached various levels of maturity and attainment and who interacted as equals. Plato was no doubt the dominant force, inspiring his colleagues by example and assisting less advanced scholars by his critical powers; but he was not above criticism, and (like the teacher in a modern graduate seminar) he may have learned as much as he taught.[4] There was undoubtedly a religious undercurrent to the enterprise; the Academy was devoted to the service of the muses, and there may have been what we would take to be religious ceremonies. However, doctrinal orthodoxy was certainly not required, and the school was (at least in principle) open to students of any persuasion. No fees were charged, and a scholar could participate in the activities of the Academy until he tired of them or his means

Fig. 4.1. The Parthenon (a temple to Athena) on the Acropolis, Athens. Built fifth century B.C.

Fig. 4.2. The schools of Hellenistic Athens. © Candace H. Smith. First published in A. A. Long and D. N. Sedley, *The Hellenistic Philosophers,* vol. 1.

of support ran out. At some point, Plato purchased a plot of land close to the Academy, which could be used for some of its activities. The possession of private property, along with Plato's provision for the selection of a successor, doubtless contributed to the longevity of the school.

Aristotle was a member of Plato's school for twenty years, until Plato's death in 348 or 347. When he returned to Athens in 335, following the imposition of Macedonian rule, Aristotle did not resume membership in the Academy, as he could perfectly well have done, but founded a rival school in another Athenian gymnasium, the Lyceum. The Lyceum, like the Academy, had long been the scene of educational activity. Aristotle and his followers were accustomed to gather in a colonnaded walk (*peripatos*) in the Lyceum and thereby acquired (or assumed) the designation "peripatetic," by which they have since been known. Aristotle's Lyceum and Plato's Academy were alike in many respects, but they differed in method and emphasis. Methodologically, Aristotle inaugurated the practice of cooperative research, exhibited in his natural history and also in the systematic collection of earlier philosophical literature. As for emphasis, Aristotle's biological interests stand in rather strong contrast to Plato's mathematical ones; and there is the obvious divergence between Platonic and Aristotelian metaphysics.[5]

Athens had by this time acquired educational leadership within the Greek world, and other teachers soon arrived to take advantage of the opportunities. Zeno of Citium arrived in Athens about 312 and subsequently began to teach in the *stoa poikilē* (painted colonnade) in a corner of the Athenian agora, thus founding a school of what came to be called "Stoic" philosophy. Epicurus, an Athenian citizen born on the island of Samos, returned to Athens about 307, purchased a house and garden, and there founded a school of "Epicurean" philosophy that survived into the Christian era.

The Academy, the Lyceum, the Stoa, and the Garden of Epicurus—the four most prominent schools in Athens—all developed institutional identities that enabled them to survive their founders. The Academy and the Lyceum seem to have had continuous existence until the beginning of the first century B.C. (perhaps until the sack of Athens by the Roman general Sulla in 86 B.C.). It is often claimed that the Academy survived until closed by the Emperor Justinian in 529 A.D. The truth seems to be that Neoplatonists *refounded* the Academy in the fifth century A.D. and managed to keep it alive until about 560 or later; however, there was no institutional continuity between this and Plato's school. The Stoa survived into the second century A.D., and the Epicurean school into the following century.[6]

Meanwhile the Athenian model had been exported to other parts of the Greek world, particularly Alexandria (in Egypt). At the death of Alexander the Great, his generals divided his empire, Egypt and Palestine falling to Ptolemy. Alexandria became Ptolemy's capital, and through his patronage and that of his successors it grew in size and magnificence and soon achieved a position of educational superiority. When Demetrius Phaleron, formerly a member of Aristotle's Lyceum, was overthrown as Dictator of Athens in 307, Ptolemy invited him to Alexandria, where he probably influenced his patron's decision to found the Museum—not a building where artifacts could be displayed, but a temple to the muses and therefore simultaneously a religious shrine and a place of learning. Connections between the Museum and the Lyceum are further illustrated by the fact that Strato, the third head of the Lyceum, spent a period at the Ptolemaic court, tutoring the royal offspring. The Museum seems to have consisted of certain buildings in the royal quarter and (since it was a temple) to have been presided over by a priest. With its associated Library (containing, by one ancient estimate, almost half a million rolls) and the generous patronage of the Ptolemaic kings, and in view of the eventual decline of Athenian schools, it became the major research institution of the Hellenistic period—one of the primary links between early Greek thought and the Roman and medieval periods.[7]

The establishment of the Museum in Alexandria is of importance not only because of the significance of the research carried out there, but also because it is the first instance of the support of advanced learning through public or royal patronage. This pattern was extended in the period 140–80 A.D. by the Roman emperors Antoninus Pius and Marcus Aurelius, who endowed imperial chairs for teachers of rhetoric and philosophy in Athens and elsewhere. Marcus Aurelius saw to the establishment in Athens of chairs for each of the major philosophical traditions—Platonic, Peripatetic, Stoic, and Epicurean—a model quickly imitated elsewhere in the Greek world. In the long run, this pattern exerted a strong influence on Roman and Christian educational practice.

THE LYCEUM AFTER ARISTOTLE

Aristotle made the acquaintance of Theophrastus (ca. 371–ca. 286) during his travels in Asia Minor, probably during his stay on the island of Lesbos (Theophrastus' birthplace) in the 340s. They became close associates, and when Aristotle returned to Athens in 335, Theophrastus joined him and participated for the next thirteen years in the activities of the Lyceum.

Upon Aristotle's death, Theophrastus assumed the headship of the Lyceum, a position that he held for thirty-six years.

Theophrastus appears to have shared Aristotle's general philosophical outlook, his methodological commitments, and his range of interests. Theophrastus continued to teach and to carry out the collaborative research projects in natural history and the history of philosophy begun during Aristotle's lifetime. He collected the opinions of the pre-Socratic philosophers in a book, which gave rise to what we now call the "doxographic" (or "opinion") tradition—a series of handbooks in which philosophical opinion on a variety of topics was collected and preserved. Most of Theophrastus' writings are now lost, but among the surviving treatises are two botanical works and a treatise on minerals that reveal a high level of commitment to the Aristotelian research program. Like Aristotle's zoological works, the botanical works contain meticulous descriptions of plant life (more than 500 varieties are mentioned), thoughtful attempts at classification, and intelligent physiological theorizing. Theophrastus accepted many of Aristotle's explanatory principles (association of life with vital heat, for example) and stressed the necessity of employing a rigorous empirical methodology. In his work *On Stones,* he followed Aristotle in dividing minerals into metals (in which the element water predominates) and "earths" (in which the element earth predominates). He proceeded to a systematic description of a wide variety of rocks and minerals.

While carrying out Aristotle's research program, Theophrastus was also prepared to question and disagree with aspects of Aristotelian natural philosophy. Three examples will serve to illustrate. Theophrastus expressed reservations about Aristotle's teleology, pointing out that not all features of the universe serve any identifiable purpose and that there is a substantial random element in the world. He reconsidered Aristotle's theory of the four elements and called into question the status of fire as an element. And he disagreed with Aristotle on light and vision, questioning Aristotle's opinion that light is the actualization of the transparency of the medium and expressing the view that the eyes of animals contain a kind of fire, the emission of which explains nocturnal vision.[8]

An achievement of quite a different sort was Theophrastus' acquisition of property for the Lyceum. Though not an Athenian citizen, Theophrastus received special dispensation to purchase a parcel of land close to the gymnasium; there, in several buildings, the school's library was presumably housed and work space provided. This real estate Theophrastus bequeathed to his scholarly colleagues in his will: "I give the garden, the *peripatos,* and all the houses along the garden to those of my friends, named herein, who wish continually to practice education and philosophy together in them . . . ; my condition is that no one alienate the property or devote it to private use but that all should hold it in common as if it were a sanctuary."⁹

The library of the peripatetic school experienced a more complicated fate. In his will Theophrastus bestowed the library (which contained not only his own books but also Aristotle's) on Neleus, whom he may have intended to succeed him. When the senior members of the community chose Strato instead, Neleus returned home to Skepsis in Asia Minor, taking the books (or many of them, at least) with him, thereby depriving the Lyceum of vital resources. This library survived more or less intact until early in the first century B.C., when (according to the historian Strabo—not to be confused with Strato) it was purchased from Neleus's heirs and returned to the peripatetic school in Athens. Shortly thereafter, Athens fell to Sulla, who shipped the books to Rome. There they came into the hands of Andronicus of Rhodes, who arranged and edited them, bringing them into wider circulation.¹⁰

Meanwhile Strato (from Lampsacus in Asia Minor) took over leadership of the Lyceum, a position that he held for eighteen years (286–268). Strato appears to have had interests almost as wide as those of Aristotle and Theophrastus. However, none of his works has survived intact, and we must be content with a fragmentary picture of his philosophical and scientific activity, reconstructed from scattered quotations and paraphrases in the works of later writers. It appears that Strato endeavored to correct and extend the work of Aristotle and Theophrastus on a variety of subjects. He certainly did not hesitate to question their views or to borrow from other philosophical traditions when there seemed good reason for doing so.

Strato's most notable contributions (as transmitted to us) were in reference to motion and the underlying structure of the physical world. Strato proposed a fundamental revision in Aristotle's theory of motion when he denied the distinction between heavy and light bodies, arguing that all bodies have weight in greater or lesser degree. Air and fire rise, then, not because they are absolutely light, but because they are displaced by heavier bodies. Strato also opposed Aristotle's theory of place and space. And he set forth observational evidence to demonstrate that heavy bodies accelerate as they descend (a feature of falling bodies not treated by Aristotle). Strato pointed to the fact that a stream of water falling from a height is continuous at the top, but discontinuous near the bottom—a fact to be explained by steadily increasing speed. In support of the same conclusion, he noted that the impact made by a falling body is a function not simply of its weight, but also of the height of its fall.¹¹

Although there is no question that Strato remained fundamentally Aristotelian in his view of the underlying structure of the corporeal world, it is also clear that he imported corpuscular notions into Peripatetic natural philosophy—possibly through the influence of Epicurus, who taught for a time in Strato's home town, Lampsacus, and also overlapped with Strato in Athens. Corpuscular ideas are most obvious in Strato's belief that light is a material emanation and that bodies are not continuous but contain interparticulate void spaces. Strato used the notion of void spaces to explain various properties of matter, including condensation, rarefaction, and elasticity. While acknowledging the existence of tiny void spaces distributed through matter, Strato denied the natural existence of continuous void space. We must be careful not to make Strato into a full-fledged atomist, for it appears that he retained a belief in the infinite divisibility of corporeal substance, thus rejecting the one absolutely essential feature of any atomistic philosophy—namely, belief in the existence of irreducible atoms.

Some of Strato's successors as head of the Lyceum are known by name, down to the end of the second century B.C. There can be no doubt that the school was the scene of regular lectures on peripatetic philosophy and continuing attempts to clarify Aristotle's philosophy and organize the materials that he had left behind. However, we have no record of fresh contributions to natural philosophy, nor of particularly sharp and telling criticism of traditional peripatetic philosophy, until after the Lyceum had ceased to function. Nonetheless, Aristotle's works continued to be known and commented upon, especially after Andronicus of Rhodes produced his new edition of the Aristotelian corpus. In the middle of the first century B.C. we find commentaries by Boethus of Sidon (a student of Andronicus) and Nicholas of Damascus (historian at the court of Herod the Great). About 200 A.D. Alexander of Aphrodisias was lecturing on peripatetic philosophy in Athens and writing important and influential commentaries on a variety of Aristotelian works. Finally, the Aristotelian commentaries of Simplicius and John Philoponus (both Neoplatonists) testify to the persistence of the Aristotelian tradition as late as the sixth century A.D. Renewed attention to this tradition in Islam and medieval Christendom would once again restore Aristotle's philosophy to a position of leadership.[12]

EPICUREANS AND STOICS

During the Hellenistic period, followers of Plato and Aristotle continued to discuss, clarify, and modify Platonic and Aristotelian philosophy. At the same time, alternative philosophical systems were developing, two of which became serious rivals. Both contained familiar elements, but both

Fig. 4.3. Epicurus, Museo Vaticano, Vatican City. Alinari/Art Resource N.Y.

were new in the prominence they gave to ethical questions. Indeed, what is striking about both is their determination to subordinate all other aspects of philosophy to ethical concerns.

The aim of philosophy, according to Epicurus (341–270 B.C.), is to secure happiness. "To say that the season for studying philosophy has not yet come, or that it is past and gone," Epicurus wrote to Menoeceus, "is like saying that the season for happiness is not yet or that it is now no more." The way to achieve happiness, Epicurus believed, was to eliminate fear of the unknown and the supernatural, and for this purpose natural philosophy was ideally suited. A maxim attributed to Epicurus reads as follows: "If we had never been molested by alarms at celestial and atmospheric phenomena, nor by the misgiving that death somehow affects us, nor by neglect of the proper limits of pains and desires, we should have had no need to study natural philosophy." Not only is natural philosophy a vehicle for the achievement of happiness; that is its sole function.[13]

Epicurus's natural philosophy borrowed many elements from ancient atomism. The universe was conceived to be eternal, consisting of an infinite void within which an infinity of atoms engage in perpetual motion, tossed about "as if in everlasting combat," like particles of dust in a beam of bright light. All things and all phenomena in our world (and in the infinity of other worlds that exist) are reducible to atoms and the void; the gods themselves must be of atomic composition. The sensible qualities of things (we now call them "secondary qualities"), such as flavor, color, and warmth, have no existence in the individual atom, the only genuine properties of which are shape, size, and weight. This is a passive, mechanistic world, in which everything is the result of mechanical causation (with one exception, to be mentioned below); there is no ruling mind, no divine providence, no destiny, no life after death. And there are no final causes: as Lucretius (d. ca. 55 B.C.) was to put it, in his account of Epicurean philosophy, "all the members [of the body] . . . existed before their use; they could not then have grown up for the sake of use."[14]

But Epicurus and his followers did more than propagate the philosophical system of the ancient atomists. They also had to adapt the atomic philosophy so that it could serve ethical functions. And they modified its content to resolve difficulties, meet objections, and generally increase its explanatory power. For example, Epicurus opposed the rationalism of Democritus, arguing that all sensation is fundamentally trustworthy.[15] From this it seemed to follow that sensible or secondary qualities possess reality at the macroscopic level, even though (as Democritus had argued) they do not exist in the atoms.

A more significant modification of the content of atomistic natural philosophy was Epicurus's doctrine of the swerve, designed not only to save atomistic cosmology from fatal objections, but also to eliminate the threat of determinism from Epicurean ethics. According to Epicurus, atoms possess shape and size (as Leucippus and Democritus had argued), but also weight. Their weight causes them to fall in the infinite void, producing what might be regarded as a primeval cosmic rain. Because none of the atoms encounter resistance, all descend at the same speed, and none is ever overtaken by another. Now this is a totally unsatisfactory cosmology, because it seems to rule out the very collisions that give atomism its explanatory power. Epicurus dealt with the difficulty by postulating an infinitesimal swerve: an atom shifts its line of descent the least possible amount, setting off a chain reaction of collisions. The most troublesome feature of this theory is that the swerve must be an uncaused event, since if it were caused, it could only be caused by collision with another atom; and the impossibility of such collisions is, of course, precisely the difficulty we are trying to escape.[16]

If we are tempted to judge Epicurus harshly for the invention of uncaused events (which are still a philosophical embarrassment, even if they do appear in some interpretations of modern quantum mechanics), we need to notice that the swerve not only accounts for the origin of the atomic maelstrom, which in turn explains the world in which we live; it also breaks the deterministic chain that would eliminate human responsibility and destroy Epicurus's ethical system. If the world is totally subject to rigid mechanical causation, then human action cannot be free; and if humans do not choose freely, they are without responsibility. The swerve introduces an element of indeterminism into the universe; and even if this does not explain how free choice is actually exercised (a question to which we still do not know the answer), by revealing a break in the chain of rigid causal necessity it makes room for the *possibility* of free human volition. This is doubtless not an entirely satisfactory solution, but to have perceived the problem of free will in a mechanical universe (Epicurus was the first to do so) is of itself a significant achievement.

The founder of Stoic philosophy was Zeno (ca. 333–262 B.C.) from Citium on the island of Cyprus. This Zeno, who is not to be confused with Parmenides' disciple of the same name, came to Athens and spent about a decade studying in a variety of Athenian schools, including the Academy, before setting up his own school in the *stoa poikilē* about 300. Zeno was succeeded by Cleanthes of Assos (331–232 B.C.) and Chrysippus of Soli (ca. 280–207 B.C.), powerful thinkers in their own right, who contributed as much as Zeno to the development of Stoicism as a systematic philosophy. Stoic philosophy, as an active scholarly tradition, survived into the second century A.D.; its influence, however, can be traced as late as the seventeenth century.[17]

Stoics and Epicureans were in radical opposition on most subjects, but they agreed on a few things. They agreed, in the first place, on the subordination of natural philosophy to ethics; in both philosophical schools, the pursuit of happiness was regarded as the goal of human existence. Stoics believed that happiness could be obtained only through living in harmony with nature and natural law; and living in harmony with nature required a knowledge of natural philosophy. In the second place, members of both philosophical schools were committed materialists, arguing strenuously that nothing exists except material stuff.

This shared materialism was important common ground; it meant that Stoics and Epicureans were allies in the battle against proponents of any nonmaterialist philosophy, such as Plato and his followers. However, once we move beyond this basic proposition, we find that Stoics and Epicureans had fundamentally different visions of the universe. Epicureans believed that matter was discontinuous and passive—consisting of discrete, unbreakable, lifeless atoms, which moved mindlessly in infinite void space.

Theirs was a mechanistic universe. Stoics, by contrast, created a model of an organic universe, characterized by continuity and activity. Let us use these contrasts (continuity-discontinuity and activity-passivity) as points of entry into Stoic natural philosophy.[18]

Matter, Stoics believed, does not present itself in the form of atoms, each with a permanent identity, but as an infinitely divisible continuum, containing no natural breaks and no void spaces. Size and shape, therefore, are not permanent attributes of matter, for matter can be chopped up into pieces of whatever size and shape we please. While allowing no void within the world, the Stoics did acknowledge an extra-cosmic void, viewing the cosmos as an island of continuous matter surrounded by an infinite void space.

Stoics followed the Epicureans in acknowledging a passive side to material things, but they were convinced that it could not be the whole story. The Epicurean position was vulnerable to the following objection. If an individual object derives all of its properties from the chance configuration of tiny lifeless pieces of matter, there can be no convincing explanation of many of the properties of the whole. The only properties possessed by Epicurean atoms are size, shape, and weight. How, then, could an Epicurean explain as simple and basic a property as cohesiveness—the fact that a rock remains a rock, resisting disintegration into its constituent particles? Where does the coldness of a block of ice come from, since coldness is not a property of its constituents? And how do we explain color, flavor, and texture? Or to turn to a far more difficult case, where do the characteristics of living things come from—the life cycle of a plant, the reproductive behavior of an insect, or the personality of a human being? If the family dog is merely a chance configuration of inert matter, how do we explain its obsession with chasing mail carriers? It would seem that besides passive matter there must be an active principle, having the capacity to organize passive matter into an organic unity and account for its characteristic behavior. There must be something that is acted upon; but there must also be something that acts, and in a materialistic world that something must be material.

The Stoics identified this active principle with breath or pneuma, the subtlest of substances, which totally interpenetrates everything, binding the recipient passive matter into unified objects and endowing these objects with their characteristic properties. But it is important to keep in mind that pneuma is more than a subtle, all-penetrating substance; it is also an active and rational substance, the source of vitality and rationality in the cosmos. Indeed, Stoics identified pneuma with divine rationality and with deity itself. The equation pneuma=reason=god may seem odd from a modern point of view, and certainly wrong from a Judeo-Christian point of view, but it was basic to Stoic cosmology. Deity had been brought down from the heavens, materialized, and made to account for activity and order in the universe.

Let us scrutinize this pneuma more closely, inquiring into its structure (if any), the source of its organizing capabilities, and its relationship to passive matter. Stoics accepted the existence of the four Aristotelian elements, but divided them into two groups on the basis of activity. They regarded earth and water, the principal ingredients of tangible objects, as passive elements, air and fire as active elements. Air and fire mix in various proportions (the Stoics had in mind a total, homogeneous blending) to produce a variety of pneumas. Thus air and fire act, while water and earth are acted upon.

Pneumas come in various grades. At the lowest level, the pneuma that accounts for the cohesion of what we would regard as inorganic bodies— rocks and minerals, for example—is called *hexis*. The pneuma of plants and animals, which gives them their vital properties, is *physis*. And the highest grade of pneuma, possessed by humans and accounting for their rationality, is *psychē*. Now Stoics identified the pneuma of an object with soul. It follows that every individual thing is permeated by soul, and this soul functions as its organizing principle. There must even be a cosmic pneuma, a world soul, since the cosmos too is an organic unity, having characteristics that require active principles for their explanation. The profoundly vitalistic character of Stoic natural philosophy should be evident.

Pneumas exist in a state of tension or elasticity. This tension accounts for that most basic property of all objects, cohesion. At higher levels, different tensions account for the variety of properties and personalities observable in the world. And finally, it may be well to reiterate that the relationship of pneuma to its host body is one of total mixture or interpenetration, both substances occupying the same space.

Stoic cosmology, like that of Plato and Aristotle, was geocentric. However, the Stoics followed the atomists and departed decisively from Aristotle, refusing to make any kind of radical distinction between the terrestrial and celestial regions; when it came to such fundamental matters as the composition and laws of nature, the Stoic cosmos was homogeneous. The Stoics agreed with Aristotle on the eternity of the universe, but in place of his belief in cosmic stability they substituted a cyclic theory inspired by pre-Socratic thought. According to a variety of Stoic thinkers, there is an eternal cosmic cycle of expansion and contraction, conflagration and regeneration. In the expansive phase, the world dissolves into fire; in the contractive phase, fire yields again to the other elements, and the world as we know it is regenerated. This cycle is repeated eternally, producing an everlasting sequence of identical worlds.[19]

Finally, we must note that the Stoic universe was conceived to be both purposeful and deterministic. Permeated as it was with mind and deity, the Stoic cosmos was inevitably suffused with purpose, rationality, and providence. At the same time, its course was rigidly determined. Stoic philosophy maintained that there are causal chains (themselves the products of divine rationality) that cannot be violated and that totally determine the sequence of events. As Cicero put it in *On Divination,* "nothing has happened which was not going to be, and likewise nothing is going to be of which nature does not contain causes working to bring that very thing about. This makes it intelligible that fate should be, not the 'fate' of superstition, but that of physics."[20]

We have seen that the natural philosophies of the Stoics and the Epicureans were, in many ways, opposites. Whereas Epicurean philosophy had as one of its principal aims the combating of Platonic and Aristotelian teleology, Stoic philosophy was directed toward the discovery of purpose and the defense of teleology. Whereas Epicureans portrayed a mechanistic universe, Stoics discovered an organic one. Whereas Epicurus struggled to introduce an element of indeterminacy into his otherwise mechanistic universe, the Stoics were content with an organic universe where rigid determinacy reigned. In the short run, the Stoic vision of the cosmos seemed the more plausible of the two and became a prominent philosophical option in late antiquity. In the long run, both Stoic and Epicurean philosophy were resurrected in the early modern period and presented as alternatives to the Platonic and Aristotelian world pictures; and each played a part in shaping the new philosophy of the seventeenth century.

FIVE

The Mathematical
Sciences in Antiquity

THE APPLICATION OF MATHEMATICS TO NATURE

Within the Western scientific tradition there has been a long debate about the applicability of mathematics to nature. The question is whether the world is fundamentally mathematical, in which case mathematical analysis is the sure route to deeper understanding, or whether mathematics is applicable only to the superficial quantifiable aspects of things, leaving the ultimate realities untouched. There can be no doubt that natural scientists in the modern period seem increasingly inclined to resolve the question in favor of the mathematical approach. However, the alternative is not without its defenders; and the debate remains lively among social scientists and historians.

The ancient Pythagoreans appear to have maintained that nature is mathematical through-and-through. If Aristotle can be trusted, the Pythagoreans went to the extreme of claiming that the ultimate reality is number (see chap. 2 for additional discussion). Plato took up with a vengeance the Pythagorean program in his theory of matter, arguing that the four elements are reducible to regular geometrical solids, which are reducible in turn to triangles. Thus for Plato the fundamental building blocks of the visible world were not material, but geometrical; moreover, what binds everything together into a unified cosmos, Plato argued, is not a physical or mechanical force, but simply geometrical proportion.[1]

There can be no question that Aristotle was mathematically informed. He modeled his theory of knowledge on mathematical demonstration, he utilized geometry in his theory of the rainbow, and he employed theory of proportion in his analysis of motion. But Aristotle was convinced that there was a difference between mathematics and natural science or physics. Physics, by his definition, considers natural things in their entirety, as sensible, changeable bodies. The mathematician, by contrast, strips away all of the sensible qualities of bodies and concentrates on the mathematical remainder:

> in his investigations [the mathematician] first abstracts every-
> thing that is sensible, such as weight and lightness, hardness
> and its contrary, and also heat and cold and all other sen-
> sible contrarieties, leaving only quantity and continuity—
> sometimes in one, sometimes in two and sometimes in three
> dimensions. . . .[2]

The mathematician is concerned only with the geometrical properties of things, and these by no means exhaust reality. Reintroduce weight, hardness, warmth, color, and the other qualities that exist in the real world, and you have moved out of the mathematical realm and back into the subject matter of physics.

So Aristotle took a middle road on the question of the applicability of mathematics to nature. He was convinced that mathematics and physics are both useful, but it was clear to him that they are not the same thing; the mathematician and the physicist may study the same object, but they concentrate on different features of it. Nonetheless, there are certain subjects—astronomy, optics, and harmonics—that exist in the borderland between mathematics and physics. In these subjects, which came to be called the "middle" or "mixed" sciences, the mathematician may be able to supply the cause or explanation of the facts established by the physicist.

Plato and Aristotle thus provided two *theories* of the relationship between mathematics and nature, and these became the poles between which natural scientists have vacillated from antiquity to the present. However, our interest is not only in theories about the applicability of mathematics to nature, but also in the ways in which these theories were *practiced.* To observe the Greeks actually at work, applying mathematics to nature, we will examine the subjects of astronomy, optics, and the balance or lever. In order to prepare ourselves for that investigation, we must first glance at the Greek achievement in pure mathematics.

GREEK MATHEMATICS

We know little about the origins of Greek mathematics. There is no question that early Greek mathematicians had access to the Egyptian, and especially the Babylonian, mathematical achievement (see chapter 1). But from the beginning Greek mathematics was different; and the difference lay chiefly in Greek geometry, with its orientation toward abstract geo-

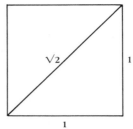

Fig. 5.1. The incommensurability of
the side and the diagonal of a square.

metrical knowledge and its formal methods of inference and proof. One reason for the Greek emphasis on geometry may have been the discovery, perhaps within the Pythagorean school, that the ratio between the side and the diagonal of a square cannot be represented by any pair of whole numbers. More technically, we refer to this as the incommensurability of the side and the diagonal; another way of expressing the same result is to speak of the irrationality of $\sqrt{2}$ (the length of the diagonal of a square whose side is 1; see fig. 5.1). It is possible that this irrationality persuaded Greek mathematicians of the unsuitability of numbers (positive integers, on the Greek view) for the representation of reality and thus encouraged the development of geometry.[3]

We have only fragmentary evidence for specific mathematical developments in the period before Euclid (who flourished about 300 B.C.), but it is universally acknowledged that those developments were codified in Euclid's own mathematical textbook, the *Elements*.[4] Here we find mathematics highly developed as an axiomatic, deductive system. The *Elements* begins with a set of definitions: of a point ("that which has no part"), a line ("length without breadth"), a straight line, a surface, a plane surface, a plane angle, right, acute, and obtuse angles, various plane figures, parallel lines, and so forth. The definitions are followed by five postulates: that a line can be drawn from a point to any other point, that a straight line can be extended continuously from either end, that a circle of any radius can be drawn about any point, that all right angles are equal, and a statement of the conditions under which straight lines will intersect. The postulates are followed by five "common notions," or axioms—self-evident truths needed for the practice of correct thinking in general and mathematics in particular. These include the claim that things equal to the same thing are equal to each other, that equals added to equals yield equal sums, and that the whole is greater than the part. These preparatory claims lay the groundwork for the propositions that fill the thirteen books that follow. A typical proposition begins with an enunciation, followed by an example, a further definition or specification of the proposition, and a construction; it ends with a proof and a conclusion. What is important to note is that the conclusion of a proper Euclidean demonstration follows necessarily from definitions, postulates, axioms, and previously proved propositions. So effectively did Euclid wield this method that, through his influence—and that of Aristotle, whose method resembled Euclid's in several critical respects—it became the standard for scientific demonstration down to the end of the seventeenth century.

We need not pause long over the content of Euclid's *Elements,* for it bears a close resemblance to the geometry taught in the modern secondary school. Books I–VI develop the elements of plane geometry; book X is devoted to a classification of incommensurable magnitudes; and books XI–XIII treat solid geometry. In books VII–IX Euclid takes up arithmetical topics, including the theory of numbers and of numerical proportion. Of the many achievements in the *Elements,* one that must be mentioned because of its future significance is the development of the method of "exhaustion"—probably borrowed by Euclid from his predecessor Eudoxus and destined to influence a variety of followers, including Archimedes. Euclid shows (XII, 2) how to "exhaust" the area of a circle by means of an inscribed polygon; if we successively double the number of sides in the polygon, we will eventually reduce the difference between the area of the polygon (known) and the area of the circle (unknown) to the point where it is smaller than any magnitude we choose (see fig. 5.2). This method made it possible to calculate the area of a circle to any desired degree of accuracy; with a little further development, it could be used to calculate the area within (or under) other curves as well. Another important feature of the *Elements* is its exploration of the properties of the five regular geometrical solids, sometimes known as the "Platonic solids," and its demonstration (XIII, 18) that there are no regular geometrical solids beyond these five.[5]

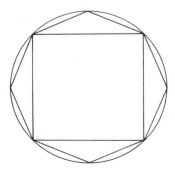

Fig. 5.2. Determining the area of a circle by the method of "exhaustion."

Euclid was followed by a series of brilliant Hellenistic mathematicians, the greatest of whom was undoubtedly Archimedes (ca. 287–212 B.C.). Archimedes contributed to both theoretical and applied mathematics, but he is especially esteemed for the elegance of his mathematical proofs. In some of his most important work, Archimedes developed the method of exhaustion and applied it to the calculation of areas and volumes, including the area enclosed within a segment of a parabola, the area bounded by certain spirals, and the surface area and volume of a sphere. He calculated an improved value for π (the ratio of the circumference to the diameter of a circle), showing that it must fall between 3 10/71 and 3 1/7. Archimedes had a profound influence on the subsequent development of mathematics and mathematical physics, particularly after his works were rediscovered and reissued in the Renaissance. We will examine his contributions to physics below.[6]

A final Greek mathematical achievement that must be mentioned is the work of Apollonius of Perga (fl. 210 B.C.) on conic sections. Apollonius studied the ellipse, parabola, and hyperbola—the plane figures formed when a circular cone is cut at various angles by a plane surface—and set out a new approach to their definition and methods of generation. His book on conic sections, like the works of Archimedes, was destined to have a major influence in the early modern period.

EARLY GREEK ASTRONOMY

Early Greek astronomy appears to have been concerned primarily with observation and mapping of the stars, with the calendar, and with the solar and lunar motions that had to be plotted before a satisfactory calendar could be constructed. The major difficulty in the calendar arose from the fact that the solar year is not an integral multiple of the lunar month. That is, in the time required for the sun to complete one circuit of the zodiac, the moon completes twelve circuits plus a fraction. Thus a calendar based on twelve months of twenty-nine or thirty days each is too short by about eleven days, and the calendar and the seasons will not be synchronized. Various schemes were developed for inserting an additional month as needed, in order to bring the calendar back into phase with the seasons. These calendric efforts culminated in the Metonic cycle, proposed by Meton (fl. 425 B.C.), based on the understanding that to a very close approximation nineteen years contain 235 months. Therefore, in a nineteen-year cycle, there will be twelve years of twelve months and seven years of thirteen months. It seems that Meton intended this as an astronomical, rather than a civil, calendar; and it was put to astronomical use for several centuries.[7]

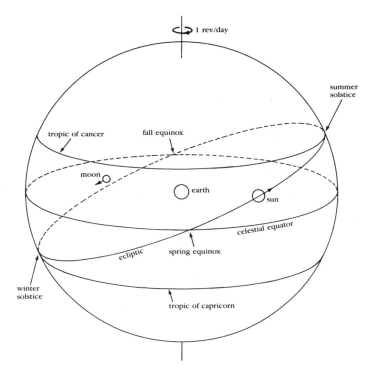

Fig. 5.3. Two-sphere model of the cosmos.

Greek astronomy took a decisive turn in the fourth century with Plato (427–348/47) and his younger contemporary Eudoxus of Cnidus (ca. 390–ca. 337 B.C.). In their work we find (1) a shift from stellar to planetary concerns, (2) the creation of a geometrical model, the "two-sphere model," for the representation of stellar and planetary phenomena, and (3) the establishment of criteria governing theories designed to account for planetary observations. Let us consider these achievements in some detail.

The two-sphere model devised by Plato and Eudoxus conceives the heavens and the earth as a pair of concentric spheres. To the celestial sphere are affixed the stars, and along its surface move the sun, the moon, and the remaining five planets. The daily rotation of the celestial sphere accounts for the observed daily rising and setting of all of the celestial bodies. Corresponding circles on the two spheres divide them into zones and mark out the motions of the wandering stars. Figure 5.3 reveals roughly what Plato and Eudoxus had in mind. The terrestrial sphere is fixed in the center, while the celestial sphere rotates daily about a vertical axis. The equator of the earth projected onto the celestial sphere defines the celestial equator. The circle along which the sun, moon, and planets move in their journey through the celestial sphere—a circle tilted approximately 23° with respect to the equator and passing through the center of the zodiac—is the ecliptic. The ecliptic and celestial equator intersect at the equinoxes; when the sun, in its annual trip around the ecliptic, reaches the fall equinox (approximately September 21), autumn begins; when it reaches the spring equinox, spring begins. The points at which the ecliptic is most distant from the equator are the solstices; when the sun reaches the summer solstice (approximately June 21), summer begins. The circles drawn parallel to the equator through the summer and winter solstice, respectively, are the Tropics of Cancer and Capricorn.[8]

The movements of the sun, moon, and remaining planets had been carefully observed and well charted by the fourth century. In the model of Plato and Eudoxus, the sun circles the ecliptic once a year, while the moon completes its circuit in a month, both moving in a west-to-east direction and with nearly uniform speed. The other planets—Mercury, Venus, Mars, Jupiter, and Saturn—also follow the ecliptic (deviating from it by no more than a few degrees), moving in the same direction as the sun and moon but with considerable variations in speed. Mars, for example, goes around the ecliptic once in about 22 months (687 days); about once every 26 months it slows to a halt, reverses itself (moving now in an east-to-west direction), halts again, and then resumes its usual west-to-east motion. This reversal of direction is called "retrograde motion," and it is exhibited by all of the planets except the sun and moon. Figure 5.4 illustrates the observed retrograde motion of Mars.

One further striking feature of planetary motion known to Plato and Eudoxus was the fact that Mercury and Venus never stray far from the sun (the maximum elongation is 23° for Mercury, 44° for Venus). Like dogs on a leash, they can run ahead of the sun or lag behind it, but they can never be more distant from it than the fixed length of the leash allows. Finally, to appreciate the achievement embodied in the two-sphere model, we

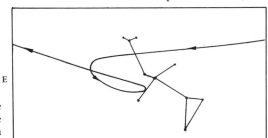

Fig. 5.4. The observed retrograde motion of Mars in the vicinity of the constellation Sagittarius, 1986. Data supplied by Jeffrey W. Percival.

E

W

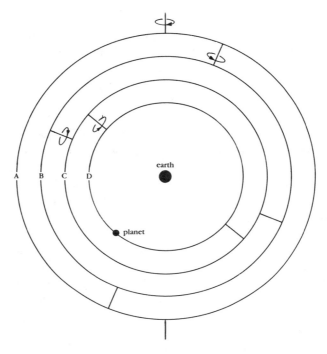

Fig. 5.5. The Eudoxan spheres for one of the planets.

must understand that all of these motions are occurring on the surface of the celestial sphere while that sphere is going through its daily rotation around the earth. The resulting motion, observed from the fixed earth, will be a combination of the irregular motion of the planet around the ecliptic and the uniform daily rotation of the celestial sphere; it is meant to capture the bewildering complexity of observed planetary positions. The two-sphere model is thus a geometrical way of conceiving and talking about planetary phenomena.

It is a fine thing to create a geometrical language for talking about planetary motion, and it is laudable to present a rough description of planetary motion around the ecliptic. But there is something more to which we might aspire: if we really wish to bring order and intelligibility to this "bewildering complexity" in the heavens, we must take the intricate, variable motion of each planet and reduce it to some combination of uniform motions. That is, we must assume that disorder conceals order, that beneath irregularity lies regularity, and that this underlying order or regularity is discoverable. A late, and possibly unreliable, account gives Plato the credit for laying out this assumption as a research program, challeng-

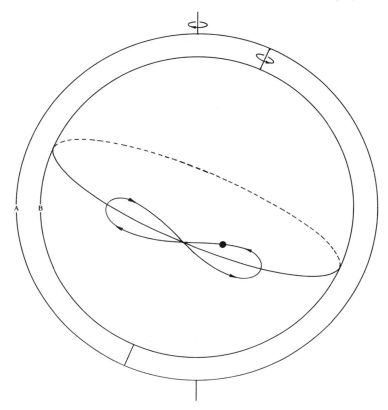

Fig. 5.6. The Eudoxan spheres and the hippopede.

ing astronomers or mathematicians to determine what combination of uniform circular motions would account for the apparent, irregular planetary motions.[9]

Whether it was indeed Plato who first posed the question, it is clear that Eudoxus first proposed an answer to it. Eudoxus's idea was ingenious, but fundamentally simple. The goal was to treat each irregular planetary motion as a composite of a series of simple uniform circular movements. To achieve this goal, Eudoxus assigned to each planet a set of nested concentric spheres, and to each sphere one component of the complex planetary motion (see fig. 5.5). Thus the outermost sphere for, say, the planet Mars rotates uniformly once a day to account for Mars' daily rising and setting; the second sphere in the series also rotates uniformly about its axis (tilted in relation to the axis of the outermost sphere), but in the opposite direction, once in 687 days, thus accounting for Mars' slow west-to-east motion around the ecliptic; and the two inner spheres account for changes in speed and latitude and for retrograde motion. Mars, then, is situated on the equator of the innermost sphere, participating not only in the motion proper to that sphere, but also in the motions transmitted downward from the three spheres above it. A similar system works for Mercury, Venus, Jupiter, and Saturn. The sun and moon, which do not undergo retrograde motion, require only three spheres apiece.[10]

The Inner Eudoxan Spheres and Retrograde Motion

If, for the sake of simplicity, we treat the interaction of spheres C and D (fig. 5.5) in isolation from the rest of the system, we find that assigning them equal and opposite rotations around axes tilted with respect to each other determines that the planet (on the equator of D) will move along a path resembling a hippopede (horse-fetter) or figure eight. We can thus visualize the motion that results from the four Eudoxan spheres by replacing C and D with a hippopede affixed to the equator of sphere B (fig. 5.6). Sphere A completes a uniform rotation once a day, carrying with it the axis of B; meanwhile, B rotates uniformly around this axis in the sidereal period of the given planet (the time required for that planet to make one complete circuit of the celestial sphere), carrying the hippopede around the ecliptic; and all the while the planet is moving around the hippopede in the direction indicated by the arrows.[11]

Eudoxus thus created the first serious geometrical model of planetary motion. Two questions naturally present themselves. First, did Eudoxus attach physical reality to the model? That is, did he conceive the spheres as physical objects? The answer seems unambiguously in the negative. There is every reason to believe that the concentric spheres of Eudoxus were intended as a purely mathematical model, with no claim to represent physical reality. Eudoxus, as far as we can tell, did not imagine that the cosmos consists of physically distinct spheres, mechanically linked to one another; rather, he was endeavoring, by means of a geometrical model, to identify the separate components of uniform motion that underlie and make sense of the complicated motions of the planets. He was searching not for physical structure, but for mathematical order.

Second, did the model work? Because no treatise by Eudoxus has survived, we have no knowledge of the geometrical details of his system. Several things can be said, however. Although Eudoxus's model is clearly mathematical, it is improbable that it was designed to yield quantitative predictions. Indeed, it is very unlikely that the notion of exact, quantitative prediction had yet entered Greek astronomy or any other Greek scientific subject; nobody aspired to more than rough qualitative agreement between theory and observations. We can, if we wish, speak of the potential of the Eudoxan model, under the assumption that the best values were chosen in every case. Under those circumstances, the system would (with one or two exceptions) have given results that correspond in a rough qualitative way, but without quantitative precision, to the astronomical observations as we know them today. Given the more limited state of astronomical knowledge and the modest aims of astronomical theory in the fourth century, this was a considerable achievement.

A generation after Eudoxus, his system was improved by Callippus of Cyzicus (b. ca. 370), who added a fourth sphere for the sun and the moon and a fifth sphere each for Mercury, Venus, and Mars. The function of the additional spheres for the sun and moon was to take into account the variations in speed of those two planetary bodies as they circle the ecliptic—the fact, for example, that the times required for the sun to pass from the summer solstice to the fall equinox and from the fall equinox to the winter solstice (see fig. 5.3, above) may differ by several days.[12]

Further development of the system of concentric spheres occurred at the hands of Aristotle (384–322). Aristotle took the Eudoxan model as modified by Callippus, but with an important difference: whereas Eudoxus appears to have considered his concentric spheres to be merely geometrical constructions, Aristotle seems to have conceived the system to be physically real and was therefore induced to think seriously about the transmission of motion from one sphere to the next. That compelled him to think about the interconnections between spheres and to realize that if all seven planets, each with its set of spheres, were nested in concentric fashion, the innermost sphere of one planet (say Saturn) would inevitably transmit its intricate motion to the outermost sphere of the planet just below it in the series (Jupiter). When the additional effect of Jupiter's own spheres was figured in, the complexity would become intolerable—and, moreover, would be at odds with the observational data. Aristotle responded to this problem by inserting a set of counteracting spheres between the innermost sphere of Saturn and the outermost sphere of Jupiter, and a similar set of counteracting spheres between the primary spheres belonging to every other pair of adjacent planets. These counteracting

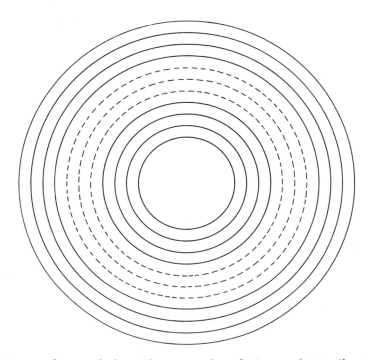

Fig. 5.7. Aristotelian nested spheres. The primary spheres for Saturn and Jupiter (four spheres apiece) are represented by solid lines. Between these two sets are three counteracting spheres (dotted lines), which counteract or "unroll" the motion of Saturn's four spheres, above, in order to transmit a simple diurnal motion to the outermost sphere for Jupiter, below.

spheres, one fewer in number than the primary planetary spheres just above them, were intended to "unroll" the system, as Aristotle put it, and restore simple diurnal motion to the outermost sphere of the next planet in the series (fig. 5.7). Aristotle left many questions of detail unanswered; his discussion of the Eudoxan system and its modifications occupies only a page or two in his work and ends with an admission of uncertainty. What is important is that Aristotle bequeathed to his successors an enormously complicated piece of celestial machinery, consisting of fifty-five planetary spheres, plus the sphere of the fixed stars.

He also bequeathed to them an important question: In astronomy, where does the balance lie between the mathematical and the physical? Is astronomy primarily a mathematical art, as Eudoxus seems to have conceived it? Or must the astronomer also concern himself with the real structure of things, as Aristotle's astronomical scheme suggests? Astronomers would ponder this question for the next two thousand years.[13]

COSMOLOGICAL DEVELOPMENTS

Contemporary with Aristotle and in the century thereafter, there were several cosmological developments of interest to astronomers. One was the proposal by Heraclides of Pontus (ca. 390–after 339), a member of Plato's Academy under Plato and Plato's successor, that the earth rotates on its axis once in twenty-four hours. This idea, which came to be fairly widely known (though it was rarely accepted as true), explains the daily rising and setting of all of the celestial bodies. Heraclides has often been credited also with the claim that the motions of Mercury and Venus are centered on the sun, but modern scholarship has revealed this interpretation to be groundless.[14]

A generation or two after Heraclides, Aristarchus of Samos (ca. 310–230 B.C.) proposed a heliocentric system, in which the sun is fixed in the center of the cosmos, while the earth circles the sun as a planet; it is usually assumed that Aristarchus also gave the other planets sun-centered orbits, although the historical evidence does not address this point. In all likelihood, Aristarchus's idea was a development of Pythagorean cosmology, which had already removed the earth from the center of the universe and put it in motion around the "central fire."[15] Aristarchus has been celebrated for his anticipation of Copernicus, and his successors have been vilified for their failure to adopt his proposal. It takes only a moment's reflection, however, to realize that we do no justice to the situation in the third century B.C. if we judge Aristarchus's hypothesis by twentieth-century evidence. The question is not whether *we* have persuasive reasons for being heliocentrists, but whether *they* had any such reasons; and the answer, of course, is that they did not. Putting the earth in motion and giving it planetary status violated ancient authority, common sense, religious belief, and Aristotelian physics; it also predicted stellar parallax (variations in the geometrical relationship between pairs of stars as the observer approaches, recedes, or otherwise changes position), which could not be observed. Meanwhile, whatever observational advantages it might have (for example, its ability to explain variations in the brightness of the planets) were available in other systems that did no violence to traditional cosmology.

There were attempts, early in the Hellenistic period, to calculate various cosmological constants. Aristarchus himself compared the earth-to-sun distance with the earth-to-moon distance, figuring the former to be about twenty times the latter (the correct ratio is about 400:1). Aristarchus's method is revealed in figure 5.8. Hipparchus (d. after 127 B.C.) calculated absolute values for the solar and lunar distances from the absence of solar

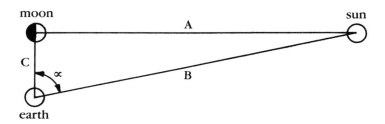

Fig. 5.8. Aristarchus's method for determining the ratio between the solar and lunar distances from the earth. The angle ∝ separating the two lines of sight is measured when the moon is at quarter-phase (A and C are then known to intersect at right angles). From this the ratio of B to C can be calculated. But the method has several drawbacks. In the first place, the exact moment when the moon's disk is half illuminated cannot be accurately determined. In the second place, a small error in the measurement of the angle ∝ (Aristarchus put it at 87°, whereas the true value is 89° 52′) leads to a very large error in the ratio of B to C.

parallax [16] and from data on solar eclipses. The assumption that solar parallax is just below the threshold of visibility gave him a value for the solar distance of 490 earth radii. From eclipse data he then obtained a value of between 59 and 67 earth radii for the lunar distance. As for the size of the earth, Eratosthenes (fl. 235 B.C.), a geographer and mathematician who headed the library in Alexandria, had calculated the earth's circumference a century earlier; his value of 252,000 stades (within 20 percent of the modern figure) became widely known and was never lost. [17]

HELLENISTIC PLANETARY ASTRONOMY

Planetary astronomy seems to have been vigorously pursued in the Hellenistic period, but we know few details because Claudius Ptolemy (at the end of the Hellenistic period) was so successful in summing up the achievements of his predecessors that their works dropped out of circulation and disappeared. We do know, because Ptolemy tells us, that Apollonius of Perga developed a new mathematical model for planetary motion in the third century B.C. And it is clear, again from comments by Ptolemy and other fragmentary remains, that Hipparchus was one of the greatest astronomers of antiquity. Hipparchus made his mark primarily in observational astronomy, producing a new and superior star map, discovering the precession of the equinoxes, developing a new astronomical sighting instrument (the diopter), and criticizing existing planetary theory. We also know that Hipparchus had access to Babylonian astronomical data, including data on planetary motions and lunar eclipses. And, what is most important, through his contact with Babylonian astronomy Hipparchus came to appreciate the goal of exact, quantitative prediction. He was the first to develop methods for assigning numerical values to geometrical models; and through his influence the demand for a quantitative match between theory and observation entered and radically transformed Greek astronomy.[18] To see the result of this transformation, we must turn to the work of Ptolemy.

Claudius Ptolemy (fl. 150 A.D.) was affiliated with the Museum in Alexandria, and its associated library. (The name "Ptolemy" offers room for confusion. It does not signify descent from the old ruling Ptolemaic dynasty, but probably represents a certain geographical sector of the city of Alexandria, used by citizens as a "tribal" name. Its importance for us is that it signifies that Claudius Ptolemy was not a recent immigrant, as many of the early Alexandrian intellectuals were, but was descended from a line of Alexandrian citizens.) It may help to guard against the foreshortening of time, seen from a distance, if we remind ourselves that Ptolemy lived some three hundred years after Hipparchus and five hundred years after Eudoxus. This means not only that he was the beneficiary of the theoretical advances made during those intervening centuries, but also that he had access to centuries of astronomical observation, both Greek and Babylonian; and even relatively crude observational data, spread over a sufficiently long span of time, are capable of yielding remarkably precise theoretical conclusions. For example, using data available in the second century B.C., Hipparchus was able to calculate the average length of the lunar month to within one second of the modern value.[19]

It would be surprising if the mathematical sophistication of Hellenistic mathematics were not reflected in Hellenistic mathematical astronomy. Ptolemy, coming near the end of that Hellenistic world, brought to planetary astronomy a level of mathematical power that Eudoxus, five hundred years earlier, could not have imagined. Ptolemy's models had the same aim as those of Eudoxus—to discover some combination of uniform circular motions that would account for the observed positions (that is, the apparent variations in speed and direction) of the planets. In addition, Ptolemy's models were capable of making accurate quantitative forecasts of future planetary positions. However, the mathematical techniques he employed were vastly different.

In the first place, instead of using spheres, he used circles. Let us see how uniform motion in a circle can be employed to give the appearance of nonuniformity. Let circle ABD (fig. 5.9) be the orbit of the planet, and allow

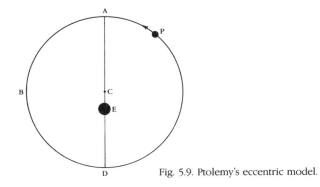

Fig. 5.9. Ptolemy's eccentric model.

the planet, P, to move uniformly around it. If the motion of P is uniform, then the planet will sweep out equal angles about the center, C, in equal times. Now if the center of uniform rotation, C, corresponds with the point of view—that is, if the earth is located at C—then the motion of P will not only *be* uniform, but also *appear* uniform. However, if the center of uniform rotation and the point of view do not coincide—if the earth, for example, is located at E—then the motion of the planet will appear nonuniform, seeming to slow as it approaches A and seeming to speed up as it approaches D. This is the eccentric model.

The eccentric model is sufficient for dealing with simple cases of nonuniform motion, such as that of the sun around the ecliptic and the resulting inequality of the seasons. For more complicated cases, Ptolemy found it necessary to introduce the epicycle-on-deferent model (fig. 5.10). Let ABD be a deferent (or carrying) circle, and draw a small circle (an epicycle) with its center on the deferent. The planet P moves uniformly around the epicycle; meanwhile the center of the epicycle moves uniformly around the deferent circle. The observer at E sees the composite of two uniformly circular motions. The exact characteristics of this composite motion will depend on the particular values chosen—the relative sizes of the two circles and the speeds and directions of motion—but it is clear that the model has considerable potential. When P is on the outside of the epicycle, as in figure 5.10, the apparent motion of the planet (as viewed from the earth) will be the sum of its motion around the epicycle and the motion of the epicycle around the deferent, and the planet will at this point have its maximum apparent speed. When P is on the inside of the epicycle, as in figure 5.11, its motion on the epicycle and the motion of the epicycle on the deferent are opposed (as viewed from the earth), and the apparent motion of the planet is determined by their difference; if the motion of P

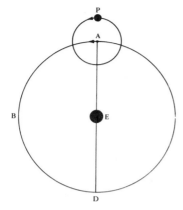

Fig. 5.10. Ptolemy's epicycle on deferent model.

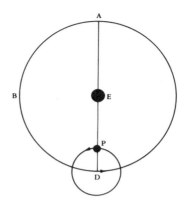

Fig. 5.11. Ptolemy's epicycle on deferent model, with the planet on the inside of the epicycle.

Fig. 5.12. Retrograde motion of a planet explained on the epicycle and deferent model. As the epicycle moves counterclockwise on the deferent, the planet moves counterclockwise on the epicycle. The actual path of the planet is represented by the heavy line.

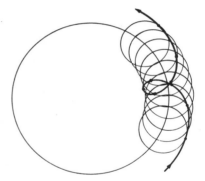

should be the greater of the two, the planet will appear to reverse itself and undergo a period of retrograde motion. This retrograde motion is illustrated in figure 5.12.

Both of these models are based firmly on the requirement that the real planetary motions—that is, the component motions—be uniform and circular. Greek astronomers have often been criticized for their "dogmatic" commitment to uniform circular motion, on the grounds that a priori assumptions (of this or any other sort) are unjustified, or at least unbecoming, in a scientist. Is such criticism justified? The truth is that the scientist, ancient or modern, begins *every* investigation with strong commitments regarding the nature of the universe and very clear ideas about which models may legitimately be employed to represent it. In Ptolemy's case, the requirement of uniform circular motion was justified above all by the nature of the inquiry; his goal was not simply to set forth the relevant observational data, in order to describe the planetary motions in all of their complexity, but to reduce the complex planetary motions to their simplest components—to discover the real order underlying apparent disorder. And the simplest motion, representing the ultimate order, is, of course, uniform circular motion.

But there were many other considerations that would have reinforced the restriction to models based on uniform motion in a circle. There was the force of common sense and the sanction of tradition, for the cyclic, repetitive nature of celestial phenomena had always suggested that celestial motion must be fundamentally uniform and circular. Furthermore, without uniform circular motion quantitative prediction would have been impossible, for the "trigonometric" methods available to Ptolemy were not readily applicable to any other kind of motion. In addition, there were aesthetic, philosophical, and religious considerations: the special character of the heavens demanded the most perfect of shapes and motions for heavenly bodies. Finally, it is instructive to note that when Copernicus broke with Ptolemy 1,400 years later, it was not because he resented the commitment to uniform circularity, but (in part) because he felt that Ptolemy had not lived up to this commitment.

In any case, the eccentric and epicycle-on-deferent models, based on uniform circular motion, were extremely powerful. But they had their limits, and their inability to account for certain planetary motions demanded the creation of yet another model, which came to be known as the equant model. Let AFB (fig. 5.13) be an eccentric circle, with center at C; and let the earth be located at E. Now, as the planet moves around the circle, in-

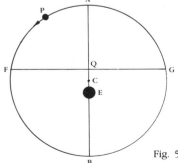

Fig. 5.13. Ptolemy's equant model.

stead of insisting that in equal times it sweep out equal angles as measured at the center (the usual definition of uniform motion), Ptolemy allowed it to move in such a way as to sweep out equal angles in equal times as measured at the equant point—a noncentral point located at Q (so chosen that QC equals CE). As the planet moves through arc AF, it sweeps out the right angle AQF. Suppose, for instance, that the planet covers this angle and arc in three years; then in the next three years, the planet must sweep out another right angle, FQB, and must therefore move through its corresponding arc, FB. In three more years, the planet moves from B to G, through right angle BQG, and so forth. If we compare the arcs traversed, it is clear that the planet has greater linear speed over FB than over AF; the planet gradually speeds up as it moves from A to B, then slows down as it moves from B to A. Viewing this variable motion from E, on the other side of the center from the equant point, will only exaggerate the apparent variability. Note, then, that in the equant model Ptolemy retained uniformity of angular motion—though not about the center—while he very definitely gave up uniformity of linear motion about the circumference. Whether this diluted version of uniformity was sufficient is a question that Copernicus raised in the sixteenth century; in the meantime, it enabled Ptolemy to complete the development of successful planetary models. Predictive success outweighed any considerations that might have dictated the stronger version of uniformity.

The three models—eccentric circle, epicycle-on-deferent, and equant—were all effective ways of employing uniform circular motion (whether that uniformity was strict or not so strict) to account for apparent irregularity in the heavens. But the models achieved their full power through combination. The eccentric circle and epicycle-on-deferent models could be easily combined by defining a deferent that was eccentric to the earth. An equant could be added, so that the center of the epicycle sweeps out equal angles in equal times as measured at a noncentral point. It was even possible to take the center of the eccentric circle and move it in a small circle around the earth—something Ptolemy had to do to make his lunar model work.[20] The most typical model for a planet (applicable to Venus, Mars, Jupiter, and Saturn) is that illustrated in figure 5.14, where ABD is an eccentric deferent, with center at C; the earth is at E and the equant point at Q. The planet moves uniformly around the epicycle; the center of the epicycle moves uniformly (as measured by the angle swept out) about the equant point, Q; and the resulting motion is viewed from the earth, E. Models such as this, developed with appropriate variations for all of the remaining planets, proved extraordinarily successful in predicting ob-

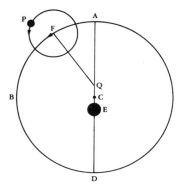

Fig. 5.14. Ptolemy's model for the superior planets. Line QF sweeps out equal angles in equal times about equant point Q.

served planetary positions. Indeed, it was precisely their degree of success that gave them such longevity and made them so difficult to displace.

It may appear that Ptolemy was engaged in a purely mathematical exercise. He entitled the treatise in which these mathematical models were presented *Mathematical Syntaxis* (or *Mathematical System*). Moreover, in the preface to this treatise he announced that speculation about divine causation of celestial motion or about the material nature of things leads only to "guesswork" and that, if the goal is to achieve certainty, the mathematical way is the only way. And at several points in the work he argued that astronomical models should be chosen on the basis of mathematical simplicity—apparently without concern for physical plausibility.

Nonetheless, if we look closely we see that nonmathematical considerations did, in fact, enter into the analysis. Ptolemy presented physical arguments for the centrality and fixity of the earth—which for him was not merely a mathematical hypothesis, but an important physical belief. He made claims about the nature of the heavens—arguing that, unlike terrestrial substance, they offer no hindrance to motion. And in another treatise, the *Planetary Hypotheses,* he tried to work out a materialized version of his mathematical models.[21] Thus, although Ptolemy was committed to a mathematical approach, his mathematical analysis did not exclude physical concerns, but functioned within the framework of traditional natural philosophy.

For all of this interest in the physics of the cosmos, the *balance* of Ptolemy's astronomical work was weighted toward mathematical analysis. And it was as a mathematician of the heavens, committed to "saving the phenomena" by mathematical means, that he influenced the Middle Ages and the Renaissance. Indeed, Aristotle and he came to symbolize the two poles of the astronomical enterprise—the former occupied especially with questions of physical structure, the latter an accomplished builder of mathematical models.

THE SCIENCE OF OPTICS

A second subject to which mathematics was successfully applied in antiquity was the science of optics. Optics includes the study of light and vision, which have been objects of investigation and speculation since earliest times. Vision has been almost universally regarded as the sense through which we learn the most about the world in which we live; and one of the most important and pleasing entities in that world proves to be light, which is not only the instrument of vision but is also connected, in the form of sunlight, with heat and life.

Any well-developed philosophy of nature must deal with vision. The atomists attributed sight to the reception in the eye of a thin film of atoms (a *simulacrum*) issuing from the surface of visible objects. According to Plato (in the *Timaeus*), fire issues from the observer's eye and coalesces with sunlight to form a medium, stretching from the visible object to the eye, through which "motions" originating in the visible object are passed to the eye and ultimately to the soul. And Aristotle argued that a potentially transparent medium is brought to a state of actual transparency when illuminated by a luminous body, such as the sun; light is simply this state of the medium. Thereafter, colored bodies in contact with this actually transparent medium produce further changes in the medium—changes that are transmitted to the observer's eye and result in visual perception of those bodies.[22]

We first see an attempt at a mathematical theory of vision in the generation after Aristotle. Euclid (fl. 300) wrote a book entitled *Optics,* in which he defined the act of vision and developed a theory of visual perspective. He argued that rectilinear rays emerge from the observer's eye in the form of a cone, with its vertex in the eye and its base on the visible object. Those things are seen on which rays fall. Having defined the visual cone, Euclid employed this geometrical entity to develop a geometrical theory of perspective. One of the postulates of the *Optics* asserts that the apparent size of an observed object is a function of the angle under which it is perceived; another maintains that the location of an observed object depends on the location within the visual cone of the ray by which it is observed (things observed by rays higher within the cone appear higher to the observer). The propositions of the book go on to analyze the appearance of an object

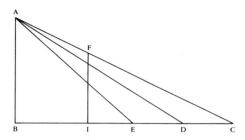

Fig. 5.15. The geometry of vision according to Euclid. A is the observer's eye, and AEC the visual cone emanating from the eye. C, the most distant of the observed points, is viewed by ray AC, which occupies a higher position within the visual cone (note where it passes through intersecting plane FI) than does ray AD, by which point D is observed.

as a function of its spatial relationship to the observer. For example, figure 5.15, shows how a more distant object intercepts a higher ray in the visual cone and thus appears higher. This is a very interesting and impressive piece of mathematical analysis, and it would prove very influential. But we should not merely be impressed by the mathematics; we should also note that the theory skips over many aspects of the process of seeing that people like Aristotle considered of fundamental importance—namely, the medium, the physical connection between the object and the eye, and the act of perception. In short, if you were willing to confine yourself to that which could be addressed geometrically, Euclid's theory was a brilliant achievement; if you were interested in any of the nongeometrical features of vision, Euclid's theory was next to useless.[23]

The greatest Hellenistic text on geometrical optics was produced 450 years after Euclid, by Ptolemy—best known as an astronomer, of course, but also author of one of the most important works on optics written before Newton. Ptolemy's *Optics* survives only in an incomplete version, but this version is sufficient to reveal the nature of Ptolemy's achievement.[24]

Ptolemy did not follow Euclid's narrow, geometrical approach to optics. Rather, he attempted to create a comprehensive theory that combined Euclid's geometrical theory of vision with a thorough analysis of the physical and psychological aspects of the visual process. Ptolemy thus presented the theory of the visual cone (applied to binocular, as well as to monocular, vision), but fleshed it out with an analysis of the radiation emanating from the eye and its interaction with visible objects. Nevertheless, the physical aspects of Ptolemy's theory did not detract from his geometrical achievement; and the geometrical portions of his text proved of major importance in teaching people how to think about vision and light geometrically.

Perhaps the most impressive aspects of Ptolemy's *Optics* from a geometrical standpoint were its theories of reflection and refraction. Oth-

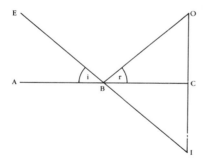

Fig. 5.16. Vision by reflected rays according to Ptolemy.

ers, including Euclid and Hero, had already written about mirrors; and Ptolemy built on their achievement. He presented a comprehensive account of reflection, which we can best explain by reference to figure 5.16. Let ABC be a plane reflecting surface, O an observed point, and E the eye. Ptolemy argued, first, that incident ray EB (remember that rays travel *from* the observer's eye *to* the observed point) and reflected ray BO define a plane that is perpendicular to the plane of the mirror; second, that the angle of incidence, *i,* is equal to the angle of reflection, *r;* and, third, that the image of O is located at I, where the extension of the ray emanating from the eye intersects the perpendicular dropped from the observed point to the reflecting surface. (In effect, the observer does not "know" that his or her visual ray has been bent by reflection at the mirror and therefore judges the object to be on the rectilinear extension of that ray.) Ptolemy applied similar rules to reflection from spherical and cylindrical mirrors, both concave and convex. He developed an impressive set of theorems dealing with the location, size, and form of images produced by reflection. It is interesting and important that he devised experiments by which to test his theory.

If Ptolemy's theory of reflection was based on the work of Euclid and Hero, his theory of refraction broke fresh ground. The basic phenomena of refraction—the illusion of the "bent" stick, half submerged in water, for example—had long been known. But Ptolemy gave refraction a thorough mathematical analysis, coupled with experimental investigation. If a ray passes obliquely from one transparent medium to another—the two media being of different optical densities—it is bent at the interface in such a way as to lie closer to the perpendicular in the denser medium. Thus, in figure 5.17, if ABC is a plane interface between air above and water below, DBF a perpendicular to that interface, E the observer's eye, and O the ob-

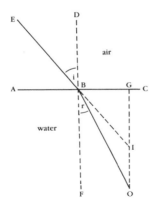

Fig. 5.17. Ptolemy's theory of refraction.

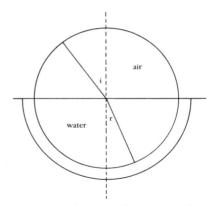

Fig. 5.18. Ptolemy's apparatus for measuring angles of incidence and refraction.

served point, the angle of incidence, EBD, is always greater than the angle of refraction, OBF. And the image of O will be located at I, where the rectilinear extension of incident ray EB intersects perpendicular OG, drawn from the observed point to the refracting surface.

Is there some fixed mathematical ratio between the angles of incidence and refraction? Ptolemy thought there must be and undertook an ingenious experimental investigation to find it. He used a bronze disk, with its circumference marked off in degrees, to measure angles of incidence and the corresponding angles of refraction in three different pairs of media (air and water, air and glass, water and glass). (See fig. 5.18.) He did not discover the desired ratio, and certainly did not discover the modern sine-law, but he did find a mathematical pattern in the data—or perhaps chose or adjusted the data to make them conform to a reasonable mathematical pattern.[25] He also passed on to future generations a thorough understanding of the basic principles of refraction, a clear and persuasive example of experimental investigation, and an important body of quantitative data.

THE SCIENCE OF WEIGHTS

The science of weights, or of the balance, was a third subject that yielded to mathematical analysis during the Hellenistic period. Indeed, it yielded more completely than the other two. In both astronomy and optics, the level of mathematization was impressive; but in both subjects there remained important physical questions for which no mathematical answer could be found. In the science of the balance beam, by contrast, the physical seemed almost completely reducible to the mathematical.[26]

The central problem was to explain the behavior of the balance beam, or lever—the fact that the beam is in balance when the weights suspended from its ends are inversely proportional to their distances (only horizontal distance counts) from the point of support or rotation. Thus a weight of 10 (fig. 5.19) on one end of a beam will balance a weight of 20 on the other end if the former is twice as far as the latter from the fulcrum. One of the earliest surviving explanations is found in a book of *Mechanical Problems,* attributed to Aristotle but actually a later product of the peripatetic school. There we find a "dynamic" account of this static phenomenon: the author explains that if a balanced beam should be set in motion, the velocities of the moving weights would be inversely proportional to the magnitude of those weights. In figure 5.20, in the time required for a weight of 20 to move distance *b,* the weight of 10 will move a distance of 2*b*. The explanatory notion operating here is that the greater velocity of the one moving body exactly compensates for the greater weight of the other.

A "static" proof of the law of the lever was produced in a treatise attributed to Euclid, and far more elegantly by Archimedes in his *On the Equilibrium of Planes*. Archimedes successfully reduced the problem to one of geometry. Except for the claim that the weights do have weight, physical considerations make no appearance. The balance beam becomes a weightless line; friction is ignored; the weights are applied to a single point of the beam and act in a direction perpendicular to it. Moreover, the demonstration based on these assumptions is Euclidean in form. Two premises provide the basis for the proof: that equal weights at equal distances from the fulcrum (and on opposite sides of it) are in equilibrium; and that equal weights situated anywhere on a lever arm may be replaced by a double weight at a point midway between them (that is, at their center

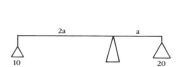

Fig. 5.19. The balance beam in a state of balance.

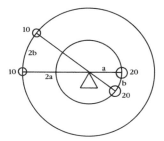

Fig. 5.20. The dynamic explanation of the balance beam.

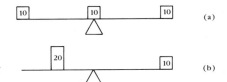

Fig. 5.21. Archimedes' static proof of
the law of the lever.

of gravity). Both premises are established by appeal to geometrical symmetry and intuition. In its simplest form, the proof asserts that the beam in
figure 5.21a, supporting three identical weights of magnitude 10, is in equilibrium by the principle of symmetry. However, we have agreed that two of
the weights can be replaced by a weight of 20 located midway between
them, as in figure 5.21b. It follows that a weight of 20 is in equilibrium with
a weight of 10 when the former is half as far as the latter from the fulcrum;
this result can be easily generalized to yield the law of the lever.

There is much more than this to Archimedes' *On the Equilibrium of
Planes;* and there are other works of his, such as *On Floating Bodies,* also
devoted to the solution of mechanical problems. But this examination of
his proof of the law of the lever reveals the thoroughness and the extraordinary skill with which he geometrized nature. Many scientific problems
continued to resist solution by mathematical methods, but Archimedes remained as a symbol of the power of mathematical analysis and a source of
inspiration for those who believed that mathematics was capable of ever
greater triumphs. His works had limited influence during the Middle Ages,
but in the Renaissance they became the basis of a powerful tradition of
mathematical science.[27]

Greek and Roman Medicine

EARLY GREEK MEDICINE

The evidence for Greek medicine is spotty, and uncertainty will always remain regarding many of the particulars of Greek healing practices. We
have only literary sources for the period prior to the fifth century. Then we
have several bodies of writings that are explicitly medical, but with severe
temporal restrictions, that inform us about medical theory and practice in
the classical and Hellenistic periods. Obviously these medical treatises
convey the views and opinions of learned physicians, many of whom were
interested in theoretical issues, such as the connection between medicine
and philosophy; but at various points they also offer revealing glimpses of
the vast substratum of popular medical belief and practice that must have
served the majority of the population. In the account that follows, we will
attempt to keep both ends of the healing spectrum in focus.

It is safe to assume that healing traditions substantially like the ones we
have seen in Egypt and Mesopotamia (chap. 1) were to be found in contemporary Greek civilization—that is, in the Bronze Age culture of the period 3000–1000 B.C. There is no question that these earliest Greeks had
contact with their Near Eastern neighbors, and we have concrete evidence
of the influence of Egyptian medical belief and practice. Thus there must
have been a wide variety of healing practices, ranging from basic surgery
and the use of internal medicines to religious incantations and dream healing. And healing practitioners of varied qualifications must have worked at
many levels, for diverse clienteles, utilizing the full range of available
medical remedies and techniques.[1]

From the ancient Greek poets Homer and Hesiod, we can glean incidental information about the nature of medical practice toward the end of this
period. In Homer's *Iliad* and *Odyssey,* the gods are implicated as causes of
plague, and they may be prayed to for healing; Hesiod, too, considers dis-

Fig. 6.1. Relief of Asclepius, the god of healing, National Archeological Museum, Athens. Alinari/Art Resource N.Y.

ease to be of divine origin.[2] Homer mentions healing incantations and pharmaceutical remedies—including some of explicit Egyptian origin. He describes a variety of wounds and, in some cases, their treatment. And he makes clear that healers were regarded as members of a distinct craft or profession—professionals in the sense that they had special skills, the exercise of which was a full-time occupation.

The religious side of healing is most clearly manifest in the cult of Asclepius, the god of healing. Asclepius, already referred to by Homer as a great physician, was subsequently deified and in the fourth and third centuries became the focus of a popular healing cult. Temples to Asclepius were built in many places—hundreds have been identified—and to these the sick would flock for cures. Central to the therapeutic process was the healing vision or dream, which was supposed to occur as the supplicant slept in a special dormitory. Healing could occur during the dream, or advice received in the dream might lead to subsequent healing. In addition, the visitor to a temple of Asclepius could expect to bathe, to offer prayers and sacrifices, and to be the recipient of purgatives, dietary restrictions, exercise, and entertainment. And, of course, it was necessary to thank the gods with a suitable offering. A number of tablets have been found at Epidaurus, the center of the cult, attesting to cures alleged to have occurred there. According to one of them, a certain Anticrates of Cnidos, who had been blinded by a spear, came to Epidaurus seeking a cure. "While sleeping he saw a vision. It seemed to him that the god pulled out the missile and then fitted into his eyelids again the so-called pupils. When day came he walked out sound."[3] Religious practices such as these remained a very significant part of ancient medicine right through the Roman period.

HIPPOCRATIC MEDICINE

In the fifth and fourth centuries, a new, more secular, and more learned medical tradition grew up alongside traditional healing practices—a medical tradition influenced by contemporary developments in philosophy and associated with the name of Hippocrates of Cos (ca. 460–ca. 370 B.C.). It is uncertain whether any of the writings (numbering between sixty and seventy) now called "Hippocratic writings" or the "Hippocratic corpus" were, in fact, written by Hippocrates. We can only affirm that they are a loosely affiliated body of medical writings, composed for the most part between about 430 and 330 B.C., later collected and attributed to Hippocrates because they shared what seemed to be "Hippocratic" traits. What were some of these traits?[4]

Fig. 6.2. The theater at Epidaurus (4th century B.C.), center of the healing cult of Asclepius. The theater seats about 14,000 people. Foto Marburg/Art Resource N.Y.

Fig. 6.3. Hippocrates (Roman copy of a Greek original). Museo di Ostia Antica.

Most prominently, the Hippocratic writings represented learned medicine. The very fact that they were "writings" already makes this point; the authors were literate. Their works were the end products of a quest for understanding. Many of the Hippocratic authors defended points of view on the nature of medicine as an art or science, on the nature and causes of disease, on the relationship of the human frame to the universe more generally, and on the principles of treatment and cure. They were engaged in what we must broadly define as natural philosophy—either as original thinkers, philosophers applying themselves to fundamental causal questions about health and disease, or as practicing physicians borrowing from the philosophical tradition. They stood at the intersection of the healing craft and the enterprise of philosophy. Hippocratic physicians may not have achieved unanimity on any of these fundamental questions, but they shared the determination to proceed in a learned manner. Even those Hippocratic authors who expressed resentment over the intrusion of philosophy into medicine did not escape its influence.[5]

What view of the medical profession do the Hippocratic writings present? We need to remember that medical practice was completely unregulated in antiquity: many kinds of healers competed for acceptance and prestige—and, of course, for business. Medicine was not studied in formal medical schools but generally learned through apprenticeship to a practicing physician. One of the concerns of the Hippocratic writings was to establish standards, drive out charlatans, and create a climate of opinion favorable to learned medicine. The stress on successful prognosis in the Hippocratic writings was designed not merely to enhance the physician's success as a healer, but also to improve his image, and thus to advance his career. Finally, the Hippocratic Oath was an attempt at self-regulation among medical practitioners.

Prominent in a number of the Hippocratic writings are theories of health and disease. What is most striking, at a general level, is the sharply reduced presence (but not total disappearance, as has sometimes been claimed) of theoretical elements of a magical and supernatural sort. The gods exist, of course, and nature itself may be regarded as divine, but intervention by the gods is ruled out as a direct cause of disease or health. We see this in various Hippocratic works, including the treatise *On the Sacred Disease* (which does not correspond exactly to any modern ailment, but includes the symptoms of epilepsy and perhaps stroke and cerebral palsy), where the author expresses the opinion

> that those who first called this disease "sacred" were the sort
> of people we now call witch-doctors, faith-healers, quacks and

charlatans. These are exactly the people who pretend to be
very pious and to be particularly wise. By invoking a divine
element they were able to screen their own failure to give
suitable treatment and so called this a "sacred" malady to conceal their ignorance of its nature.[6]

The author proceeds to offer his own naturalistic account, based on the
blockage of "veins" by phlegm from the brain. What is important here is
that nature is assumed to act uniformly; whatever the causes may be, they
are not capricious, but uniform and universal.

The Hippocratic treatises often associate disease with some imbalance
in the body or interference with its natural state. In several of the treatises,
disease is associated with the bodily humors or fluids. One version of this
theory is worked out in *On the Nature of Man,* where it is argued that four
humors—blood, phlegm, yellow bile, and black bile—are the basic constituents of the body and that imbalances among these humors are responsible for disease:

> The human body contains blood, phlegm, yellow bile and
> black bile. These are the things that make up its constitution
> and cause its pains and health. Health is primarily that state in
> which these constituent substances are in the correct propor-
> tion to each other, both in strength and quantity, and are well
> mixed. Pain occurs when one of the substances presents ei-
> ther a deficiency or an excess, or is separated in the body and
> not mixed with the others.[7]

Each of these humors was associated with a pair of the basic qualities: hot,
cold, moist, and dry. This scheme linked disease to excess or deficiency of
warmth and moisture and led to the conclusion that different humors tend
to predominate during different seasons. Phlegm, for example, which is
cold, increases in quantity during the winter; and therefore during the
winter phlegmatic ailments are particularly common. Blood predominates
in the spring, yellow bile in the summer, and black bile in the autumn.
Seasonal factors are, of course, not the only causes of disease: food, water,
air, and exercise also contribute to one's state of health.

If disease is associated with imbalance, then therapy must be directed
toward the restoration of balance. Diet and exercise (which together make
up what were called "regimen") were among the most common therapies.
Purging the body—through blood-letting, emetics, laxatives, diuretics, and
enemas—was another way of redressing an imbalance of bodily fluids.

Careful attention to seasonal and climatic factors, and to the natural disposition of the patient, was also part of successful therapy. And through it all, the physician was to keep in mind that nature has its own healing power and that the physician's most basic task is to assist the natural healing process. A considerable part of the physician's responsibility was preventive—the giving of advice on how to regulate diet, exercise, bathing, sexual activity, and other factors that would contribute to the patient's health.

But the learned physician did not merely give advice. He also engaged in what we might regard as the "clinical" side of medical practice. Various Hippocratic treatises offer instruction on examination procedures, diagnosis, and prognosis (prediction of the probable future course of the disease). We are told what symptoms to look for and how to interpret them; the physician is to examine the patient's face, eyes, hands, posture, breathing, sleep, stool, urine, vomit, and sputum; he is to be alert to coughing, sneezing, hiccuping, flatulence, fever, convulsions, pustules, tumors, and lesions. Case histories, which reveal the typical course of a given disease, are supplied. Many of these are remarkable for their precision and clarity. Consider, for example, the following description of what must have been an epidemic of mumps:

> Many people suffered from swellings near the ears, in some
> cases on one side only; in others both sides were involved.
> Usually there was no fever, and the patient was not confined
> to bed. In a few cases there was slight fever. In all cases the
> swellings subsided without harm, and none suppurated [i.e.,
> discharged pus], as do swellings caused by other disorders.
> The swellings were soft, large and spread widely; they were
> unaccompanied by inflammation or pain, and they disap-
> peared leaving no trace. Boys, young men and male adults in
> the prime of life were chiefly affected, and . . . those given to
> wrestling and gymnastics were specially liable.[8]

On the basis of the symptoms observed, the physician rendered a diagnosis and a prognosis. And finally, if the case was treatable, he prescribed treatment. Treatment, as we have noted, was often dietary or directed toward the regulation of exercise and sleep; it might also include bathing and massage. But there were many specific ailments thought to respond to certain internal or external medicines; hundreds of the latter, mostly herbal, are mentioned in the Hippocratic writings: laxatives, purges, emetics (to induce vomiting), narcotics, expectorants (to promote coughing), salves, plasters, and powders. And finally, the treatment of wounds, fractures, and dislocations is also covered in the Hippocratic writings—with a level of skill that has elicited admiration from modern physicians.

Finally, we must say a word about the principles of inquiry embodied in this medical literature. Once we get beyond the commitment to a critical approach to the healing enterprise and the resolve to employ naturalistic principles of explanation and therapy, unanimity disappears. Some treatises display a strong inclination toward philosophical speculation. The author of *The Nature of Man,* for example, offers a speculative theory of human nature and of health and disease, and from this theory draws out several therapeutic principles. However, other treatises within the Hippocratic corpus attack the speculative approach. The author of *On Ancient Medicine* expresses skepticism about the use of hypotheses in medicine, especially the hypothesis that disease is the result of imbalance among the four qualities; he argues that this theory does not lead to remedies that differ in any significant way from remedies prescribed by other physicians, but simply surrounds them with a fog of "technical gibberish."[9] He, and other Hippocratic authors of skeptical bent, preferred to have physicians proceed cautiously, on the basis of accumulated experience, accepting causal theories only when they were supported by overwhelming evidence. As we have seen, the admonition to proceed experientially bore fruit in the careful diagnostic procedures and the impressive case histories of the Hippocratic corpus. Occasionally we even find a specific observation made to confirm a theoretical conclusion—as in *On the Sacred Disease,* where the author proposed dissection of a goat that had the disease, in order to demonstrate that the ailment results from accumulation of phlegm in the brain.[10]

We must conclude this discussion of Hippocratic medicine with two cautionary reminders. First, when learned medicine appeared, it did not drive out its rivals. Learned medicine was never the only kind of medicine, nor even the most popular, but functioned alongside traditional forms of healing belief and practice. Throughout Greek antiquity (from the fifth century B.C. onward) the sick could turn for help to learned physicians, priestly healers in the temples of Asclepius, midwives, herb-gatherers, and bonesetters. Moreover, there can be no doubt that the lines demarcating these various types of healers were vague—so that, for example, temple healing might be closely associated with learned medicine. Furthermore, there seems no question that the sick sometimes experimented with alternative types of healing simultaneously or sequentially.

Second, if traditional medical practices continued to exist alongside learned medicine, they were also, to some extent, incorporated within it. That is, we must not inflate learned Greek medicine into an early version of modern medicine. Greek medicine was . . . well, Greek. It had to be fitted into the worldview and philosophical outlook of the ancient Greeks; and that means it did not exclude all medical beliefs and practices that the modern Western physician would find bizarre or repugnant. Thus dream healings remained a part of medicine, including Hippocratic medicine, throughout antiquity.[11] And although divine interference was ruled out, religious elements did not altogether disappear. To offer the simplest example, in the opening lines of the Hippocratic oath, the physician swears by Apollo and Asclepius and calls on the gods and goddesses to witness his oath. If we are tempted to dismiss this example on the grounds that it may represent empty ritual (comparable to atheists or agnostics swearing on the Bible in a courtroom), a more persuasive case may be found in the Hippocratic author who recommends prayers along with regimen.[12] Or to consider a subtler case of religious presence, when the author of *On the Sacred Disease* denies that disease is the result of divine intervention, he is only arguing that every disease has a natural cause; he is not opposed to the view that this natural cause is itself an aspect or a manifestation of divine agency. Most Hippocratic physicians undoubtedly continued to believe that natural things partake of divinity, and that disease is simultaneously natural and divine.

HELLENISTIC ANATOMY AND PHYSIOLOGY

Our sources for Greek medicine are strangely bifurcated. We have the Hippocratic writings, which teach us much about early Greek medicine; and we have a variety of sources from the early Christian era, which give us a good picture of medicine under the Roman Empire. But there is an intervening period of four hundred to five hundred years for which we have only fragmentary remains of medical literature. The explanation is not that medicine ceased to be practiced or that medical treatises ceased to be written (although the production of medical treatises no doubt had its ups and downs); it is rather that, for reasons unknown to us, the medical writings of this intervening period did not survive. Developments in medicine during this period must, therefore, be reconstructed from fragmentary descriptions in the work of later authors.[13]

Knowledge of human anatomy and physiology among Hippocratic physicians seems to have been quite limited. There is little evidence of the systematic dissection of human bodies during, or prior to, the period in which the Hippocratic treatises were produced—owing, no doubt, to traditional taboos regarding proper burial of the dead and perhaps also to the absence of any good reason for supposing that human dissection could provide beneficial knowledge. Such anatomical knowledge as existed was undoubtedly acquired in the course of surgery or the treatment of wounds or by analogy with animal anatomy (well understood, thanks to Aristotle).

It was an event of great significance, therefore, when the practice of human dissection began in Alexandria in the third century B.C.[14] How this extraordinary innovation came about, we do not exactly know. It was undoubtedly connected with royal patronage from the Ptolemaic dynasty, which was powerful enough to violate traditional burial taboos if it so desired; it may have had something to do with medical developments, which elevated the significance of anatomical knowledge, or with the transplantation of Greek medicine into a new social and religious setting; it appears to have occurred within a philosophical context in which new questions coming to the fore may have called for new methods of investigation. Whatever the reasons, ancient testimony is virtually unanimous in maintaining that Herophilus of Chalcedon and Erasistratus of Ceos were the first to engage in systematic dissection of the human body; if we are to believe the Roman encyclopedist Celsus and the church father Tertullian, they even engaged in the vivisection of prisoners.

What did they learn? Herophilus (d. ca. 260 or 250 B.C.), a native of Asia Minor, studied medicine under Praxagoras of Cos before migrating to Alexandria, where he worked under the patronage of the first two Ptolemaic rulers. His pathological theory and therapeutic practice, insofar as we can tell, seem to have been Hippocratic in character; it was as an anatomist that he broke fresh ground.[15] Herophilus investigated the anatomy of the brain and nervous system, identifying two of the brain's membranes (the dura mater and pia mater) and tracing the connections between the nerves, the spinal cord, and the brain. His distinction between sensory and motor nerves reveals his understanding of the functions of the nervous system. He examined the eye with great care, identifying its principal humors and tunics and creating a technical nomenclature that has survived to the present; he traced the optic nerve from the eye to the brain and argued that it was filled with a subtle pneuma.

Herophilus also explored the organs of the abdominal cavity. He presented careful descriptions of the liver, the pancreas, the intestines, the reproductive organs, and the heart. He distinguished veins from arteries by the thickness of their walls. He examined the valves in the heart. He studied the arterial pulse—though he did not understand it as a simple mechanical response to the pumping action of the heart—and employed variations in pulse rhythms as a diagnostic and prognostic tool. He described the ovaries and the Fallopian tubes and wrote a treatise on obstetrics. Even this brief sketch reveals Herophilus's remarkable achievements as a student of human anatomy and physiology.

His work was continued by his approximate contemporary, Erasistratus (b. ca. 304 B.C.), from the island of Ceos, who had studied medicine in Athens within the peripatetic school and at Cos.[16] Erasistratus followed up and improved on Herophilus's investigations of the structure of the brain and the heart. He supplied an excellent description (which Galen quotes for us) of the bicuspid and tricuspid valves and their function in determining one-way flow through the heart; the heart, in Erasistratus' view, functioned as a bellows, expanding to draw blood or pneuma into it, contracting to expel blood into the veins and pneuma into the arteries. The expansion and contraction of the heart, Erasistratus held, were the result of an innate faculty residing in the heart; he argued, correctly, that expansion of the artery during the arterial pulse was simply a passive response to the expansion and contraction of the heart.

Although we have important pieces of physiological theory from Herophilus (his theory of the pulse, for example), he seems to have been more interested in structure than in function. We find a great deal more physiology in the work of Erasistratus. Apparently influenced by the peripatetic school, Strato in particular, Erasistratus argued that matter consists of tiny particles separated by minute void spaces; he combined this corpuscularianism with the theory of pneuma, to explain a variety of physiological processes. Let us use his explanations of digestion, respiration, and the vascular systems (of particular importance because of their subsequent influence on Galen) for illustrative purposes.

Erasistratus believed that all tissue in the body contains veins, arteries, and nerves, and that these vessels serve as channels by which various substances fundamental to the functioning of the body are conducted to its various organs. Food enters the stomach, where it is mechanically reduced to juice, which passes through tiny pores in the walls of the stomach and intestines to the liver, there to be converted into blood. Blood is then sent through the veins to all parts of the body, where it is responsible for nourishment and growth. The arteries, by contrast, contain only pneuma, inhaled from the atmosphere in respiration and drawn down to the left side of the heart through the "vein-like artery" (our pulmonary vein); from the heart, pneuma is distributed through the arteries to all parts of the body, endowing these parts with their vital capacities. Finally, the nerves contain a finer form of pneuma, "psychic" pneuma, produced from arterial pneuma by refinement in the brain and responsible for sensation and motor functions. To explain the movement of these substances throughout the body, Erasistratus appealed to nature's abhorrence of a vacuum: the pumping action of the heart or the consumption or wastage of matter in an organ requires that blood or pneuma be immediately drawn in to occupy the newly created or newly vacated space.

This is a very impressive theory, portions of which were to survive within Western physiological thought for nearly two thousand years. But even in Erasistratus' day, an apparently fatal objection was posed—namely, that when an artery (the channel by which *pneuma* is conducted to all parts of the body) is cut, *blood* flows from it. Erasistratus responded to the objection by arguing that the veins and arteries do not, under normal circumstances, communicate; however, when an artery is opened, the escaping pneuma creates, or threatens to create, a vacuum; this potential vacuum, in turn, opens up tiny channels (anastomoses) between the veins and arteries, allowing blood to pass temporarily from the veins to the arteries and follow the escaping pneuma out through the wound.

Erasistratus' theory of nutrition and vascular flow led easily to a theory of disease. Erasistratus held that disease is caused principally by the flooding of veins with surplus blood, owing to excessive eating. If, for example, the veins are sufficiently charged with blood, the normally closed channels between the venous and arterial systems may be forced open; blood may then pass into the arteries and be sent through the arterial system to the extremities, where it causes inflammation and fever. It follows, from such a theory of disease, that therapy must be directed toward reducing the quantity of blood. This may be accomplished through limiting the intake of food or (less commonly) through blood-letting.

HELLENISTIC MEDICAL SECTS

Herophilus and Erasistratus attracted substantial attention within the medical world and drew leading physicians and medical theorists into their orbits. These students and observers, though no doubt inspired by the example and the teachings of the two men, seem not to have considered themselves bound to any sort of doctrinal orthodoxy. After all, Herophilus and Erasistratus had themselves disagreed on many issues. A student of Herophilus, Philinus of Cos, wrote a book against certain teachings of Herophilus and the Herophileans, which set off a round of attack and counterattack. Over the next several centuries a substantial polemical literature was produced by the Herophileans and their critics (who became known

Fig. 6.4. Greek physician, grave relief, 480 B.C. Antikenmuseum Basel, Inv. no. BS 236.

as "empiricists"). Hellenistic medicine was beginning to break up into rival medical sects, each with its own medical theories and its own methodological program.

In the long run, several groups emerged.[17] One family of sects, descended in part from the Herophileans and Erasistrateans, was already lumped together in antiquity under the rubric "rationalists" or "dogmatists." It must be stressed that the "rationalists" or "dogmatists" did not constitute a unified or coherent movement but were divided on many issues; if they were united by anything, it was their general commitment to speculative, theoretical medicine—the attempt to apply, to the medical realm, the methods of natural philosophy that we have observed in the principal philosophical schools. *Some* "rationalists" continued to defend human dissection as a valuable methodological tool, capable of contributing to the formulation of hypotheses regarding the hidden causes of disease; what *all* would have agreed on is the value of physiological theory for the practice of medicine.

Their principal rivals and detractors, the "empiricists," adopted the radically opposite opinion that theoretical speculation, including the quest for physiological knowledge and the hidden causes of disease, was a waste of time; and, especially, that human dissection made no useful contribution to medical knowledge and should be forbidden. The "empiricists" maintained, in short, that the anatomical and physiological tradition developed by Herophilus, Erasistratus, and their theoretically oriented followers was a medical dead end, which was to be avoided. The successful physician should concentrate on visible symptoms and visible causes and recommend therapy on the basis of past experience (his own and that of his predecessors) of the efficacy of various remedies.

In the first century A.D., there appeared in Rome a third group of physicians, known as "methodists," whose basic claim was that the "rationalists" and "empiricists" had made medicine unnecessarily complicated—that the intricacies of learned medicine, including anatomy and physiology and the quest for the causes of disease (both hidden and visible), could be dispensed with. The core of "methodist" teaching was that disease depends on tenseness and laxness of the body and that the prescribed treatment follows directly and "methodically" from this premise. Such teaching proved very popular among Roman aristocrats, whose support made "methodism" a powerful medical force in Rome and throughout the Hellenistic world. A fourth doctrinal school was that of the "pneumatists," who built a medical philosophy on Stoic principles. And finally, we must mention Asclepiades of Bithynia (fl. 90–75 B.C.), an influential Roman physician, who repudiated humoral theories in favor of atomistic doctrines.

GALEN AND THE CULMINATION OF HELLENISTIC MEDICINE

It was this medical world that Galen entered when, at age sixteen, he decided on a medical career. Born in Pergamum (one of the leading intellectual centers of Asia Minor and of the entire Hellenistic world) in 129 A.D., Galen studied philosophy and mathematics before turning to medicine.[18] His travels, in pursuit first of education and later of patronage, illustrate the high level of mobility enjoyed by scholars in the ancient world. Galen studied medicine in Pergamum and Smyrna (both in Asia Minor), then in Corinth on the Greek mainland, and finally in Alexandria. From Alexandria, he returned to Pergamum as physician to the gladiators, then moved to Rome in search of patronage, returned to Pergamum, went back to Italy, and eventually settled in Rome, where he enjoyed the friendship and served the medical needs of the rich and powerful, including the Emperors Marcus Aurelius, Commodus, and Septimius Severus. He died after 210. Galen produced an enormous body of writings, the surviving portions of which occupy twenty-two volumes in the standard nineteenth-century edition. These writings, which summed up the knowledge and adjudicated the principal disputes of the ancient tradition of learned medicine, established Galen as the leading medical authority of antiquity—rivaled only by Hippocrates—and gave him unparalleled influence well into the modern period.

Galen was a broadly educated philosopher, informed on all of the major philosophical controversies of antiquity and committed to the integration of medicine and philosophy. He was powerfully influenced by the Hippocratic corpus, by Plato, Aristotle, and the Stoics, by the anatomical and physiological works of Herophilus and Erasistratus and the medical controversies of the Hellenistic period. He has been described as an eclectic rationalist,[19] more interested in the disease than in the patient, viewing the latter as a vehicle by which to gain understanding of the former. Central to his medical aims was the need to classify diseases—to discover the universals that lay behind the particulars—and to search for their hidden causes. And he was convinced that anatomical and physiological knowledge was essential to the success of this venture.

Hippocratic influence was of critical importance in shaping Galen's medical philosophy (though he felt free to borrow selectively and to interpret the borrowed elements loosely). It furnished his view of the human frame and the task of the physician, his stress on the importance of clinical observation and case histories, his concern with diagnosis and prognosis, and his general therapeutic notions. From the Hippocratic treatise *On the Nature of Man,* Galen took the doctrine of the four humors—the view that the fundamental constituents of the human body are blood, phlegm, yellow bile, and black bile, reducible in turn to the basic qualities hot, cold, wet, and dry. The four humors, he argued, come together to form tissues; tissues combine to form organs; organs unite to make up the body.

Disease may be connected either with disequilibrium among the humors and their constituent qualities or with the specific state of particular organs; one of Galen's principal innovations in the art of diagnosis was to localize disease by identifying specific afflicted organs. Galen's discussion of fevers illustrates both aspects of his theory of disease. Generalized fevers, he argues, are produced throughout the body by the heat of putrefying humors; localized fevers result from noxious or toxic humors within a specific organ, leading to changes, such as hardening or swelling, and to pain. For purposes of diagnosis, Galen leaned especially on pulse and examination of the urine; but he also perceived the need to examine all of the other signs stressed in the Hippocratic corpus. In his *On the Art of Healing,* he wrote:

> When you meet the patient, you study the most important symptoms without forgetting the most trivial. What the most important tell us is corroborated by the others. One generally obtains the major indications in fevers from the pulse and the urine. It is essential to add to these the other signs, as Hippocratics taught, such as those that appear in the face, the posture the patient adopts in bed, the breathing, the nature of the upper and lower excretions, . . . presence or absence of headache, . . . prostration or good spirits in the patient, . . . [and] the appearance of the body.[20]

Essential to the successful practice of medicine, Galen believed, was knowledge of the structure and functioning of the individual organs. He preached the importance of anatomical knowledge, but acknowledged that in his day dissection of humans was no longer possible. He urged his reader to be alert to the possibility of fortuitous anatomical observation, through the disintegration of a tomb or the discovery of a skeleton along a roadside; for those who could manage it, he recommended a visit to Alexandria, where the skeleton could still be examined firsthand; but he acknowledged that for the most part human anatomy would have to be inferred by analogy, from the dissection of animals whose anatomy resembles that of humans. Galen himself dissected a variety of animals, including a small monkey known as the barbary ape (the macaque). His skill as an anatomist is obvious from several anatomical works, including a guide to dissection, *On Anatomical Procedures*. He supplied excellent descriptions of the bones, the muscles, the brain and nervous system, the eyes, the veins and arteries, and the heart. He borrowed, of course, from the work of Herophilus and Erasistratus; but he did not hesitate to correct these predecessors when he felt they were in error. Unfortunately, Galen's dissection of animals led to the mistaken attribution to humans of certain anatomical features found only in his animal sources; the most notorious case is that of the *rete mirabile,* to which we will return. Nonetheless, it was Galen's anatomical works, rather than those of Herophilus and Erasistratus, that survived; and thus it was Galen who supplied Europe with its only systematic account of human anatomy until the Renaissance.

Galen's physiological system had even more complex roots. Plato had argued for a tripartite soul, consisting of a superior ("rational") part and two inferior parts (associated with the passions and appetites), lodged, respectively, in the brain, the chest, and the abdominal cavity. Galen adopted this scheme and proceeded to correlate the three faculties of the soul identified by Plato with the three basic physiological functions defined by Erasistratus; the result was a tripartite organizational framework for physiology. In this scheme, the brain (seat of the soul's rational faculties) was identified as the source of the nerves. Following Erasistratus, Galen argued that the nerves contain psychic pneuma, which accounts for sensation and motor functions. The heart (seat of the passions) became for Galen the source of the arteries, which convey life-giving arterial blood (and vital pneuma) to all parts of the body. And the liver (seat of desire or appetite) was made the source of the veins, which nourish the body with venous blood.[21]

These three physiological systems, as Galen developed them, were not totally independent but had interconnections. It may be helpful, therefore, if we work through them from beginning to end—from the initial intake of food to the final distribution of psychic pneuma through the nerves. Food arrives in the stomach, where it is reduced to juice (*chyle* is the Greek term)—not merely through mechanical action, as Erasistratus believed, but through cooking by the body's vital heat. *Chyle* passes through the walls of the stomach and intestines into the surrounding mesentery veins, which convey it to the liver. In the liver, *chyle* undergoes further refinement and cooking, to yield venous blood. This venous blood, which is nourishment for the body, moves slowly outward through the veins to the various tissues and organs, where it is consumed. Thus the venous system has its origin in the liver; it conveys venous blood to all parts of the body; and it is responsible for nourishment.[22]

Venous blood reaches the right side of the heart through the *vena cava*. A major blood vessel (Galen's artery-like vein, our pulmonary artery) carries some of this venous blood to the lungs, which, like all other organs, require nourishment. The remainder of the venous blood seeps slowly through tiny pores in the heavy muscle (the interventricular septum) that divides the right ventricle of the heart from the left. Galen acknowledged that these pores are too small to be visible, but he argued that since the incoming *vena cava* is larger than the outgoing artery-like vein, some venous blood must go elsewhere; moreover, the disparity in size is too great to be accounted for by the fact that the heart (like every other organ) consumes a certain amount of venous blood as nourishment; and finally, the principle that nature does nothing in vain guarantees that the small pits in the surface of the interventricular septum must lead somewhere. It follows that

> the thinnest portion of the blood is drawn from the right ven-
> tricle into the left, owing to there being perforations in the
> septum between them: these [perforations] can be seen for
> a great part [of their length]; they are like pits with wide
> mouths, and they get constantly narrower; it is not possible,
> however, actually to observe their extreme terminations,
> owing both to the smallness of these and to the fact that when
> the animal is dead all the parts are chilled and shrunken.[23]

What happens when venous blood reaches the left side of the heart? Here we must introduce Galen's theories of vitality and respiration.[24] Galen joined Plato, Aristotle, and the anonymous author of *On the Heart* (a treatise formerly considered Hippocratic, but probably Hellenistic) in identifying life with innate heat; moreover, he shared their view that the principal seat of this life-giving heat is the heart. Maintaining the right degree of vital heat is critical, of course, and it is the lungs and respiration that perform this function. On the one hand, the lungs surround the heart and diminish or moderate its heat. On the other, they nourish the "fire" within the heart by sending air to it through the vein-like artery (our pulmonary vein); and by the same mechanism they provide a way for the heart to rid itself of the waste products of burning. As the heart expands, air is drawn from the lungs into the left ventricle of the heart; as the heart contracts, soot and smoky vapors are sent in the other direction and exhaled into the atmosphere. Air reaching the left ventricle of the heart during the expansive phase mixes with venous blood that has passed through the interventricular septum—blood that has also been heated and thus vivified by the innate heat in the heart. The product is a finer, purer, and warmer arterial blood, now charged with vital spirit or pneuma, which is conveyed throughout the body by the arteries. In the course of defending this theory, Galen devoted considerable attention to proving, contrary to Erasistratus, that the arteries do indeed contain blood. Thus we have Galen's second major physiological system—the arterial system, rooted in the heart, conveying arterial blood through the arteries, imparting life to the tissues and organs of the body.

The brain, like every other organ, is the recipient of arterial blood. Some of this arterial blood passes into the *rete mirabile*—a network of fine arteries found in certain ungulates (where it serves cooling functions) and mistakenly attributed by Galen to humans. In passing through the arteries of the *rete mirabile* the arterial blood is refined and emerges as the finest grade of spirit or pneuma—psychic pneuma. This pneuma is sent to all parts of the body through the nerves and accounts for sensation and motor functions; and thus we have the third of Galen's major physiological systems.

Before leaving Galen's physiology, it is necessary to make one more point. Galen found Erasistratus' attempt to mechanize physiology unpersuasive. In particular, he did not believe that the movement of fluids through the body could be sufficiently explained in terms of pumping action or nature's abhorrence of a vacuum. He accepted that the heart functions as a bellows, drawing air from the lungs during expansion and propelling arterial blood into the arteries during contraction, and that the arteries themselves have active motions that move fluids along. But in addition, he was persuaded that all organs possess nonmechanical faculties by which they attract, retain, and repel fluids on the basis of need. Thus the liver has the capacity to draw to itself the *chyle* that it requires. Similarly, venous blood moves through the body not because it is pumped, but because the organs of the body attract, retain, and repel it according to their need for nourishment.

Galen's medical system proved exceedingly persuasive, dominating medical thought and teaching throughout the Middle Ages and into the early modern period. Part of its persuasive appeal lay in its comprehensiveness. Galen addressed all of the major medical issues of the day. He could be both practical, as in his pharmacology, and theoretical, as in his physiology. He was philosophically informed and methodologically sophisticated.[25] His work embodied the best of Greek pathological and therapeutic theory. It contained an impressive account of human anatomy and a brilliant synthesis of Greek physiological thought. In short, Galen offered a complete medical philosophy, which made excellent sense of the phenomena of health, disease, and healing.

But there was another reason for Galen's popularity. Galen introduced into his anatomy and physiology a massive dose of teleology, which would endear him to Islamic and Christian readers. Galen was not himself a Christian, and his teleological approach had no Christian roots but was inspired by Plato's *Timaeus* and Aristotle's *The Parts of Animals* and by Stoic thought. Like Aristotle—indeed, more than Aristotle—Galen found evidence of intelligent design in the animal and human frame, and his *On the Usefulness of the Parts of the Body* is a litany of praise to the wisdom and providence of the Demiurge (a term and a conception obviously borrowed from Plato). In this book Galen wrote:

> And I consider that I am really showing him [the Demiurge]
> reverence not when I offer him [sacrifices] . . . of bulls and
> burn incense . . . , but when I myself first learn to know his
> wisdom, power, and goodness, and then make them known to
> others. I regard it as proof of perfect goodness that one
> should will to order everything in the best possible way, not
> grudging benefits to any creature, and therefore we must
> praise him as good. But to have discovered how everything
> should best be ordered is the height of wisdom, and to have
> accomplished his will in all things is proof of his invincible
> power.[26]

Galen argued that Nature (or the Demiurge) does nothing in vain; that the structure of the human body is perfectly adapted to its functions, unable to be improved even in imagination. Galen even presented the beginnings of a natural theology—that is, a theory of god or the gods based on evidence found in nature. At the conclusion of *The Usefulness of Parts,* he called attention to lessons that can be learned about the world soul from the investigation of animal anatomy:

> For when in mud and slime, in marshes, and in rotting plants
> and fruits, animals are engendered which yet bear a mar-
> velous indication of the intelligence constructing them, what
> must we think of the bodies above [i.e., the celestial bodies]?
> . . . Thus, when anyone looking at the facts with an open
> mind sees that in such slime of fleshes and juices there is yet
> an indwelling intelligence and sees too the structure of any
> animal whatsoever—for they all give evidence of a wise Cre-

ator—he will understand the excellence of the intelligence in the heavens.[27]

As the reader can easily imagine, Galen's teleology, as well as his desire to fit humans and their diseases into a complete and satisfying (and, of course, ancient) worldview, have not always gone down well with modern scholars. Indeed, Galen has been the target of a certain amount of abuse from medical historians, angry at him for not being modern.[28] Of course, Galen was merely being a second-century Greco-Roman; and if we concentrate on his deficiencies from a modern standpoint, we are apt to miss the opportunity to learn from his life and thought what it meant to be a physician in the declining years of Greco-Roman civilization. Galen pulled together several strands of ancient thought: he summed up more than six hundred years of Greek and Roman medicine; at the same time, he fitted that medicine into an ancient philosophical and theological framework. That teleology pervades Galen's work is a useful reminder that the question of order and organization in the universe remained a problem of central importance, which every major thinker still felt obliged to address and on which the last word had yet to be said—a problem, indeed, on which the last word has yet to be said today. That the gods figure into Galen's worldview, and even into his medical system, is not a feature to be regretted, but one to be understood as typical of ancient medicine and philosophy. Galen did not differ substantially from the Hippocratic writers, or from his principal philosophical guides, in his view of the gods. Although he admitted divinity into the medical realm, as in his acknowledgement of the healing power of Asclepius,[29] he did not permit this belief to interfere with the formulation of a medical philosophy restricted to natural causation. Galen certainly believed that behind the admirable design found in living things could be discerned a designer; but this belief had no major influence on his analysis of disease or on his diagnostic and therapeutic procedures.

SEVEN

Roman and Early Medieval Science

GREEKS AND ROMANS

Galen's career (examined in the previous chapter) nicely illustrates the interpenetration of Greek and Roman intellectual life. Born and raised in Pergamum in Asia Minor—well within the boundaries of the Roman Empire, but still a stronghold of Greek culture—Galen continued his education in Corinth (on the Greek mainland) and in Alexandria. It was a Greek education that he received—delivered in Greek and based on the Greek classics—and it was the Greek intellectual tradition that he thus joined. But Galen finished his career in Rome, serving Roman emperors and lecturing to Roman audiences. His biography thus raises the question to which the opening sections of this chapter will be devoted: what was the political, cultural, intellectual, and especially scientific relationship between Greece and Rome?

The autonomy and the dynamic political life of the Greek city-states ended with the conquests of Alexander the Great (334–323 B.C.) and the establishment of a Greek empire. However, intellectual life in the successor states, after the division of Alexander's empire among his generals, was the object of sporadic, sometimes generous, patronage and remained vigorous, at least for a time. Meanwhile, Rome grew from an insignificant Etruscan town in the seventh century B.C. to a thriving republic in the fifth and fourth centuries. By 265 B.C. it controlled the Italian peninsula, and by 200 it had enough of a foreign policy and sufficient military might to intervene in Greek affairs during the Second Macedonian War (200–197). Rome gradually extended its influence over Greek lands in the next 150 years; by the death of Julius Caesar in 44 B.C., Rome was in control of virtually the entire Mediterranean basin, including Greece, Asia Minor, and North Africa (see map 3).

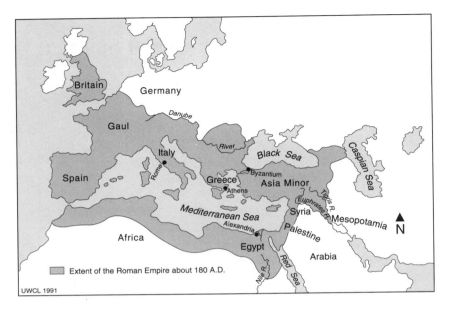

Map 3. The Roman Empire

In the Greek provinces, Roman control did not lead to the collapse of culture and learning. On the contrary, as the Roman writer Horace (d. 8 B.C.) pointed out so famously, while Rome captured Greece militarily and politically, the artistic and intellectual conquest belonged to the Greeks.[1] As Rome's power and prosperity increased, its leisured class began to appreciate Greek achievements in literature, philosophy, politics, and the arts. Any Roman wishing to acquire sophistication in such matters could do no better than to imitate the Greek achievement—to borrow from a culture of superior accomplishment in these areas.

The linguistic and geographical barriers that one might expect to have hindered such borrowing did not, in fact, prove to be a serious problem in the early years of cultural contact. The ability to read and speak Greek was common in Italy, which had Greek settlements going back many centuries: recall, for example, Parmenides and Zeno from the city of Elea and the Pythagoreans of southern Italy visited by Plato. Rome itself had a community of Greeks by the second century B.C., and bilingualism (at some level) flourished among the Roman upper classes. With increasing frequency Greek scholars settled in Rome, voluntarily or as slaves; and it was easy to find Greek teachers willing to expound the content of Greek literature and philosophy. Study abroad in the Greek provinces was another alternative,

Fig. 7.1. The ancient forum in Rome. Alinari/Art Resource N.Y.

which, for a young Roman with serious scholarly aspirations, became almost obligatory. By such mechanisms, Rome and its environs acquired a substantial circle of Greek and Roman scholars—all of them in touch with the Greek learned tradition. Eventually, Roman scholars began to convey aspects of the Greek intellectual achievement to a Latin readership. In a few cases, texts were even translated from Greek to Latin.[2]

A number of these circumstances are illustrated by the career of Cicero (106–43 B.C.), a highly educated Roman statesman and man of letters. Cicero studied with Greek teachers, first in Rome, subsequently in Athens and on the island of Rhodes; he learned Greek, of course, and also mastered significant portions of Greek philosophy and was strongly influenced by Stoicism and the epistemological theories that developed within the Platonic school in the third century.[3] Cicero wrote Latin treatises on a variety of topics and produced a Latin translation of Plato's *Timaeus* (which has not survived).

Support for scholarship was, at the beginning, entirely private. A member of the upper class might devote some of his leisure time to reading and learned discussion; he might have a library, possibly even a substantial one. But anybody who lacked means of his own would have to find a patron. The possible arrangements actually covered a wide spectrum, from distinguished scholars attached to the households of the wealthy to educated, Greek-speaking slaves. The obligations of the scholar who made it to the top of this ladder might be to advise or provide intellectual companionship for his patron, or to care for his patron's library; if less fortunate or less able, he would likely be charged with the education of his patron's children and perhaps be given menial tasks in addition.

The level of discourse in these settings varied. The Roman scholar who wished to proceed at the highest level would do it in Greek. It follows that such scholarly discourse as occurred in Latin (whether spoken or written) fell somewhat below the highest levels of Greek scholarship, which have thus far occupied our attention. Latin was employed when the linguistic limitations of the audience demanded; and it was a lighter, more popular version of Greek learning that appealed to that audience. Certain prominent historians of science, sneering at popularization, as though nothing but "cutting-edge" research matters, have been very critical of the Greeks for having developed a popular level of learning, and of the Romans for drawing on it.[4] But this reflects a very narrow outlook. In fact, there must be multiple levels of knowledge and expertise within any scholarly tradition. For every Aristotle, capable of confronting perplexing philosophical or scientific problems in original ways, there were thousands of educated Greeks and Romans whose aspirations did not and could not go beyond grasping what Aristotle had achieved or reconciling Aristotle's views with those of other acknowledged authorities. Inevitably, any program of creative research is accompanied by other movements directed toward preservation, commentary, education, popularization, and transmission. We can see this in our own school system.

Given these circumstances, it was natural for scholars setting out to sample and interpret the Greek intellectual achievement for a Roman audience to concentrate on that which interested their Roman patrons—not the subtleties of Greek metaphysics and epistemology, nor the technical details of Greek mathematics, astronomy, and anatomy, but subjects of practical value and intrinsic appeal. A certain amount of mathematics would be included for utilitarian reasons, or as training for the intellect. Medicine hardly needed justification, though Romans were initially suspicious of certain aspects of Greek medicine. Logic and rhetoric were important in the law courts and the political arena. And Epicurean and Stoic philosophy addressed pressing ethical and religious concerns. But science or natural philosophy, beyond the basics, was rarely valued except as an amusement. This state of affairs is vividly illustrated by the fact that for Romans the most celebrated astronomical authority was Aratus of Soli (d. 240 B.C.), whose poem on the constellations and weather prognostication (*The Phaenomena*) was translated into Latin at least four times, while the technical works of Eudoxus and Hipparchus remained unavailable or unknown.[5]

Such science or natural philosophy as Romans knew, then, tended to be a limited, popularized version of the Greek achievement. Generations of historians have sought to explain the Roman failure to master the more abstruse or technical aspects of Greek learning in terms of intellectual inferiority, moral weakness, or temperamental defect. It is often said that Romans simply did not have theoretical minds—though it is then quickly added (because everybody has to be good at something) that they made up for this deficiency with administrative and engineering talent.[6] In fact, there is no mystery about the level or the degree of Roman intellectual effort and no reason to be surprised or critical. We need always to remember that the Roman aristocracy regarded learning, except for clearly utilitarian matters, as a leisure-time pursuit. Romans, then, did the obvious thing: they borrowed what seemed interesting or useful. If certain Greeks had devoted their lives to subjects that were abstract, technical, impractical, and (as some no doubt judged) boring, that was no reason for large numbers of Romans to make the same mistake. Members of the Roman upper class had about the same level of interest in the fine points of Greek natural philosophy as the average American politician has in metaphysics and epistemology. At best, their desire was, as the Roman playwright Ennius put it, "to study philosophy, but in moderation."[7] The only surprise is that historians should have expected it to be otherwise.

POPULARIZERS AND ENCYCLOPEDISTS

I have described the circumstances under which science and natural philosophy were pursued by members of the Roman upper classes, and the factors that motivated them. I must now illustrate the resulting intellectual tradition, examining specific genres of Latin literature that dealt with scientific subjects or shaped the intellectual environment within which science was pursued, and surveying a few of the most influential works.

One of the best known and perhaps most influential of the early popularizers was the Stoic Posidonius (ca. 135–51 B.C.). Born in Syria of Greek parents, Posidonius studied in Athens and subsequently became head of the Stoic school on the island of Rhodes. He exercised a strong indirect influence on Roman intellectual life through his many pupils, one of whom was Cicero; but he also traveled to Rome and impressed the Romans in person. Posidonius was the closest thing to a universal scholar that we can find in the first century B.C. He was interested in history, geography, moral philosophy, and natural philosophy and wrote voluminously on all of these subjects. Among his works (all written in Greek) were commentaries on Plato's *Timaeus* and Aristotle's *Meteorology;* Lucretius borrowed heavily from the latter in writing his *On the Nature of Things.*

Posidonius's works have not survived, and our knowledge of them is therefore secondhand; but one of his most influential investigations seems to have been the determination of the earth's circumference—which he put first at 240,000 stades (a little smaller than Eratosthenes' estimate), later at 180,000 stades. The importance of this smaller figure is that it was picked up by Ptolemy, passed on to the readers of Ptolemy's *Geography,* and used in the fifteenth century by Christopher Columbus as the basis for his calculation of the distance between Spain and the Indies.

Posidonius exercised substantial influence on Latin writers, such as Varro (116–27 B.C.), thereby helping to determine the form and content of education and scholarship in Latin. Varro, regarded by his Roman admirers as a phenomenal scholar, studied in Rome and Athens; he proved to be a prolific writer, turning out Latin works on a variety of topics (about seventy-five titles, almost all of them now lost). The most important of these was an encyclopedia, *Nine Books of Disciplines,* which was to become a model and source for subsequent Roman encyclopedists. One notable feature of the *Disciplines* was its use of the liberal arts (the subjects deemed suitable for the education of a Roman gentleman) as organizing principles. Varro identified and gave a basic account of nine such arts: grammar, rhetoric, logic, arithmetic, geometry, astronomy, musical theory, medicine, and architecture. Varro's list, narrowed by subsequent writers who omitted the last two arts, came to define the classical seven liberal arts of the medieval schools—the first three arts coming to be known as the trivium, the remaining four as the quadrivium.[8]

Varro's contemporary and friend, Cicero, had quite a sophisticated

Fig. 7.2. Cicero, Museo Vaticano, Vatican City. Alinari/Art Resource N.Y.

knowledge of Greek philosophy—having studied with the Stoic Posidonius, the Epicurean Phaedrus, and the Platonists Philo of Larissa and Antiochus of Ascalon.[9] Cicero was heavily influenced in his intellectual method by the skeptical tendencies that had developed within the Platonic school; in particular, he became convinced that probability was the most anybody could achieve in philosophical matters and, consequently, that the best way of discovering truth was through the critical sifting of past opinion. The product of this belief was a series of dialogues in which Cicero reported the opinions of his teachers, friends, and earlier writers on a variety of philosophical topics. For the opinions of his predecessors, especially those from the more distant past, Cicero borrowed from existing handbook literature, including the "doxographic" (or "opinion") tradition initiated by Theophrastus.

Cicero thus simultaneously drew on, and contributed to, the popularization movement. He provided his readers with an account of recent and contemporary controversy on the major philosophical issues, including some of the questions that have occupied us in previous chapters—the nature of the underlying reality, the source of order in the universe, the role of the gods, the nature of the soul, and the process of knowing. His own worldview was constructed from a combination of Platonic and Stoic elements, and Cicero became one of the major sources of Stoic philosophy for the Middle Ages and the early modern period. He identified God with nature, nature with fire, and all three (God, nature, and fire) with the active force responsible for the existence, activity, and rationality of the universe. He described the Stoic cosmological cycle of successive conflagration and regeneration. And he advocated the idea of a close parallelism between the macrocosm (god and the universe) and the microcosm (the individual human), arguing that god bears the same relationship to matter in the universe as the human soul bears to the human body. The macrocosm-microcosm analogy was to become a staple of medieval and Renaissance thought and a central theme of astrological writing. Cicero did not devote much attention to the mathematical sciences, which he considered of value principally for their ability to sharpen the wits of young men; however, his discussion of planetary motion in the heavens and his translation of Aratus's astronomical poem, *The Phaenomena,* reveal that he was not totally disinterested or uninformed on such matters.

One of Varro's and Cicero's contemporaries, Lucretius (d. 55 B.C.), wrote a long philosophical poem, *On the Nature of Things.* At one level, this work is a defense of Epicurean natural philosophy, which aims to overcome the fear of death by touting the explanatory power of atoms and the void. However, within this basic Epicurean framework, *On the Nature of Things* is encyclopedic in its scope and popular in its level of presentation and choice of detail. Lucretius discusses the infinity of worlds, their creation and destruction, and such basic astronomical data as the path of the sun around the zodiac, the resulting inequality of days, and the phases of the moon; the mortality of the soul; sense perception, including delusions of the senses; sleep, dreams, and love; mirrors and the reflection of light; the origins of plant and animal life, including a denunciation of teleological explanation in biology; the origins and history of the human race; and extraordinary meteorological and geological phenomena, such as lightning, thunder, earthquakes, rainbows, volcanoes, and magnetic attraction. Lucretius concludes with an account of the great Athenian plague.[10]

Varro, Cicero, and Lucretius represent the flowering of Roman intellectual life in the latter days of the Roman Republic. Others who contributed to this intellectual enterprise were Vitruvius (d. 25 B.C.), a contemporary who wrote on architecture, and several writers of the early imperial period: Celsus (fl. 25 A.D.), author of an influential medical encyclopedia, and the Stoic Seneca (d. 65 A.D.), who wrote on natural philosophy, including meteorology (this portion of his work being heavily dependent on Posidonius).[11]

However, the man universally regarded as the pinnacle of the popularization movement is Pliny the Elder (23/24–79 A.D.). Pliny is the central figure in most accounts of Roman science, and we too must briefly examine his work. He was born in northern Italy, into the provincial nobility, and educated in Rome. After a successful military career (the route to advancement for a man of Pliny's social standing), he turned to literary endeavors and ended in the service of the Emperors Vespasian and Titus. He wrote several books on the history of Rome and its wars, a book on grammar, and the work on which his fame now rests, the *Natural History,* dedicated to Titus.

The *Natural History* is a remarkable work, which resists easy characterization and really must be read to be appreciated.[12] Pliny had a voracious appetite for information. In the preface to the *Natural History,* he reports that he and his assistants perused two thousand volumes by some one hundred authors and extracted from them twenty thousand facts. It appears that Pliny worked out a system of note cards, so that he could manually sort his twenty thousand pieces of information; the cards were organized by subject and assembled to produce the *Natural History.*[13] The energy with which Pliny proceeded is astounding. His nephew tells us that he rose as early as midnight and worked nearly round the clock, reading or being read to, making or dictating notes. If we are to understand Pliny's achievement, it is important for us to grasp his fascination with factual data. Although in the *Natural History* Pliny sometimes offers explanations of natural phenomena, his goal was not to produce a comprehensive, carefully reasoned natural philosophy, but to create a vast storehouse of interesting and entertaining information—a book, his nephew tells us, "no less varied than nature herself."[14]

Pliny's purpose, then, was to survey the universe and the natural objects that populate it. He devoted seventy-two pages (in a modern English translation) simply to a list of the contents of the *Natural History* and the authorities consulted. Among the subjects treated were cosmology, astronomy, geography, anthropology, zoology, botany, and mineralogy. Pliny had a flair for picking out matters of unusual interest, and he has often been described primarily as a purveyor of marvels; and, to be sure, natural marvels are not scarce in the pages of the *Natural History.* Pliny reports a series of celestial portents (including multiple suns and moons), thunderbolts called forth by prayers and rituals, the greatest earthquake in human memory (which demolished twelve cities in Asia), human sacrifice among transalpine tribes, a boy said to have been regularly transported to and from school on the back of a dolphin, and exotic races of monsters (including the Arimaspi, who have one eye in the center of their foreheads,

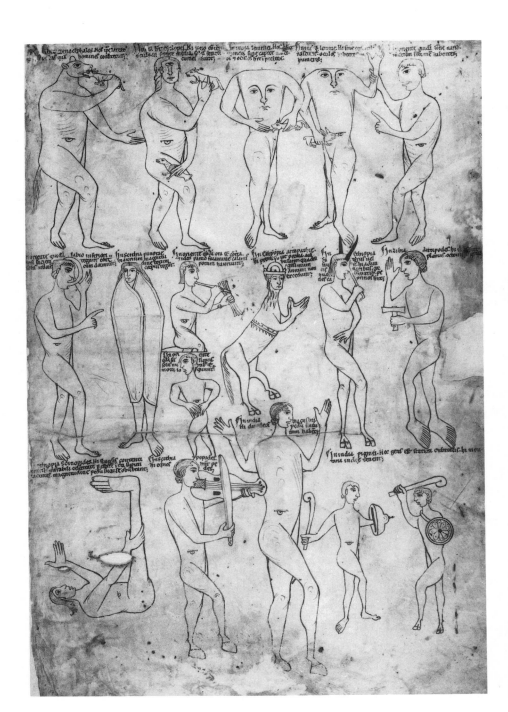

Fig. 7.3. Pliny's monstrous races, British Library, MS Harley 2799, fol. 243r (12th c.). By permission of the British Library.

Fig. 7.4. Macrobius on rainfall. A thirteenth-century scribe's attempt to illustrate Macrobius's argument that if we do not assume that all rain falls toward the center of the earth along a radius, we must accept the ridiculous consequence that the portion of it that misses the earth would find itself ascending toward the other hemisphere of the heavens. British Library, MS Egerton 2976, fol. 49v (13th c.). By permission of the British Library. On this illustration and the accompanying argument, see John E. Murdoch, *Album of Science: Antiquity and the Middle Ages,* pp. 282–83.

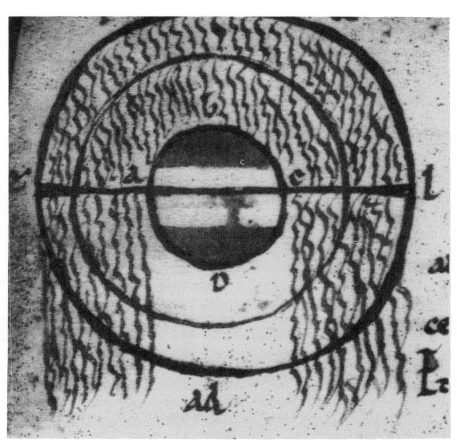

the Illyrians, who kill with a glance of the evil eye, and the Monocoli, who have only a single leg but manage nonetheless to hop with remarkable swiftness).[15]

Just as it would be a mistake to overlook the marvelous element in Pliny's *Natural History,* so it would be a mistake to disregard the more prosaic and commonplace. Pliny's account of astronomy and cosmology is a good example of the latter.[16] He describes the celestial and terrestrial spheres and the circles used to map them. He knows that the planets move through the band of the zodiac in a west-to-east direction, and he knows the approximate periods with which they do so; he described planetary retrogressions and reports that Mercury and Venus remain within 22° and 46°, respectively, of the sun. He discusses the motion, phases, and eclipses of the moon; and he understands solar and lunar eclipses as a function of the relative dimensions of the bodies involved and the shadows thus cast. With regard to the dimensions of the earth, Pliny reports Eratosthenes' value of 252,000 stades for its circumference. Pliny thus communicates substantial pieces of cosmological and astronomical knowledge, though not always reliably and certainly not up to the standards of the mathematical astronomer. He was neither borrowing from the tradition of mathematical astronomy (for example, the astronomical sections of the *Natural History* do not reveal the influence of Hipparchus) nor writing for an audience of astronomical specialists. He was merely endeavoring to communicate the bare essentials to a public not interested in, or equipped to deal with, observational or mathematical complexity.

Pliny was not a typical Roman scholar. Most obviously, nobody matched him for energy and dedication to the task of collecting information. Moreover, his coverage was considerably wider than that of any Roman predecessor (including Varro, who had confined himself to nine arts); in the preface to the *Natural History,* Pliny points out, correctly, that he is the first to attempt to deal with the whole of the natural world in a single work. And finally, Pliny surpassed most of his predecessors and contemporaries in the superficiality of his treatment. Nonetheless, he does serve as a useful measure of what the educated Roman might be expected to know—after Pliny, if not before. And the fact that the *Natural History* survived, while many other popularizing works did not, helped to determine the level and content of early medieval learning.

Thus far we have concentrated on Roman literature of an encyclopedic nature—attempts to collect in a single work large quantities of information drawn from many different sources. But Rome also saw the development of a commentary tradition, in which the narrative took its structure and a good part of its content from a single authoritative text. This tradition illustrates the ancient tendency to identify certain venerable or privileged texts as the repositories of knowledge and to measure learning by the ability to read and interpret those texts. An important example of the Roman commentary tradition is the *Commentary on the Dream of Scipio,* by Macrobius (who flourished in the first half of the fifth century, some 350 years after Pliny). This work, which employs Cicero's *Dream of Scipio* as the occasion for an exposition of Neoplatonic philosophy, enjoyed extremely wide circulation during the early Middle ages. We will not explore its contents except to note that in it Macrobius set out a comprehensive philosophy of nature, largely Platonic in inspiration, which included substantial sections on arithmetic, astronomy, and cosmology.[17]

One last Roman compiler must be considered because of the window he offers us on the mathematical arts at their best in the schools of the later Roman Empire—and also because his book became one of the most popular school texts of the Middle Ages. Martianus Capella was probably a North African from the city of Carthage; he thus serves, in addition, to remind us of the strength of the scholarly tradition in the Roman provinces, especially those of North Africa, during the later years of the Empire. Martianus is customarily dated to the period 410–39, but on the basis of slender evidence. The book of his that proved so influential was an allegory, entitled *The Marriage of Philology and Mercury,* in which seven bridesmaids offer surveys of their respective liberal arts to the wedding guests at a celestial marriage ceremony.[18]

The first of the mathematical arts to be presented is geometry. Through the mouth of the bridesmaid Geometry, Martianus briefly surveys the highlights of Euclid's *Elements,* including most of the definitions, all of the postulates, and three of the five axioms with which that work begins (see chap. 5, above). He discusses and classifies plane and solid figures, including Plato's five regular solids; he defines right, acute, and obtuse angles; and he touches on proportionality, commensurability, and incommensurability. But the bulk of this chapter is devoted to a discourse on geography, derived from Pliny and others. Martianus begins with proofs of the earth's sphericity; a report of Eratosthenes' figure for its circumference, accompanied by a faulty account of Eratosthenes' method of calculation; and arguments for the centrality of the earth within the universe. He discusses the five climatic zones and the division of the habitable world into three continents (Europe, Asia, and Africa), and proceeds to offer an extremely swift tour of the known world (basically a boiled-down version of Pliny's similar discussion).

Arithmetic comes next. Martianus begins with an account, heavily Pythagorean in tone, of the first ten numbers, explaining the virtues and associations of each, the deities with which they are connected, and their interrelationships. For example, three

> is the first odd number [one doesn't count as odd for Martianus], and must be regarded as perfect. It is the first to admit of a beginning, a middle, and an end, and it associates a central mean with the initial and final extremes, with equal intervals of separation. The number three represents the Fates and the sisterly Graces; and a certain Virgin who, as they say, "is the ruler of heaven and hell," is identified with this number. Further indication of its perfection is that the number begets the perfect numbers six and nine. Another token of its respect is that prayers and libations are offered three times. Concepts of time have three aspects; consequently, divinations are expressed in threes. The number three also represents the perfection of the universe. . . .[19]

Martianus proceeds to the classification of numbers and a discussion of what we would regard as their purely mathematical properties. He defines numbers as prime (integrally divisible by no number except the number 1) or composite; even or odd; plane and solid; perfect, deficient, or superabundant. Perfect numbers, for example, are those for which the sum of the factors equals the number ($1+2+3=6$); deficient numbers are those for which the sum of the factors is less than the number ($1+2+7<14$). Martianus also defines and classifies various ratios or proportions. For example, the ratio of 8 to 6 is supertertius, because the first number is a third again larger than the second; and the ratio of 6 to 8 is subtertius by similar reasoning.

Martianus begins his account of astronomy with a reference to Eratosthenes, Hipparchus, and Ptolemy—men whose reputation he knew, but whose works he had undoubtedly never seen. His chapter on astronomy includes basic cosmological and astronomical information, probably drawn from Varro, Pliny, and other sources.[20] He defines the celestial sphere and its major circles. He describes the zodiac, which he breaks into twelve signs of 30° each. He names and catalogues the major constellations. He identifies the traditional seven planets and describes their principal motions with more sophistication than was typical of handbook literature. For example, he reveals accurate knowledge of the approximate periods of their west-to-east motion around the ecliptic and a good grasp

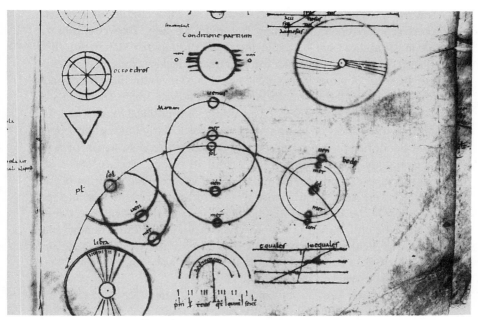

Fig. 7.5. Martianus Capella on the motions of Venus and Mercury. Several attempts to capture Martianus's theory of the motions of Venus and Mercury in relation to that of the sun. The drawing at center right places Venus and Mercury on sun-centered orbits. From a ninth-century copy of Martianus's *Marriage of Philology and Mercury,* Paris, Bibliothèque Nationale, MS Lat. 8671, fol. 84r.

of the retrograde motions of the superior planets. One of the most interesting and influential features of this chapter is Martianus's discussion of the inferior planets, Mercury and Venus, which he believes move through orbits centered on the sun (see fig 7.5). Eleven hundred years later, Copernicus would cite Martianus in support of this feature of his own system.[21]

TRANSLATIONS

In the early years of cultural contact between Rome and its Greek neighbors (soon to become subjects), there was no problem of scholarly access. Widespread bilingualism, ample opportunity for travel or study abroad, and the easy availability of Greek teachers provided educated Romans with the means by which to participate in the Greek intellectual tradition. For those of lesser linguistic accomplishment or more modest aspirations, there were popularizing works in Latin and a few translations. Among the latter we have already noted Cicero's translations of Plato's *Timaeus* and *The Phaenomena* of Aratus.

Toward the end of the second century A.D., the conditions that had favored scholarship and learning began to weaken. Two centuries of peace and stability gave way, after the death of Emperor Marcus Aurelius (180 A.D.), to political turmoil, civil war, urban decline, and eventual economic disaster. Beginning about 250, attack and invasion from barbarians on the frontiers of the Empire became an additional threat. The results of these events included the loss of economic and political vitality and the general deterioration of living conditions, particularly among the upper classes. Economic problems, exacerbated by an inadequate supply of slave labor and general depopulation (as a result of plague, warfare, and a declining birth rate), contributed to the loss of leisure—the absolute prerequisite for serious scholarly endeavor. A further problem that affected scholarship in the West was diminishing communication with the Greek East. Near the end of the third century, and again in the fourth century, the Empire was divided administratively into eastern and western halves. Increasingly, those two halves went their separate ways, and the Latin West gradually lost its vital contact with the Greek East.

Under such circumstances, intellectual continuity between East and West was broken. Bilingualism in the western regions of the Roman Empire declined, as did basic literacy, and the problem of access to Greek learning began to be felt. This is not to suggest that there was a complete break, but only that the connection became thinner and more precarious. Several men in the later years of the Roman Empire, aware of the growing threat, attempted to lessen its impact by translating some of the more fundamental Greek philosophical literature into Latin. Two of these men are of particular importance for the history of science.[22]

About Calcidius, the first of the two, we know virtually nothing. Even his dates are uncertain, although several lines of argument suggest that he may have lived late in the fourth century.[23] At any rate, he translated Plato's *Timaeus* from Greek to Latin; and it was his version of this treatise, rather than Cicero's, that survived into the Middle Ages and became identified with medieval Platonism. Accompanying the translation was a long commentary, in which Calcidius drew on the doxographic tradition and a variety of late antique philosophers to explain and elaborate Plato's cosmological ideas.

The other translator, Boethius (480–524), lived well over a century later, after Rome had fallen under barbarian rule. Born into the Roman aristocracy, Boethius was active in affairs of state and was appointed to high political office in the regime of Theodoric the Ostrogoth; he was subsequently accused of treason and put to death. We do not know anything about

Boethius's education, but his career bears testimony to the continuing existence of at least fragments of the Greek intellectual tradition within the Roman senatorial class. Boethius set out, as he tells us, to make available to the Latins as much of Plato and Aristotle as he could lay his hands on, and also to reconcile their philosophies. He succeeded in translating a number of Aristotle's logical works (which became known collectively as the "old logic"), Euclid's *Elements,* and Porphyry's *Introduction to Aristotle's Logic.* In addition, Boethius wrote handbooks, based on Greek sources, on several of the liberal arts, including arithmetic and music.[24]

By the time Boethius was put to death in 524, the West had been largely cut off from original Greek science and natural philosophy. It possessed Plato's *Timaeus,* some of Aristotle's logical works, and a few other bits and pieces—none of which, in all probability, had a very wide circulation. Beyond that, its knowledge of the Greek achievement was limited to commentaries, handbooks, compendia, and encyclopedias. Rome had managed to preserve and transmit the Greek intellectual tradition only in a thin and limited version.

THE ROLE OF CHRISTIANITY

There is one piece of the picture that we have thus far omitted from consideration. Christianity grew from a small Jewish sect in a remote corner of the Roman Empire into a major religious force in the third century and the state religion by the end of the fourth. This book is not the appropriate place to investigate the details of that extraordinary development. What is important for our purposes is the fact that Christianity came to play a powerful religious role in the late Roman Empire. From this fact follows the question that we must now take up—namely, how did the dominance of Christianity affect knowledge of, and attitudes toward, nature? The standard answer, developed in the eighteenth and nineteenth centuries and widely propagated in the twentieth, maintains that Christianity presented serious obstacles to the advancement of science and, indeed, sent the scientific enterprise into a tailspin from which it did not recover for more than a thousand years. The truth, as we shall see, is far different and much more complicated.[25]

One of the charges frequently leveled against the church is that it was broadly anti-intellectual—that the leaders of the church preferred faith to reason and ignorance to education. In fact, this is a considerable distortion. Although Christianity seems at first to have appealed to the poor and disenfranchised, it soon reached out to the upper classes, including the educated. Christians quickly recognized that if the Bible was to be read, literacy would have to be encouraged; and in the long run Christianity became the major patron of European education and a major borrower from the classical intellectual tradition. Naturally enough, the kind and level of education and intellectual effort favored by the church fathers was that which supported the mission of the church as they perceived it.

When the church developed a serious intellectual tradition, as it did in the second and third centuries, the driving forces were defense of the Christian faith against learned opponents (an enterprise known as "apologetics") and the development of Christian doctrine. For such purposes, the logical tools developed within Greek philosophy proved indispensable. Furthermore, aspects of Platonic philosophy seemed to correlate nicely with, and therefore to support, Christian teaching. For example, Plato had staunchly defended divine providence and the immortality of the soul; better yet, Plato's Demiurge looked very much like a monotheistic answer to the multiple gods of pagan polytheism; and this Demiurge could, with only a little stretching, be viewed as the Christian creator-God. Thus in the second and third centuries we find a series of Christian apologists putting Greek philosophy, especially Platonism, to Christian use.[26]

But this development did not please everybody. Some Christians regarded the Greek philosophical tradition as a source of error rather than truth. For every Plato, author of a philosophy compatible with Christian theology, there were an Aristotle and an Epicurus, whose worldviews were diametrically opposed to Christian doctrine on points of critical importance. Tertullian (ca. 155–ca. 230), a native of Carthage in Roman Africa, denounced philosophy as a source of heresy and warned against those who try to construct Christian doctrine out of Stoic and Platonic materials. However, a more typical attitude was that of Augustine (354–430), another North African, who accepted Greek philosophy as a useful, if not perfectly reliable, instrument. Philosophy, in Augustine's influential view, was to be the handmaiden of religion—not to be stamped out, but to be cultivated, disciplined, and put to use.

Now natural philosophy could not be separated from the rest of philosophy and, therefore, shared the fate of the larger whole of which it was a part. Like philosophy generally, it received mixed reviews from the intellectual leaders of the early church, ranging from mistrust and dislike to appreciation and enthusiasm—the same spectrum of opinion as we find in pagan circles. Augustine, who did so much to determine medieval attitudes, admonished his readers to set their hearts on the celestial and eternal, rather than the earthly and temporal. Nonetheless, he acknowledged that the temporal could serve the eternal by supplying knowledge about nature that would contribute to the proper interpretation of Scripture and the development of Christian doctrine. And in his own works, including his theological works, Augustine displayed a sophisticated knowledge of Greek natural philosophy. Natural philosophy, like philosophy more generally, was to serve handmaiden functions.[27]

Whether this represents a blow against the scientific enterprise or modest, but welcome, support for it depends largely on the attitudes and expectations that one brings to the question. If we compare the early church with a modern research university or the National Science Foundation, the church will prove to have failed abysmally as a supporter of science and natural philosophy. But such a comparison is obviously unfair. If, instead, we compare the support given to the study of nature by the early church with the support available from any other contemporary social institution, it will become apparent that the church was one of the major patrons— perhaps *the* major patron—of scientific learning. Its patronage may have been limited and selective, but limited and selective patronage is better than no patronage at all.

But a critic determined to view the early church as an obstacle to scientific progress might argue that the handmaiden status accorded natural philosophy is inconsistent with the existence of genuine science. True science, this critic would maintain, cannot be the handmaiden of anything, but must possess total autonomy; consequently, the "disciplined" science that Augustine sought is no science at all. The appropriate response is that totally autonomous science is an attractive ideal, but we do not live in an ideal world. Many of the most important developments in the history of science have been produced by people committed not to autonomous science, but to science in the service of some ideology, social program, or practical end; for much of its history, the question has not been *whether* science will function as handmaiden, but *which* mistress it will serve.

ROMAN AND EARLY MEDIEVAL EDUCATION

One of the ways in which the church became a patron of learning was through the creation and support of schools. We have already touched upon education in Rome; let us look more closely at the Roman schools, then at the early medieval schools that replaced them.[28]

Elementary education in Rome was generally a function carried on in the home, presided over by a parent or a tutor, who taught the child (beginning at about age seven) to read, write, and calculate. Organized primary schools also existed for those who needed or preferred them. The education of girls ceased at this point; if a boy were destined for additional education, he would be sent at about age twelve to study Latin grammar and literature (especially poetry) with a grammarian. The study of literature imparted not only writing skills and a knowledge of literary forms, but also, through the content of the works studied, a broad cultural education. Subsequent study, at about age fifteen, called for the skills of the rhetorician in a school of rhetoric. Here the student prepared for a career in politics or law by mastering the theory and techniques of public speaking. To go beyond this level of education was to engage in advanced study with a philosopher; this was possible for those of exceptional means or ambition, but it was done exclusively in Greek. Natural philosophy and the mathematical sciences would receive only limited attention in these educational settings: they would probably put in an appearance in the teaching of the grammarian or rhetorician; they might figure a little more prominently in the teaching of the philosopher. Seldom would instruction surpass the level achieved in Martianus Capella's *Marriage of Philology and Mercury.*

Roman education began as private enterprise, dependent on the initiative of parents and teachers. Schools were held in a variety of physical settings, including homes, rented shops, public buildings, and the open air. Eventually, a system of municipal and imperial support developed, which established paid positions for teachers in most of the major cities, not only in Italy but also in provinces such as Spain, Gaul, and North Africa. Paid positions were provided for grammarians and rhetoricians, and on occasion for philosophers as well. At its zenith, Rome boasted an educational system that provided an impressive measure of educational opportunity for members of the upper class throughout the Empire.

As the Empire declined, so did its educational program. Invasion, civil disorder, and economic collapse brought deterioration of the conditions that had favored schools and education. Particularly critical were the loss of urban vitality and decline in the size, affluence, and influence of the upper classes from which the schools had always drawn their support. Disinterest and neglect by the Germanic tribes that overran the Empire in the fourth and fifth centuries were significant contributing factors. However, deterioration was gradual rather than precipitous, particularly in regions bordering the Mediterranean. Roman Britain and northern Gaul quickly lost contact with the classical tradition, but schools and intellectual life continued to exist (if not to thrive) in Rome, northern Italy, southern Gaul, Spain, and North Africa.

The relationship of Christianity to the demise of classical education is an extremely difficult and complex problem. As we have seen, there were leaders of the church who were profoundly worried about the pagan content of classical education and who denounced the schools as a threat. The literature studied in the schools was frequently polytheistic and, by Christian standards, immoral; it certainly did not have the edifying qualities of, say, the Psalms or Jesus' sermon on the mount. We might, therefore, expect the church to have moved quickly to establish an alternative, Christian educational system; or if it did not, then when Christianity became the state religion we would expect the pagan schools to have been radically transformed into Christian institutions. However, neither event occurred. The fact is that the majority of the early church fathers valued their own classical education and, while recognizing its deficiencies and dangers, could conceive of no viable alternative to it; consequently, instead of repudiating the classical culture of the schools, they endeavored to appropriate it and build upon it. Substantial numbers of Christians continued to send their children to Roman schools; and educated Christians participated in those schools as teachers of grammar, rhetoric, and philosophy (much as religious people participate in modern secular education), no doubt allowing their Christian beliefs and sentiments to influence the curriculum to a degree, but not departing fundamentally from the classical tradition. As for the clergy, they were drawn from people who had already completed grammatical, and perhaps rhetorical, studies; their theological and doctrinal education would then take place informally, through a process of apprenticeship, or possibly in an episcopal school run by a bishop for the training of converts and prospective clergy.

But participation in the schools was not the same as unqualified enthusiasm and wholehearted support. The church remained deeply ambivalent and divided on the value and appropriateness of classical education and, though willing to use the schools, was not likely to go out of its way to save them from the various forces pushing them towards extinction—especially if an acceptable alternative should present itself. Such an alternative emerged in the fifth century as a byproduct of monasticism.

Christian monasticism appeared in the West during the fourth century. Monasteries spread rapidly, providing retreats for Christians who wished to withdraw from the world in the pursuit of holiness. In the sixth century St. Benedict (d. ca. 550) established a monastery at Monte Cassino, to the south of Rome, and drew up rules governing the lives of the monks who settled there—rules that came to be widely adopted within Western monasticism. The Benedictine Rule dictated all aspects of the life of the monk and nun, obliging them to devote the major share of their waking hours to

Fig. 7.6. A monk in his study. Florence, Biblioteca Medicea Laurenziana, Codex Amiatinus (7th–8th c.).

worship, contemplation, and manual labor. Worship included reading the Bible and devotional literature, and that required literacy. The Benedictine Rule also mandated books, tablets, and writing implements for all monks and nuns. Since monasteries accepted young children (committed by their parents to the monastic life), the monastery was obliged to teach them to read—though in the early centuries of monasticism this rarely, if ever, occurred in a formal monastery school. Monasteries also developed libraries and scriptoria (rooms where books needed by the monastic community were produced by copyists).[29]

Initially the education that occurred within the monastery was intended solely to meet the internal needs of the monastic community. It was directed by the abbot or abbess or an educated monk or nun, and it was designed to provide the literacy required for the religious life and thus, ultimately, to promote spirituality. It has often been claimed that as the classical schools disappeared, monasteries felt growing pressure from the local gentry and nobility to provide education for their young—that is, for children who were not destined to become monks or nuns—and that monasteries established "external schools" for this purpose. In fact, there is no evidence before the ninth century for the existence of external schools in the monasteries; and thereafter the practice seems to have been exceedingly rare. If we find men with a monastic education performing administrative functions in church and state, this is not because monasteries set out to educate the lay public through external schools; it is because lay students were sometimes admitted to the internal monastic schools, but more especially because the monasteries contained a pool of talent (educated for monastic purposes) that could be tapped for service outside the monastery.[30]

There is dispute among historians about the degree to which classical learning entered the monasteries—dispute deriving, perhaps, from differences between monasteries or between medieval writers who addressed the subject of monastic learning. What seems clear is that the stress was on spiritual development and whatever was thought to contribute to that. The Bible was the central core of the educational program; biblical commentaries and devotional writings supplemented the biblical text. Classical pagan literature, widely judged to be irrelevant or dangerous, was not prominent. But there were many exceptions; indeed, we often find the use of pagan sources by the very people who were denouncing them. Augustine's admonition for Christians to borrow what is true and useful from pagan literature seems often to have been heeded, and an examination of writings emanating from the monasteries reveals a surprisingly extensive, if selective, knowledge of ancient sources. The mathematical arts of the quadrivium were rarely pursued beyond the most elementary level, but there were exceptions to this generalization.

A good illustration of the penetration of classical learning into the monasteries is to be found in Ireland from the sixth century onward (a circumstance for which we have no adequate historical explanation). Here we find significant attention given to the classical pagan authors. Some Greek was known, and the mathematical arts of the quadrivium (particularly as applied to the calendar) were well developed.[31]

Another impressive exception to the monastic apathy toward classical education was the monastery of Vivarium, founded by a member of the Roman senatorial class, Cassiodorus (ca. 480–ca. 575), upon his retirement from public life. Cassiodorus established a scriptorium in his monastery, arranged for the translation of Greek works into Latin, and made study an essential part of the routine of his monks. He also wrote a handbook of monastic studies, in which he recommended a surprisingly numerous collection of pagan authors. In this handbook he briefly discussed each of the seven liberal arts. That this was more than lip service is revealed by a treatise on the calendar (still in existence) that appears to have been written at Vivarium during Cassiodorus's lifetime. It is clear that Cassiodorus shared the universal monastic view that secular studies were to be pursued only insofar as they served sacred purposes; where he differed from other leaders of the monastic movement was in his opinion of the range of secular studies capable of making such a contribution.[32]

These exceptions are important; but they do nothing to overturn the generalization that monasteries were dedicated to spiritual pursuits. Learning was cultivated, but only insofar as it contributed to religious ends. Science and natural philosophy were extremely marginal to this enterprise—although not entirely absent. What, then, is the significance of monasticism for the history of science, and why are we devoting space to it in this book? Was this not the "dark age" of the history of science—an age during which nothing of significance transpired?

There is no question that knowledge of Greek natural philosophy and mathematical science had fallen off precipitously, and few original contributions to it appeared in Western Europe during the early centuries of the medieval period (roughly 400–1000). If we are looking for new observational data or telling criticism of existing theory, we will find little of it here. Creativity was not lacking, but it was directed toward other tasks— survival, the pursuit of religious values in a barbaric and inhospitable world, and even (on occasion) exploration of the extent to which knowl-

Fig. 7.7. A medieval scribe. Oxford, Bodleian Library, MS Bodley 602, fol. 36r (13th c.).

edge about nature was applicable to biblical studies and the religious life. The contribution of the religious culture of the early Middle Ages to the scientific movement was thus one of preservation and transmission. The monasteries served as the transmitters of literacy and a thin version of the classical tradition (including science or natural philosophy) through a period when literacy and scholarship were severely threatened. Without them, Western Europe would not have had more science, but less.

TWO EARLY MEDIEVAL NATURAL PHILOSOPHERS

It may be worthwhile to conclude with a couple of illustrations of the early medieval contribution to science or natural philosophy—more specifically, to call attention to two men whose names have become synonymous with early medieval natural philosophy and the medieval worldview.

Isidore of Seville (ca. 560–636) was raised in Spain, then under Visigothic rule, and educated by his older brother (perhaps in a monastic or an episcopal school), before succeeding his brother as archbishop of Seville in 600. He was the outstanding scholar of the late sixth and early seventh centuries and illustrates the relatively high level of learning and culture available (but certainly not common) in Visigothic Spain during his lifetime. Isidore's works range widely over biblical studies, theology, liturgy, and history. He wrote two books of special interest to the historian of science: *On the Nature of Things* and *Etymologies.* These works, based on both pagan and Christian sources (including Lucretius, Martianus Capella, and Cassiodorus), communicate a brief, superficial version of Greek natural philosophy. The *Etymologies,* which exists in more than a thousand manuscripts (one of the most popular books of the entire Middle Ages), offers an encyclopedic account of things by way of an etymological analysis of their names. It covers the seven liberal arts, medicine, law, timekeeping and the calendar, theology, anthropology (including monstrous races), geography, cosmology, mineralogy, and agriculture. Isidore's cosmos is geocentric, composed of the four elements. He believes in a spherical earth and reveals an elementary understanding of the planetary motions. He gives an account of the zones of the celestial sphere, the seasons, the nature and size of the sun and moon, and the cause of eclipses. One of the notable features of his natural philosophy is his vigorous attack on astrology.[33]

If there is a certain vagueness regarding Isidore's intellectual formation, that of the Venerable Bede (d. 735) is known in considerably greater detail. At the age of seven, Bede entered the monastery of Wearmouth in Northumbria (northeastern England, near modern Newcastle) and there spent the remainder of his life studying and teaching, first as a student in the monastic school and eventually as a monastic schoolmaster. The monasteries of Northumbria were the direct offspring of Irish monasticism and thus inherited the Irish concern for quadrivial studies and the classics, but they were also in touch with the best of contemporary Continental scholarship. Bede, doubtless the most accomplished scholar of the eighth century, wrote on the whole range of monastic concerns, including a series of textbooks for monks. He is best known for his *Ecclesiastical History of the English People*. He also wrote a book, *On the Nature of Things* (based especially on Pliny and Isidore), and two textbooks on timekeeping and the calendar. In the latter, designed to regulate the daily routine of the monks and teach them how to determine the religious calendar, Bede made the most of the limited astronomical knowledge at his disposal and existing treatises on calendrics, to lay a solid foundation for what came to be called the science of "computus," establishing principles of time-keeping and calendar control that were eventually adopted throughout Christendom.[34]

Isidore and Bede are fitting representatives of the tradition of popularization and preservation that we have been tracing in this chapter—men who struggled to preserve the remnants of classical learning and pass them on, in usable form, to the Christian world of the Middle Ages. But is this tradition worthy of the attention we have given it? Does it merit a chapter in a book on the history of early science? If the history of science were simply the chronicle of great scientific discoveries or monumental scientific thoughts, Isidore and Bede would have no place in it; no scientific principles circulate today under their names. However, if the history of science is the investigation of the historical currents that converged to bring us to the present scientific moment—the strands that must be grasped if we are to understand where we came from and how we got here—then the enterprise in which Isidore and Bede were engaged is an important part of the story. Neither Isidore nor Bede was a creator of new scientific knowledge, but both restated existing scientific knowledge in an age when the study of nature was a marginal activity. They provided continuity through a dangerous and difficult period; in so doing, they powerfully influenced for centuries what Europeans knew about nature and how Europeans thought about nature. Such an achievement may lack the drama of, say, discovering the law of gravitation or devising the theory of natural selection, but to affect the subsequent course of European history is no mean contribution.

Science in Islam

LEARNING AND SCIENCE IN BYZANTIUM

While the classical tradition was slowly declining in the Latin West, and natural philosophy was being transformed into the handmaiden of theology and religion, what was happening in the Greek-speaking East? Although the East experienced many of the same misfortunes as the West—invasion, economic decline, and social upheaval—the consequences were less severe. A higher level of political stability was maintained, as the eastern half of the old Roman Empire gradually separated itself from the West, giving rise to what we now call Byzantium or the Byzantine Empire, with its capital in Constantinople (present-day Istanbul). That the city of Constantinople did not fall to invaders before 1203, while Rome was sacked as early as the fifth century, tells us something about the relative levels of stability. Greater social and political stability meant greater continuity in the schools; the tradition of classical studies thus waned more slowly in Byzantium and never entirely disappeared; and, of course, the East never found itself separated from the original sources of Greek scholarship by a linguistic barrier.[1]

But it does not follow that natural philosophy and mathematical science flourished. The study of nature was as impractical in the East as it was in the West; the fathers of the Greek church had the same ambivalence toward it as did their Western counterparts, and shared the same determination to subordinate it to theology and the religious life. Scholarly interests in the East were generally theological or literary. Authors felt obliged to limit themselves to the structure and vocabulary of the classical period; this led to imitative tendencies that (it is often claimed) stifled creativity. Insofar as philosophical labors were undertaken, they tended toward commentary on the classical authors; such commentary inevitably included a small amount of natural philosophy, mathematical science, and medicine.

These are sweeping generalizations, of course, and we must be careful not to leave the impression that scholarly achievement was absent or limited. The Platonic tradition (more accurately called the "Neoplatonic tradition," since it departed from Plato on many important matters) was represented by a series of distinguished scholars. Although there was no longer a living peripatetic tradition, there were attempts to assimilate Aristotelian to Platonic philosophy; and certain philosophers of the Byzantine period wrote important commentaries on Aristotle, in which they explained, embellished, or criticized his philosophy of nature, addressing the Aristotelian texts with a level of sophistication unmatched by any Latin-speaking contemporary.

Themistius (d. ca. 385), who taught philosophy in Constantinople and served as tutor to the imperial offspring, wrote influential paraphrases and summaries of a variety of Aristotelian works, including the *Physics, On the Heavens,* and *On the Soul.* Simplicius (d. after 533), an Athenian Neoplatonist determined to reconcile Platonism and Aristotelianism, wrote intelligent commentaries on these same three works. And John Philoponus (d. ca. 570), a Christian Neoplatonist who taught in Alexandria, wrote commentaries on Aristotle's *Physics, Meteorology, On Generation and Corruption,* and *On the Soul.* In these commentaries he attempted, in conscious opposition to Simplicius, to demonstrate the profound errors propagated by Aristotle, including the celestial-terrestrial dichotomy and the notion of an eternal universe. He also offered a systematic and original refutation of Aristotle's theory of motion, denying Aristotle's explanation of projectile motion and the claim that heavy bodies fall through a medium with speeds proportional to their weights. Through the eventual translation of their works into Arabic and Latin, all three men—Themistius, Simplicius, and Philoponus—helped to determine the subsequent course of Aristotelian natural philosophy.[2]

The argument, then, is that Byzantine intellectual life was in decline, like that in the West, but less precipitously; and we can find examples of sophisticated scholarship within the Byzantine Empire that cannot be matched anywhere in the Latin-speaking world. But this was not the only difference. The East also participated in a critically important process of cultural diffusion by which Greek learning was transmitted to far-flung regions of Asia and North Africa, where it would subsequently be assimilated by non-Greeks. This process of diffusion and assimilation is the real subject of the present chapter.

THE EASTWARD DIFFUSION OF GREEK SCIENCE

Although Greek influence had long extended beyond the Greek home-land, cultural diffusion as a conscious policy began with the military campaigns of Alexander the Great.[3] When Alexander conquered Asia and North Africa (334–323 B.C.), he not only acquired territory but also established beachheads of Greek civilization. His campaigns took him as far south as Egypt, as far east as Bactria (in Central Asia near the present-day Soviet-Afghan border) and beyond the Indus River into the northwestern corner of India (see map 2). Behind him he left garrisons and a batch of cities named Alexandria (at least eleven); successful efforts at colonization enlarged the Greek presence, and in the long run these Greek cities became centers of Greek culture, from which Hellenism could emanate into the surrounding regions. The most notable centers of Greek culture thus established were Alexandria in Egypt and the Kingdom of Bactria in Central Asia.

But conquest and colonization were not the only mechanisms of diffusion. Religion also played a decisive role in the spread of Greek learning. Many of the details are obscure, but for our purposes a sketch may be sufficient. In the millennium after Alexander's conquests, his Asian territories (especially, present-day Syria, Iraq, and Iran) proved to be fertile ground for a variety of major religious movements. At one time or another Zoroastrianism, Christianity, and Manicheism contended with each other for converts; all three were based on sacred books and thus, of necessity, cultivated at least a measure of learning. Christianity and Manicheism, in particular, had acquired Greek philosophical underpinnings and thus contributed to the Hellenization of the region. Let us concentrate our attention for a moment on the Christian contribution.

There was a strong Christian presence in Syria from the beginning; and in the first few centuries of the Christian era missionary activity led to the establishment of Christian churches through a wide region of western Asia. In the fifth and sixth centuries, reinforcements arrived in the form of dissident Christian sects seeking refuge from persecution. The Christianization of the Byzantine Empire in the fourth century had led to a series of bitter theological disputes and rifts within the Byzantine church. The most critical of the disputes for our purposes concerned Christ's nature—specifically, the relationship between Christ's humanity and divinity. The extreme positions—that of the Nestorians, who emphasized Christ's humanity over his divinity, and that of the Monophysites, who leaned in the other direction—were condemned in church councils held in 431 and 451.[4] During the ensuing conflict, Nestorian leaders established themselves in the school at Edessa in Syria (then the eastern limit of the Byzantine Empire). Struggle with the Monophysites (who were strong in Syria) and the eventual closing of the school by order of the emperor in 489 caused the Nestorians to seek refuge in the city of Nisibis, to the east, just over the Persian border. There, with the encouragement of the local bishop, they created a center of Nestorian higher education. Biblical studies and theology were the focus of attention, of course, but Aristotelian logic (one of the necessities of serious theology) was also taught, along with other aspects of Greek philosophy. Nisibis may also have developed a program of medical instruction.

From this foothold in Persia, the Nestorians managed, in the next century, not only to shape Persian Christianity, but also to exercise a broad influence on Persian intellectual life. By steps that we only dimly understand, Nestorians managed to insinuate themselves into positions of power and influence and to impart a taste for Greek culture in the Persian ruling class. We see the results in the invitation issued by the Persian king Khusraw I about 531, to the philosophers from the Academy in Athens (expelled by a decree of the Byzantine Emperor Justinian), to settle in Persia. This same Khusraw is reputed to have been knowledgeable in Platonic and Aristotelian philosophy and to have had Greek philosophical works translated for his use; Nestorian connections are revealed in his treatment by a Nestorian physician. Khusraw II (590–628) had two Christian wives—one of them, at least, a Nestorian before her conversion to Monophysitism—and an influential physician-advisor who also vacillated between the Nestorian and Monophysite sects.[5]

An influential mythology has developed around Nestorian activity in the city of Jundishapur in southwestern Persia. According to the often-repeated legend, the Nestorians turned Jundishapur into a major intellectual center by the sixth century, establishing what some have chosen to call a university, where instruction in all of the Greek disciplines could be obtained. There is alleged to have been a medical school, with a curriculum based on Alexandrian textbooks, and a hospital modeled on those that had developed within the Byzantine Empire, which kept the realm supplied with physicians trained in Greek medicine. Moreover, Jundishapur is held to have played a critical role in the translation of Greek scholarship into Near Eastern languages and, indeed, to have been the single most important channel by which Greek science passed to the Arabs.[6]

Recent research has revealed a considerably less dramatic reality. There is no persuasive evidence for the existence of a medical school or a hospital at Jundishapur, although there seems to have been a theological school and perhaps an attached infirmary. No doubt Jundishapur was the scene of serious intellectual endeavor and a certain amount of medical practice—it furnished several physicians for the Islamic court at Baghdad in the eighth century—but it is doubtful that it ever became a major center of medical education or of translating activity. If the story of Jundishapur is unreliable in its details, the lesson it was meant to teach is nonetheless a valid one. Nestorian influence, though not focused on Jundishapur, did play a vital role in the transmission of Greek learning to Persia and ultimately to the Arabs. There is no question that Nestorians were foremost among the early translators; and as late as the ninth century, long after Persia had fallen to Islamic armies, the practice of medicine in Baghdad seems to have been monopolized by Christian (probably Nestorian) physicians.[7]

But there is a linguistic shift here, which we must also take into account. Although the content of the education available in Nisibis, Jundishapur, and other Nestorian centers was predominantly Greek, the language of instruction was not. Teaching was in Syriac, a Semitic tongue (a dialect of Aramaic) widespread in the Near East; this, along with Greek, was the language of culture in Persia, and it was adopted by the Nestorians as their literary and liturgical language. The teaching program, then, required the translation of Greek texts into Syriac. Such translations were made at Nisibis and other locations, beginning as early as 450. Again we are short on detail, but logical works by Aristotle and Porphyry appear to have been among the earliest translated. Medical literature, works on mathematics and astronomy, and various philosophical treatises were eventually rendered as well.

Several points merit special emphasis. First, let us be clear that this is a story about the *transmission* of learning. Our subject (in the early sections of this chapter) is not original contributions to natural philosophy, but the preservation and eastward diffusion of the Greek heritage into Asia, where it would subsequently be absorbed into Islamic culture. Second, this process of cultural diffusion was quite slow, but also of very long duration—occupying a period of nearly a thousand years, from the Asian conquests of Alexander the Great (about 325 B.C.) to the founding of Islam in the seventh century A.D. Third, the story must not be oversimplified to the point where the diffusion of Greek learning is viewed as hanging on the slender thread of Nestorian activity in the city of Jundishapur or any other specific location. Rather, we must see this as a wide movement of cultural diffusion, whereby the aristocracies of western Asia assimilated broadly and deeply, and by a variety of mechanisms, the fruits of Greek culture. We must now consider the further transmission of those fruits to Islam.

THE BIRTH, EXPANSION, AND HELLENIZATION OF ISLAM

The Arabian peninsula, wedged between Persia to the north and east and Egypt to the west, had been untouched by Alexander's military campaigns and not much affected by Byzantine territorial ambitions. Jewish and Christian communities had flourished for a time in the south, but by the seventh century their influence had diminished to a modest level. Except on the southern and northern edges, the population was largely nomadic, although cities had been established around pilgrimage sites and along the major trading routes. It was in one of these cities, Mecca, that Muḥammad was born late in the sixth century and from which he preached the new religion of Islam. Muḥammad had a series of revelations in which the Koran (or Qurʾān, the holy book of Islam) was dictated to him by the angel Gabriel. The central theme of these revelations was the existence of a single omnipotent, omniscient god, Allah, creator of the universe, to whom the faithful (called "Muslims" or "Moslems") must submit. This book came to define all aspects of Islamic faith and practice; it was the source of Islamic theology, morality, law, and cosmology, and thus the centerpiece of Islamic education; it served to codify Arabic as a written language, and it remains the principal model for Arabic literary style.[8]

Muḥammad both practiced and taught the necessity of holy war and compulsory conversion. Before his death in 632, his band of followers had overrun the Arabian peninsula and conducted successful raids to the north; after his death Muslim forces emerged from their homeland and rapidly put both Byzantine and Persian armies to flight, thus gaining control of major portions of the Near East. In twenty-five years of stunning military success, Islam subjugated almost the whole of Alexander's Asian and North African possessions, including Syria, Palestine, Persia, and Egypt. Within a century, the remainder of North Africa and almost the whole of Spain fell to Muslim arms.

Muḥammad left no male heir or designated successor; consequently leadership of the developing Islamic Empire became a matter of bloody dispute. The first caliphs ("successors" of Muḥammad) were chosen from Muḥammad's early followers. In 644 ʿUthman of the Umayyad family became caliph, and in 661 his cousin Muʿāwiyah, who had been governor of Syria. In the interests of security, Muʿāwiyah and his successors ruled from

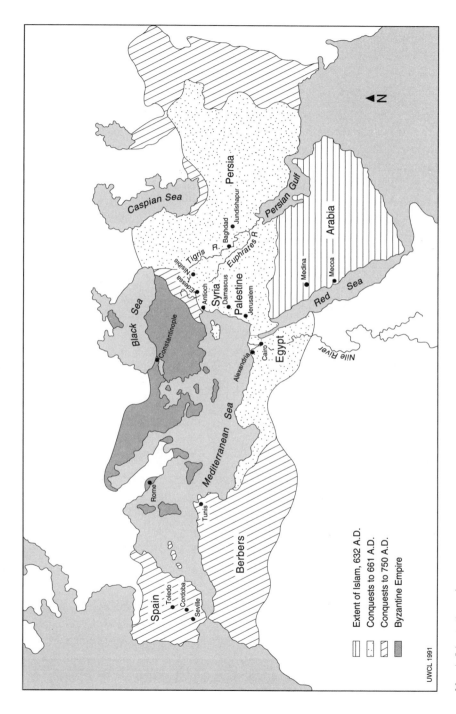

UWCL 1991

Map 4. Islamic Expansion

Legend:
- Extent of Islam, 632 A.D.
- Conquests to 661 A.D.
- Conquests to 750 A.D.
- Byzantine Empire

Damascus in Syria, where Umayyad strength was concentrated. Here the Umayyad dynasty, which held power for about a century, came into contact with educated Syrians and Persians, whom it used as secretaries and bureaucrats; and thus on a small scale began the Hellenization of Islam.

The process of Hellenization accelerated after 749. In that year a new dynasty, the ʿAbbasids (descended from Muḥammad's uncle, al-ʿAbbās), came to power. The ʿAbbasid caliphs had no intention of remaining in Damascus: like the Umayyads a century earlier, they wanted their capital situated in friendly territory. In 762, al-Manṣūr (754–75) built a new capital, the city of Baghdad, on the Tigris River. Al-Manṣūr's court in Baghdad was not famous for piety but cultivated a religious climate that was relatively intellectual, secularized, and tolerant. More importantly, the Islamic Empire was being transformed from a warrior aristocracy into a centralized state, which called for a much more substantial administrative bureaucracy than anything Muḥammad, his immediate successors, or the early Umayyads could have imagined. The staffing of this bureaucracy could hardly be accomplished from among the warriors who made up the conquering armies, and the caliphs had no reasonable alternative but to make use of educated Persians (generally recent converts to Islam, though the use of Christians was not unknown).

The Persian influence is especially apparent in the powerful royal advisors from the Barmak family—formerly from the province of Bactria and recent converts to Islam. Khālid ibn Barmak served al-Manṣūr; and his son Yaḥya became vizier (chief advisor and tutor of the caliph's heirs) under al-Manṣūr's grandson, Hārūn ar-Rashīd (786–809). The Christian influence is most clearly evident in the practice of medicine at court. In 765, al-Manṣūr was treated by a Nestorian physician from Jundishapur, Jūrjīs ibn Bakhtīshuʿ. Jūrjīs was apparently successful, for he remained in Bagdad as the caliph's personal physician, becoming a powerful court figure; his son succeeded him, and for several generations the Bakhtīshuʿ family held the post of court physician. Finally, it is important to note that there were also influences emanating from India in the east; some of these were the long-term result of the earlier Hellenization of India.

TRANSLATION OF GREEK SCIENCE INTO ARABIC

The translation of Greek and Syriac works into Arabic began under al-Manṣūr, but became serious business under Hārūn ar-Rashīd, who sent agents to Byzantium in search of manuscripts. Al-Maʾmūn (813–833), Hārūn's son, founded a research institute, the House of Wisdom, in

Baghdad; and here translation reached its peak. At the head of the House of Wisdom was Ḥunayn ibn Isḥāq (808–73)—a Nestorian Christian and an Arab, descended from an Arab tribe that had converted to Christianity long before the religion of Islam existed. Ḥunayn, who studied medicine with the distinguished physician Ibn Māsawaih, was bilingual from childhood in Arabic and Syriac; as a young man he went to the "land of the Greeks" (perhaps Alexandria), where he acquired a thorough mastery of Greek. Returning to Baghdad, he attracted the notice of a member of the Bakhtīshuᶜ family and a set of wealthy brothers (the "sons of Mūsa"), and through these patrons he was introduced to al-Ma'mūn. At some point Ḥunayn accompanied an expedition to Byzantium, in search of manuscripts. He served as translator under several caliphs and finished his career as chief royal physician, replacing one of the Bakhtīshuᶜ.[9]

Ḥunayn's translating activity is of such critical importance as to deserve our careful attention. Ḥunayn was assisted by his son Isḥāq ibn Ḥunayn, his nephew Ḥubaysh, and others. Many of their translations were collaborative efforts. For example, Ḥunayn might translate a work from Greek to Syriac, whereafter his nephew would render the Syriac text into Arabic. Ḥunayn's son Isḥāq translated from both Greek and Syriac into Arabic, as well as revising the translations of his colleagues. And Ḥunayn, besides producing his own translations from Greek to Syriac or Arabic, seems to have insisted on checking the translations of his charges. Ḥunayn and his co-workers were extremely sophisticated in their methods. They understood the need to compare manuscripts whenever possible, in order to weed out errors. And instead of following the common translating practice of mechanical word-for-word substitution (which suffers from the severe disadvantage that not every Greek word has a counterpart in Arabic or Syriac, while failing also to take into account syntactical differences between the languages), Ḥunayn grasped the meaning of a sentence in the original Greek and rendered it by a sentence of equivalent meaning in Arabic or Syriac.

The bulk of Ḥunayn's translations were medical, with special emphasis on Galen and Hippocrates. He rendered about ninety of Galen's works from Greek to Syriac and about forty from Greek to Arabic. He translated some fifteen Hippocratic works. Ḥunayn also translated (or corrected) three of Plato's dialogues, including the *Timaeus;* translated various Aristotelian works (in most cases from Greek to Syriac), including the *Metaphysics, On the Soul, On Generation and Corruption,* and part of the *Physics;* rendered a variety of other works on logic, mathematics, and astrology; and produced a Syriac version of the Old Testament. Ḥunayn's son Isḥāq translated more Aristotle, as well as Euclid's *Elements* and Ptolemy's *Almagest.*

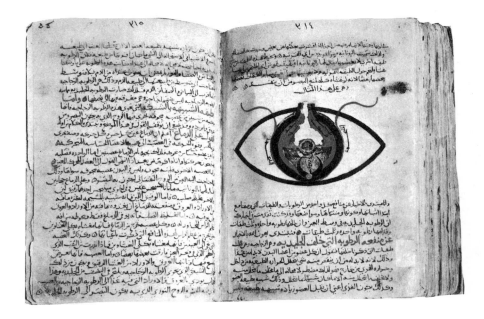

Fig. 8.1. Ḥunayn ibn Isḥāq on the anatomy of the eye, from a thirteenth-century copy of Ḥunayn's *Book of the Ten Treatises of the Eye,* Cairo, National Library.

Their co-workers in Baghdad and their contemporaries elsewhere added to these translations; for example, Thābit ibn Qurra (836–901), a trilingual pagan (that is, neither a Christian nor a Muslim) who spent most of his career in Baghdad, translated mathematical and astronomical treatises, including works of Archimedes. Translation activity continued at a high level for more than a century after Ḥunayn and Thābit. By the year 1000 A.D., almost the entire corpus of Greek medicine, natural philosophy, and mathematical science had been rendered into usable Arabic versions.

THE ISLAMIC RESPONSE TO GREEK SCIENCE

But the question arises: usable for what? What did members of the Muslim ruling caste see in Greek science that made them willing to pay for translations and support scholarship in Greek scientific disciplines? How were the translated works received by these patrons and by literate Muslims more generally? What functions did Greek science serve in the Islamic world, and how well did it blend with other aspects of Islamic culture? In particular, was there a religious price that had to be paid for the acceptance of Greek science?

We know in general what was translated, and in many cases we know whom to thank. But rarely do we have any exact knowledge of the motivations that lay behind a particular translation. One factor that must have been almost universal is that the patrons of translation were literate, or aspired to literacy, or at least wished to be associated with literacy (if only for the prestige it bestowed); they were people who wished to participate, in one way or another, in the most advanced intellectual culture available. But an explanation in terms of the cultural level of the patrons and recipients seems insufficient. These cultured Muslims were willing to invest in Greek science because they believed (rightly or wrongly) that it had value—that it contributed to the achievement of some worthwhile end. The pursuit of knowledge for its own sake was never endorsed by Islamic religious ideology, nor by any other thread in the cultural fabric. As in medieval Christendom, science was justified by virtue of its utility.[10]

Medicine is a science of apparent utility, and it may well be medicine that first attracted Muslim patronage; certainly medical translations were among the earliest. Medicine, in turn, called for philosophical equipment—or so the reader of Galen's works would gather. Indeed, Galen himself had written on logic and used natural philosophy in his medical writings; and it must have been clear to the translators and their patrons that a full grasp of Galen's medical philosophy demanded a broad knowledge of Greek thought, including Platonic and Aristotelian philosophy.[11] The utility of astronomy, astrology, mathematics, alchemy, and a certain amount of natural history must also have been evident. And finally, in Islam there were successful attempts to create a scholastic theology, imbued with Greek logic and metaphysics. It would appear, then, that the translation of almost any Greek medical, mathematical, or philosophical work could (with only a little stretching) be justified on utilitarian grounds: some were of critical importance; the rest must have seemed at least vaguely useful.

There was no necessary connection between the translation of a book into Arabic and its wide dissemination in Islam or the assimilation of its contents into Islamic culture. After all, translation requires only a translator and perhaps a patron, while dissemination and assimilation are broad cultural phenomena. Once the linguistic barrier had been surmounted through translation, formidable obstacles remained. One of them was the

Fig. 8.2. Ibn Ṭūlūn Mosque (9th century), Cairo. Foto Marburg/Art Resource N.Y.

lingering question of utility, which could not be answered for a culture, as it might for a patron, with a wave of the hand. To the strict Muslim, knowledge was always a means rather than an end, subordinated to the achievement of personal salvation, the acquisition of wisdom (defined in religious terms), the government of the Islamic commonwealth, or some other manifestly practical purpose.

Another obstacle that Greek science had to overcome was its foreign origin and its rational character. Muslims themselves divided learning into two categories: traditional, on the one hand; foreign or rational, on the other. The traditional disciplines were those based on the Koran: grammar, poetry, history, theology, and law. These rested on divine authority and were often taught orally (reflecting the oral nature of Muḥammad's revelations and his own teaching); the obligation of the practitioner of such disciplines was completeness and fidelity of transmission. By contrast, the foreign disciplines obtained from the Greeks were of human rather than divine origin; they were to be apprehended by reason, rather than accepted on the basis of authority or tradition; their transmission was primarily by means of the written word, and they were subject to critical commentary and correction. Any attempt to apply the methodology of the foreign sciences to the traditional disciplines would run obvious risks; and it was inevitable, therefore, that the foreign sciences should be seen as a threat by people of conservative bent.

What, then, was the fate of the foreign sciences in Islam? No simple answer, applicable to all times and places, is possible. Indeed, the historical situation was so complex that historians who specialize on Islam cannot agree on how to characterize it. Two quite different interpretations are currently in circulation. According to one of them, the foreign sciences never ceased to be viewed by the great majority of Muslims as useless, alien, and perhaps dangerous. They went against the grain of orthodox thought, met no fundamental need, and were excluded from the developing educational system. As a result, the foreign sciences were never deeply integrated into Islamic culture, but survived on the margins. The undeniably great achievements of Islamic scientists and natural philosophers, therefore, must have emanated from isolated enclaves of scholars protected from the pressures of orthodoxy (as at a royal court during a period of unusual tolerance) or willing, for reasons known only to themselves, to swim against the cultural stream. This has been called the "marginality thesis," because of its claim that science in Islam was never more than a marginal pursuit.[12]

The alternative theory views the Islamic encounter with Greek learning in a quite different light. While acknowledging that suspicion and hostility existed, this theory maintains that on the whole Greek science and natural philosophy enjoyed a reasonably hospitable reception in Islam. After all, Islam did not reject the fruits of foreign learning but, despite conservative opposition, undertook a remarkable program of recovery and cultivation. Moreover, one can point to many examples of the integration of Greek disciplines into traditional learning and Islamic culture more generally. Thus logic became incorporated into theology and law; astronomy became an indispensable tool for the *muwaqqit,* who was responsible for determining the times of daily prayer in his locale; and mathematics became essential for a wide variety of commercial, legal, and governmental purposes. That mathematics and astronomy were occasionally taught in the most highly developed of the Muslim schools, the *madrasahs* or colleges of law, testifies to the high level of acceptance and integration. According to this interpretation, Islam successfully appropriated large portions of foreign learning, despite opposition; let us call this the "appropriation thesis." On this view, the foreign sciences did not conquer the traditional disciplines, but made peace with them by agreeing to serve as their handmaidens.[13]

The gap between these two interpretations is substantial; and, given the current state of research on the history of Islamic science, the dispute does not seem likely to be soon resolved. But several things can be said, which may help to mediate between the two positions. First, we must acknowledge that the marginality thesis in its strong form is untenable. The cultivation of Greek natural philosophy and mathematical science was far too widespread and successful to be viewed as a marginal product of Islamic culture. But while granting this to the "appropriationists," we must go on to point out that science was far from central to Islamic culture and that there were forces within Islam tending to marginalize the foreign sciences—which is to say that the "marginalists" have their eye on some genuine feature of Islamic culture. To be specific, Greek learning never found a secure institutional home in Islam, as it was eventually to do in the universities of medieval Christendom. One reason why this was so, is that Islamic schools lacked the structure and uniformity of those in the West, particularly at the higher levels.[14] This lack of structure offered freedom to the individual scholar to pursue whatever specialty he wished. Freedom insured diversity and created room for the practitioner of Greek philosophy and science; but it also insured that Islamic schools would never develop a curriculum that systematically taught the foreign sciences. In short, Islamic education did nothing to prohibit the foreign sciences; but neither did it do much to support them. This fact may help us to understand the decline of Islamic science in the thirteenth and fourteenth centuries.

THE ISLAMIC SCIENTIFIC ACHIEVEMENT

Early in the twentieth century, the distinguished physicist-philosopher-historian Pierre Duhem threw out a challenge for historians of Islamic science when he wrote: "There is no Arabian [read "Islamic"] science. The wise men of Mohammedanism were always the more or less faithful disciples of the Greeks, but were themselves destitute of all originality."[15] Duhem was clearly wrong, but his statement is useful nonetheless as a means of focusing our attention on the critical issue: by seeing precisely what is wrong with Duhem's claim, we stand to learn something important about the character of the Islamic scientific achievement.

It is simply not true that Muslim practitioners of Greek science were "destitute of all originality"; and one possible response to Duhem, therefore, is to demonstrate this by enumerating the many original contributions of Islamic physicians, mathematicians, and natural philosophers. To offer but a single example, the eleventh-century Muslim Ibn al-Haytham turned his critical powers on nearly the whole of the Greek scientific achievement and made contributions of the utmost importance and originality to astronomy, mathematics, and optics. Unfortunately, to carry out this program of recounting Muslim contributions to the various sciences would require volumes, and we must be content with more modest goals—though we will, below and in subsequent chapters, deal with Islamic contributions to certain specific areas of scientific discourse.[16]

But Duhem's statement offers us another point of entry into the problem, which may lead us to the central issue. Duhem maintains that Muslim scholars interested in the foreign sciences "were always the more or less faithful disciples of the Greeks." He says this with derogatory intent, as proof that the Muslims were not genuine scientists; that is, he associates discipleship with an unscientific attitude (which tells us something about his definition of science). However, we can turn Duhem's point around and argue that it was precisely by becoming the disciples of the Greeks that Muslims entered the Western scientific tradition and became scientists or natural philosophers. Discipleship, on this view, is essential, rather than antipathetic, to the scientific enterprise; and Muslims became scientists not by repudiating the existing scientific tradition, but by joining it—by becoming disciples of the most advanced scientific tradition that had ever existed.

What does it mean to be a disciple? For would-be Muslim scientists, it meant adopting both the methodology and the content of Greek science. By and large, Islamic science was built on a Greek foundation and carried out according to Greek architectural principles; Muslims did not attempt to pull down the Greek edifice and begin from the ground up, but applied themselves to completing the Greek project. This does not mean that originality and innovation were absent; it means that Muslim scientists expressed originality and innovation in the correction, extension, articulation, and application of the existing framework, rather than in the creation of a new one. If this seems to be a damning admission, let it be understood that the great bulk of modern science consists in the correction, extension, and application of inherited scientific principles; a fundamental break with the past is approximately as exceptional today as it was in medieval Islam.

Muslim scientists were aware of this relationship to the past. An early Muslim scientist, al-Kindī (d. ca. 866), who pursued the mathematical sciences under several early ʿAbbasid caliphs in Baghdad, acknowledged his debt to ancient predecessors and his membership in an ongoing tradition. Had it not been for the ancients, al-Kindī wrote,

> it would have been impossible for us, despite all our zeal,
> during the whole of our lifetime, to assemble these principles
> of truth which form the basis of the final inferences of our
> research. The assembling of all these elements has been
> effected century by century, in past ages down to our own time.

Al-Kindī conceived his obligation to be the completion, correction, and communication of this body of ancient learning. He continued:

> It is fitting then [for us] to remain faithful to the principle that
> we have followed in all our works, which is first to record in
> complete quotations all that the Ancients have said on the
> subject, secondly to complete what the Ancients have not fully
> expressed, and this according to the usage of our Arabic lan-
> guage, the customs of our age, and our own ability.

Two hundred years later al-Bīrūnī (d. after 1050) could still judge that the task facing Muslim scientists was "to confine ourselves to what the ancients have dealt with and endeavor to perfect what can be perfected."[17]

Islamic astronomy is a good illustration of the relationship between Islamic and Greek science. Muslim astronomers produced a great deal of very sophisticated astronomical work. This work was carried out largely within the Ptolemaic framework (though we must acknowledge early Hindu influences on Islamic astronomy, largely displaced by subsequent access to Ptolemy's *Almagest* and other Greek astronomical works). Muslim astronomers sought to articulate and correct the Ptolemaic system, improve the measurement of Ptolemaic constants, compile planetary tables based on Ptolemaic models, and devise instruments that could be used for the extension and improvement of Ptolemaic astronomy in general.

To give but a few examples, al-Farghānī (d. after 861), an astronomer employed at the court of al-Ma'mūn, wrote an elementary, nonmathematical textbook of Ptolemaic astronomy, which had wide circulation in Islam and (after translation into Latin) in medieval Christendom. Thābit ibn Qurra (d. 901), another court astronomer in Baghdad, studied the apparent motions of the sun and moon on Ptolemaic principles; he concluded that the precession of the equinoxes is nonuniform and devised a theory of variable precession (called "trepidation") to account for it. Al-Battānī (d. 929) introduced mathematical improvements into Ptolemaic astronomy, studied the motion of the sun and moon, calculated new values for solar and lunar motions and the inclination of the ecliptic, discovered the movement of the line of apsides of the sun (the shifting of the sun's perigee, or closest approach to the earth, in the heavens), drew up a corrected star catalogue, and gave directions for the construction of astronomical instruments, including a sundial and a mural quadrant. The fact that al-Battānī was still being cited in the sixteenth and seventeenth centuries (by Copernicus and Kepler, among others) testifies to the quality of his astronomical work. Finally, Islam saw a debate between defenders of the physically oriented concentric spheres of Aristotle and the mathematically oriented Ptolemaic system; this debate, conducted mainly in twelfth-century Spain, ended indecisively.[18]

Optics is another example of distinguished scientific achievement in Islam. Here we find innovations at least as fundamental as those in astronomy—innovations, nonetheless, that grew out of the reconciliation and consolidation of a variety of ancient traditions. To be specific, Ibn al-Haytham (d. ca. 1040), who served at the court in Cairo (where a separatist Muslim dynasty had established its own caliphate), followed Ptolemy's lead in combining what had originally been separate Greek approaches to optical phenomena—mathematical, physical, and medical. In the case of Ibn

في صفه حركات عطارد ونقسها النصفها النبلنا اول اسنه واسنع ماسه
لبرزجور اما الوسط اصله ط د طه د طه وهومثل وسط الشمس والزهره والاوج
رب نسبه و اصل الحاصه المناح الدكور د د د ٥ وحركة الوسط مثل حركه
وسط الشمس وهى فى السنه با مطبه مد وحركه حا مه عطارد فى عسرى
سنه فارسيه با مط ٢٢ د وفى سنه احج مرا وفى شهرد حد ر د د اوفى بوم
د د و كله ١٤ الى لركط وفى ساعه د ٢٢ رمو و كون واناا سعلنا حركه الاوج
من حركه الوسط لحمل حركة الميكر وطريق هومن هنا الكوكب كطريق نعوم زحل

Fig. 8.3. The motion of Mercury according to Ibn ash-Shātir (14th c.). Oxford, Bodleian Library, MS Marsh 139, fol. 29r.

Fig. 8.4. The eyes and visual system according to Ibn al-Haytham. From a copy of al-Haytham's *Book of Optics* made in 1083 A.D., Istanbul, Süleimaniye Library, MS Fatih 3212, vol. 1, fol. 81v.

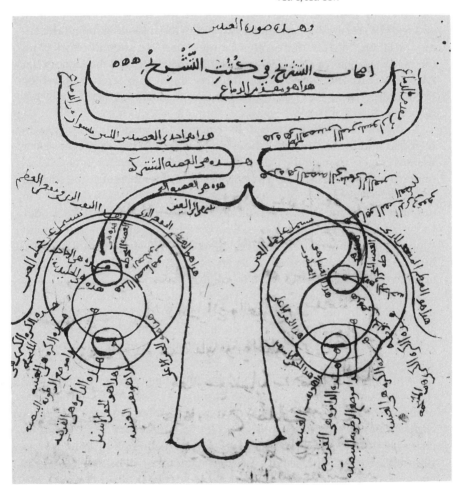

al-Haytham, this synthesis produced a new theory of vision, built on the idea that light is transmitted from the visual object to the eye, which was to prevail first in Islam and then in the West (see chap. 12, below) until Kepler devised the theory of the retinal image in the seventeenth century.[19]

THE DECLINE OF ISLAMIC SCIENCE

The scientific movement in Islam was both distinguished and durable. Translation of Greek works into Arabic began in the second half of the eighth century; by end of the ninth century translation activity had crested, and serious scholarship was under way. From the middle of the ninth century until well into the thirteenth, we find impressive scientific work in all the main branches of Greek science being carried forward throughout the Islamic world. The period of Muslim preeminence in science lasted for five hundred years—a longer period of time than has intervened between Copernicus and ourselves.

The scientific movement had its origins, for practical purposes, in Baghdad under the ʿAbbasids, though there came to be many other Near Eastern centers of scientific patronage. Early in the eleventh century, Cairo, under the Fatimids, came to rival Baghdad. In the meantime, the foreign sciences had made their way to Spain, where the Umayyads, displaced in the Near East by the ʿAbbasids, built a magnificent court at Cordoba. Under Umayyad patronage, the sciences flourished in the eleventh and twelfth centuries. Instrumental in this development was al-Ḥakam (d. 976), who built and stocked an impressive library in Cordoba. Another large collection of scientific books was to be found in Toledo.

But during the thirteenth and fourteenth centuries, Islamic science went into decline; by the fifteenth century, little was left. How did this come about? Not enough research has been done to permit us to trace these developments with confidence, or to offer a satisfactory explanation, but several causal factors can be identified. First, conservative religious forces made themselves increasingly felt. Sometimes this took the form of outright opposition, as in the notorious burning of books on the foreign sciences in Cordoba late in the tenth century. More often, however, the effect was subtler—not the extinction of scientific activity, but alteration of its character, by the imposition of a very narrow definition of utility. Or to reformulate the point, science became naturalized in Islam—losing its alien quality and finally becoming Islamic science, instead of Greek science practiced on Islamic soil—by accepting a greatly restricted handmaiden role. This meant a loss of attention to many problems that had once seemed important.

Fig. 8.5. Interior of the Great Mosque of Cordoba, built in the middle of the 8th century A.D.
Foto Marburg/Art Resource NY.

Second, a flourishing scientific enterprise requires peace, prosperity, and patronage. All three began to disappear in late medieval Islam as a result of continuous, disastrous warfare among factions and petty states within Islam and attack from without. In the West, the Christian reconquest of Spain began to make serious, if sporadic, headway after about 1065 and continued until the entire peninsula was in Christian hands two centuries later. Toledo fell to Christian arms in 1085, Cordoba in 1236, and Seville in 1248. In the east, the Mongols began to apply pressure on the borders of Islam early in the thirteenth century; in 1258 they took Baghdad, thus bringing the ʿAbbasid caliphate to an end. In the face of debilitating warfare, economic failure, and the resulting loss of patronage, the sciences were unable to sustain themselves. In assessing this collapse, we must remember that at an advanced level the foreign sciences had never found a stable institutional home in Islam, that they continued to be viewed with suspicion in conservative religious quarters, and that their utility (especially as advanced disciplines) may not have seemed overpowering. Fortunately, before the products of Islamic science could be lost, contact was made with Christendom, and the process of cultural transmission began anew.

NINE

The Revival of Learning in the West

THE MIDDLE AGES

To this point, I have employed the expression "Middle Ages" without definition and without specifying exact chronological limits. This may be a case where inexactness is a virtue, since historians themselves do not agree on what the expression means; but the time has come to be a little more definite. The idea of the Middle Ages (or medieval period) first arose in the fourteenth and fifteenth centuries among Italian humanist scholars, who detected a dark middle period between the bright achievements of antiquity and the enlightenment of their own age. This derogatory opinion (captured in the familiar epithet "dark ages") has now been almost totally abandoned by professional historians in favor of the neutral view that takes "Middle Ages" simply as the name of a period in Western history, during which distinctive and important contributions to Western culture were made—contributions that deserve fair and unbiased investigation and appraisal.

The chronological limits of the Middle Ages are necessarily blurred, because medieval culture (whatever exactly we take it to be) appeared and disappeared gradually, and at different times in different regions. If we must have dates, then the Middle Ages may be taken to cover the period from the end of Roman civilization in the Latin West (500 is a good round number) to 1450, when the artistic and literary revival commonly known as the Renaissance was unmistakably under way. It will be convenient for our purposes to subdivide this period into the early Middle Ages (approximately 500 to 1000), a transition period (1000 to 1200), and the high or late Middle Ages (1200 to about 1450). These are not exactly the standard subdivisions (the "high" and "late" Middle ages are frequently differentiated), but they will serve our purposes.

CAROLINGIAN REFORMS

We have observed (chap. 7) the declining fortunes of the Roman Empire in the Latin West and the appearance of socioreligious structures, such as monasticism, that we think of as characteristically medieval. Western Europe went through a process of de-urbanization; the classical schools deteriorated, and leadership in the promotion of literacy and learning passed to monasteries, where a thin version of the classical tradition survived as the handmaiden of religion and theology. This is not to suggest that monastic education totally obliterated the alternatives. Some municipal schools survived, especially in Italy; palace and episcopal schools never completely disappeared; and some of the great households always managed to arrange for private tutoring. The claim is simply that the monasteries became the dominant educational force.

Did this spell the end of serious scholarship? Some, who have chosen to define "scholarship" as a continuation of Greek and Roman scholarship, have judged so. But this seems a serious mistake. There is no question that scholarship declined in quantity and quality; but the notion that it disappeared as a productive enterprise is an illusion fostered by failure to look for the right thing or to look for it in the right places. In fact, scholarship continued, but in new forms and with changed focus.

The new focus was religious or ecclesiastical: what came to occupy the best scholarly minds was biblical interpretation, religious history, church governance, and the development of Christian doctrine. Boethius (480–524) not only translated parts of Aristotle's logic and composed handbooks on the liberal arts, as we have seen; he also wrote a group of short treatises that addressed contemporary theological controversies. Isidore of Seville (ca. 560–636) wrote not only the *Etymologies* and *On the Nature of Things,* the encyclopedic works that contain his natural philosophy, but also manuals for instructing the clergy on matters of history, theology, biblical interpretation, and liturgy. Gregory of Tours (d. 595) wrote a *History of the Franks,* which documents the spread of Christianity in Frankish territory. Gregory the Great (ca. 550–604), who became pope in 590, produced an influential corpus of sermons, lectures, dialogues, and biblical commentaries. And Bede (d. 735) left us biblical commentaries, sermons, and hagiography (saints' lives), along with his books on timekeeping and the calendar.

There is virtually no science or natural philosophy in these religious and theological works, but Greek logic and metaphysics do put in an appearance. Boethius set the pattern in his determined effort to think through such problems as divine foreknowledge and the nature of the divine trinity with the help of Aristotelian logic and Platonic and Aristotelian metaphysics. Isidore attempted to explain the origin of various Christian heresies by their parallels within the philosophical tradition. Even Gregory the Great, an outspoken critic of pagan learning, revealed at many points the implicit or explicit philosophical underpinnings of his theology.[1]

Late in the eighth century there was a burst of scholarly activity associated with the court of Charlemagne (Charles the Great). In 768 Charlemagne inherited a Frankish kingdom that encompassed portions of modern Germany and most of France, Belgium, and Holland. By the time of his death in 814, Charlemagne had enlarged the kingdom (which we know as the Carolingian Empire) to include more German territory, Switzerland, part of Austria, and more than half of Italy—the first serious attempt at centralized government in Western Europe since the disappearance of the Roman Empire (see map 5). As part of a program to strengthen church and state (and at the same time to act the part of emperor), Charlemagne undertook educational reforms, importing scholars from abroad to staff a palace school and ordering the establishment of monastery and cathedral schools throughout the realm. Charlemagne persuaded Alcuin, headmaster of the cathedral school at York in northern England, to come and direct this educational enterprise.

Alcuin (ca. 730–804), a beneficiary of the tradition of Irish scholarship (see chap. 7) who could trace his intellectual lineage directly back to Bede, created a thriving palace school, which educated the royal family and supplied the realm with educated religious and political functionaries. We know little about the curriculum, but it is clear that the seven liberal arts were included and that even astronomy was taught at some level; Alcuin himself wrote textbooks on the trivium. Alcuin's pupils were appointed as bishops and abbots, and through his efforts and theirs the average educational level of the clergy was raised. Around Alcuin there formed a circle of scholars interested in contemporary theological controversies and capable of contributing to them. Under his leadership books were collected, corrected, and copied—including the works of the church fathers and an occasional classical author. Finally, one of the most important and long-lasting steps taken by Charlemagne and Alcuin was the imperial edict mandating the establishment of cathedral and monastery schools, which contributed to a wider dissemination of education (directed toward the clergy, of course) than the Latin West had seen for several centuries and laid a foundation for future scholarship.[2]

The benefits from these educational reforms can be seen in the careers

Map 5. Carolingian Empire about 814

of two scholars, one from the ninth century and one from the tenth. John Scotus Eriugena (fl. 850–75), an Irishman attached to the court of Charlemagne's grandson Charles the Bald, was undoubtedly the ablest scholar of the ninth century in the Latin West. Eriugena had many gifts, including a keen, original mind and rare linguistic talent. He had an excellent command of Greek, probably first acquired in some Irish monastic school but improved after his arrival on the Continent, which he put to use in the translation of several Greek theological treatises into Latin: first, at the re-

Fig. 9.1. Personification of the quadrivium. From left to right: music, arithmetic, geometry, and astronomy. From a ninth-century copy of Boethius's *Arithmetic,* Bamberg, MS Class. 5 (HJ.IV.12), fol. 9v.

quest of Charles the Bald, the works of pseudo-Dionysius (an anonymous Christian Neoplatonist of about 500 A.D.); later, works of several Greek church fathers. Eriugena also produced original and sophisticated theological treatises, in which he developed the Neoplatonism of pseudo-Dionysius and attempted a synthesis of Christian theology (with a Greek slant) and Neoplatonic philosophy; his *On Nature,* an attempt at a comprehensive account of created things, contains a well-articulated (and, of course, thoroughly Christian) natural philosophy. Finally, he wrote a commentary on Martianus Capella's widely influential textbook of the liberal arts, *The Marriage of Philology and Mercury,* presumably in connection with his teaching responsibilities. Eriugena had an immediate impact on an entourage of disciples, and through them a continuing influence on Western thought.[3]

Another beneficiary of the Carolingian educational reforms emerged a century later from the monastery school at Aurillac in south-central France. Gerbert (ca. 945–1003) enjoyed a meteoric career, explained by a combination of intellectual talent and political opportunism. Born in humble circumstances, he received an impressive education at Aurillac and later in northern Spain, where he studied for a period of time. He went from Spain to the important cathedral school at Reims in northern France, first as a student of logic, subsequently as headmaster. From Reims he moved to northern Italy as abbot of the monastery at Bobbio, back to Reims as archbishop, and back to Italy as archbishop of Ravenna. Finally, in 999 his patron Otto III (Saxon emperor) saw to his election as Pope Sylvester II.

It has been customary to concentrate on Gerbert's role as one of the initiators of fruitful intellectual contact between Islam and Latin Christendom. But before examining that aspect of his achievement, we must note that he also contributed to an older tradition of scholarship—the recovery and dissemination of the classical liberal arts, especially Aristotelian logic as transmitted through Boethius and other Latin sources. At Reims Gerbert lectured on various logical works of Aristotle, Cicero, Porphyry, and Boethius; he also composed at least one logical treatise of his own. Gerbert's fame, however, rests on his contribution to the mathematical quadrivium, and here the connection with Islam was critical. When Gerbert crossed the Pyrenees into the northeastern corner of Spain in 967 to study with Atto, bishop of Vich, there can be no doubt that his purpose was to master the mathematical sciences, which were apparently more highly cultivated there (by virtue of the proximity of Islam) than anywhere north of the Pyrenees.

We know nothing in detail about Gerbert's studies, but his subsequent

Fig. 9.2. The cloister, Santa Maria de Ripoll in northeastern Spain. Courtesy of William J. Courtenay.

career provides eloquent testimony to his mastery of the mathematical sciences—unequalled in the Latin West in many centuries, though still very modest by comparison with the best of Greek mathematics—and to his familiarity with Islamic achievements in mathematics and astronomy. His correspondence, despite the turbulent political and religious context within which much of it was written, is laced with references to mathematics, astronomy, manuscripts to be copied or corrected (including Pliny's *Natural History*), translated books, and works to be obtained (including those of Boethius and Cicero). In one letter, Gerbert requests a book on multiplication and division by Joseph the Spaniard (an Arabic-speaking Christian); in another, he solicits a book on astronomy translated from the Arabic by Lupitus (archdeacon of the cathedral of Barcelona); in yet another, he announces the discovery of an astronomical work that he believes to have been written by Boethius. He praises his patron, Otto III, for his interest in numbers. He instructs friends and associates on how to solve various arithmetical and geometrical problems. He presents instructions on the construction of an astronomical model (a hemisphere on which the principal celestial circles and constellations were marked) and on the use of the abacus for multiplication and division (employing Arabic numerals).

Finally, it is probable that Gerbert, during his three years with Atto in Vich, had some connection with the nearby monastery of Santa Maria de Ripoll. Exactly how close the connection was, we do not know; but at the time, Ripoll seems to have been a center of quadrivial studies based on Arabic sources. A surviving Latin manuscript from the monastery library (dating from about the time of Gerbert's visit to Spain) contains translated versions of a number of important Arabic treatises on mathematics and the astrolabe (an instrument for making astronomical observations and calculations). It is possible that Gerbert carried a copy of one or more of these treatises over the Pyrenees upon his return; we know that fifty years later one of them could be found in the monastery at Reichenau in southern Germany. Gerbert may also have written his own treatise on the astrolabe. What is clear is that Gerbert used his influential posts as teacher and church dignitary to advance the cause of mathematical science in the West.[4]

THE SCHOOLS OF THE ELEVENTH AND TWELFTH CENTURIES

When Gerbert died in 1003, Western Europe was on the eve of political, social, and economic renewal. The causes of this renewal were numerous and complex. One of them was the emergence of stronger monarchies, capable of administering justice and reducing the level of internal disorder and violence. At the same time, secure borders were restored, after the Viking and Magyar invasions of the ninth and tenth centuries. Indeed, having been on the receiving end of aggression for so many centuries, Europeans were about to turn around and become the aggressors, driving the Muslims out of Spain and dispatching armies of crusaders to rescue the Holy Land.

Political stability led to the growth of commerce and increased affluence. The extension of a money economy to the countryside enhanced trade in agricultural products. Technological developments played a critical role in supplying necessities and producing sources of wealth. The refinement and spread of the water wheel, for example, gave rise to a minor industrial revolution; and agricultural innovations, such as crop rotation and the invention of the horse collar and the wheeled plow (combined, possibly, with improved climatic conditions), led to a major increase in the food supply.[5] One of the most dramatic results of these changes was a population explosion; exact figures are not available, but between 1000 and 1200 the population of Europe may have doubled, tripled, or even quadrupled, while the city-dwelling portion of this population increased even more rapidly.[6] Urbanization, in turn, provided economic opportunity, allowed for the concentration of wealth, and encouraged the growth of schools and intellectual culture.

It is widely agreed that a close relationship exists between education and urbanization. The disappearance of the ancient schools was associated with the decline of the ancient city; and educational invigoration followed quickly upon the re-urbanization of Europe in the eleventh and twelfth centuries. The prototypical school of the early Middle Ages was a monastic school—rural, isolated from the secular world, and dedicated to narrow educational objectives (even if those objectives were sometimes stretched by external pressures). With the shift of population to the cities in the eleventh and twelfth centuries, urban schools of various sorts, which to this point had been minor contributors to the educational enterprise, emerged from the shadow of the monastery schools and became major educational forces. This development was assisted by reform movements within monasticism, aimed at reducing monastic involvement in the world and reemphasizing the spiritual nature of the monk's calling. Among the urban schools that came into prominence at this time were cathedral schools; also schools run by parish clergy and a wide variety of public schools, both primary and secondary, not directly linked to ecclesiastical needs but open to anybody who could afford them.[7]

The educational aims of the new urban schools were far broader than those of the monastery schools. The emphasis of the teaching program varied from one school to another, according to the vision and specialty of the master who directed it, but in general urban schools expanded and reoriented the curriculum to meet the practical needs of a diverse and ambitious clientele, which would go forth to occupy positions of leadership in church and state. Even the cathedral schools, which resembled the monastery schools in having exclusively religious aims, based their curriculum on a broader conception of the range of studies that would contribute to religious ends. And if the educational ambitions of a master or his students went beyond what could be sustained within the framework of the cathedral school, they might detach themselves from the cathedral and operate independently of its authority. Indeed, it was quite possible for "schools" to be migratory, rather than geographically fixed, and to follow the itinerary of a charismatic master whose teaching, wherever it might take place, held the students together.[8] The product of these new arrangements was a rapid broadening of the curriculum: logic, the quadrivial arts, theology,

Fig. 9.3. A grammar school scene. The master threatens his pupils with a club. Paris, Bibliothèque National, MS Fr. 574, fol. 27r (14th c.).

law, and medicine came to be cultivated in the urban schools to a degree unheard of within the monastic tradition. The new schools multiplied in number and size; at their best, they emitted an aura of intellectual excitement that attracted the ablest masters and students into their orbits.

In France, some of the most robust schools were attached to (or operated in the shadows of) cathedrals in regions influenced by the Carolingian reforms of the ninth century. Laon was an early leader, with a significant cathedral school by 850 and a strong reputation in theology as late as the eleventh and twelfth centuries. In the tenth century Gerbert was attracted to the cathedral school at Reims as student and master. In the twelfth century, schools at Chartres, Orleans, and Paris emerged as leading centers for the liberal arts. The most famous of the twelfth-century schools is the cathedral school of Chartres—though the degree and duration of its preeminence have recently been called into question.[9] Schools in nearby Paris certainly flourished about the same time, offering instruction in a wide range of subjects, including the liberal arts. Outside of France, the leading schools were less apt to have any connection with cathedrals: Bologna became renowned for advanced legal instruction (by private teachers) early in the twelfth century, and by the end of the century Oxford (which had no cathedral) gained a reputation for studies in law, theology, and the liberal arts.

Several characteristics of these schools are important for our purposes. First, they witnessed a determined effort to recover and master the Latin classics (or Greek classics available in ancient Latin translations), surpassing anyting seen in the early Middle Ages. Bernard of Chartres spoke for his age when he pictured his generation as dwarfs standing on the shoulders of giants, able to see farther not by virtue of individual brilliance but through mastery of the classics. Among the favorite Roman authors were the poets Virgil, Ovid, Lucan, and Horace. Cicero and Seneca were valued as moralists, and Cicero and Quintilian as models of eloquence. The logical works of Aristotle and his commentators (especially Boethius) were carefully studied and applied to all manner of subjects. Legal studies were critically dependent on the recovery of the *Digest,* a summary of Roman legal thought. And Martianus Capella, Macrobius, and Plato (through Calcidius's translation of the *Timaeus* and accompanying commentary) served as the principal sources for cosmology and natural philosophy. None of this is to suggest that the pagan classics displaced the Christian sources that had formed the core of monastic education; rather, the newly recovered sources took their place alongside the Bible and the writings of the church fathers; it was assumed that these bodies of literature were mutually compatible and that recovery of the ancient classics was simply a matter of expanding the sources from which one might legitimately learn.[10]

Second, the urban schools, like European society more generally, saw a marked "rationalistic" turn—that is, an attempt to apply intellect and reason to many areas of human enterprise. There were attempts, for example, to rationalize commercial practices and the administration of church and state through record-keeping and the development of accounting and auditing procedures. One historian has described this as a "managerial revolution."[11] The same confidence in human intellectual capacity pervaded the schools, where phlosophical method was applied with increasing zeal to the whole of the curriculum, including biblical studies and theology.

The application of reason to theology was not new. As we have seen, the earliest Christian apologists undertook a reasoned defense of the faith; and scholars of the early Middle Ages (inspired by the example of Boethius) made a persistent effort to apply Aristotelian logic to knotty theological problems. The difference in the eleventh and twelfth centuries was in the lengths to which theologians were willing to go in the application of philosophical method. Anselm of Bec and Canterbury (1033–1109) is an ex-

Fig. 9.4. The west facade (12th c.) of Chartres Cathedral.

cellent example.[12] Though perfectly orthodox in his theological beliefs, Anselm was prepared to stretch the limits of theological methodology: to explore what unaided reason could achieve in the theological realm, to ask whether certain fundamental theological doctrines were true as judged by rational or philosophical criteria. His best-known piece of theological argumentation is a proof of God's existence (known as the "ontological

Fig. 9.5. The chained library of Hereford Cathedral.

proof") in which he places no reliance on biblical authority. Anselm's purpose was entirely constructive; clearly he applied philosophical method
to doctrines about the existence and attributes of God not because he
doubted the doctrines but in order to buttress them and make them evident to nonbelievers. At first glance this may not seem particularly daring,
but in fact the risks were serious: if reason can prove theological claims,
presumably it can also disprove them. This is not a problem as long as
reason arrives at the "right" answer; but what shall we do if, having committed ourselves to reason as the arbiter of truth, we find reason and faith
in opposition?[13]

A generation after Anselm, Peter Abelard (ca. 1079–ca. 1142), a brilliant,
restless, and abrasive student and teacher in the schools of northern
France (including Paris and Laon), extended the rationalist program begun
by Anselm. In various works, he defended theological positions considered dangerous by his contemporaries and was twice condemned by the
religious authorities. Abelard's best known book was entitled *Sic et non*

Fig. 9.6. Hugh of St. Victor teaching in Paris. Oxford, Bodleian Library, MS Laud. Misc. 409, fol. 3v (late 12th c.).

(roughly translatable as *Yes and No* or *Pro and Con*); in this sourcebook for students, he assembled conflicting opinions of the church fathers on a series of theological questions. He was using conflicting opinions to pose problems, which must then become the objects of philosophical investigation; in his view, the road to belief passes through doubt. There can be no question that Abelard intended to reason about, and in support of, the faith: he wrote on one occasion that he did not "wish to be a philosopher if it meant rebeling against [the Apostle] Paul, nor an Aristotle if it meant cutting [himself] off from Christ."[14] There is also no question that he was perceived by those of more conservative outlook, such as the monastic reformer Bernard of Clairvaux, who thundered against him, as a dangerous champion of philosophical method. The fact that Abelard attracted a following of enthusiastic students must have confirmed Bernard's worst fears.

In the work of Anselm and Abelard and like-minded contemporaries, we can see the makings of a confrontation between faith and reason. Anselm and Abelard raised in a compelling way such questions as: How does one "know" in the theological realm? Are the rational methods employed in other school subjects (logic, natural philosophy, and law) applicable also to theology, or does theology submit to some other master? How are conflicts between reason (Greek philosophy) and revelation (the truths revealed in the Bible) to be resolved? Worries about questions such as these jeopardized the intellectual revival and established an agenda for philosophers and theologians of the thirteenth and fourteenth centuries. The wholesale translation of Greek and Islamic philosophical and scientific literature, which was about to begin, would only intensify the problem. We will return to this subject below (chap. 10).

NATURAL PHILOSOPHY IN THE TWELFTH-CENTURY SCHOOLS

Natural philosophy did not hold center stage in the twelfth-century schools, but it did benefit from the general intellectual ferment. The determination among scholars to master the Latin classics extended to the classics of natural philosophy—Plato's *Timaeus* with Calcidius's commentary, Martianus Capella's *Marriage of Philology and Mercury,* Macrobius's *Dream of Scipio,* Seneca's *Natural Questions,* Cicero's *On the Nature of the Gods,* and the works of Augustine, Boethius, and John Scotus Eriugena. Most of these texts have a Platonic tilt, and the scholars who read and analyzed them were sharply drawn to the Platonic conception of the cosmos. Plato's *Timaeus,* source of the most coherent discussion of cosmological and physical problems then available, and also the repository of Plato's own words, became the central text. That position of centrality, in turn, gave the *Timaeus* the power to shape the agenda and content of twelfth-century natural philosophy. This does not mean that the Platonism of the twelfth century was pure or entirely without rival: certain Stoic ideas managed to elbow their way into the Platonic milieu; toward the end of the century Aristotle's physical and metaphysical works began to make their presence felt; and in the thirteenth century Platonic philosophy would retreat before an Aristotelian onslaught. But for the time being, Plato held the position of leadership.[15]

But Plato was a versatile guide, and Platonic leadership could mean many different things. The *Timaeus* is first and foremost an account of the formation of the cosmos by the divine craftsman. The obvious and most pressing task, therefore, was to reconcile Platonic cosmology (or the aspect of cosmology that deals with origins, known as "cosmogony") with the account of creation in the book of Genesis, as explained over the centuries by the fathers of the church. Or to put it a little differently, the task was to apply all of the cosmology and physics that could be learned from Plato and the other ancients to the elucidation of the Genesis account of creation. Science, it should be noticed, was still expected to function as handmaiden.

A number of excellent scholars applied themselves to this project in the twelfth century. One of them was Thierry of Chartres (d. after 1156), a teacher of international renown at Chartres and (perhaps) Paris. Thierry wrote a commentary on the six days of creation, in which he managed to read the content of Platonic cosmology (along with pieces of Aristotelian and Stoic natural philosophy) into the biblical text. One of the major needs was to explain the specific sequence of God's creative activity as described in Genesis. According to Thierry, the four elements were created by God in the first instant of time; everything that followed was a natural unfolding of the order inherent in that initial creative act. Once created, fire immediately began to rotate (owing to its lightness, which forbids rest), while also illuminating the air, thus accounting for day and night (the first day of creation). During the second rotation of the fiery heaven, the fire heated the waters below, causing them to ascend as vapors until suspended above the air, forming what the biblical text refers to as the "waters above the firmament" (second day). Reduction in the quantity of water below, owing to vaporization, caused dry land to emerge from the seas (third day). Further heating of the waters above the firmament led to the formation of the celestial bodies, composed of water (fourth day). Finally, heating of the land and the lower waters brought forth plant, animal, and human life (fifth and sixth days).[16]

This is a very brief and incomplete survey of Thierry's commentary, but it is enough to reveal the nature of the philosophical program on which he and various contemporaries had, under Platonic inspiration, embarked. Thierry's cosmology may not be sophisticated by modern standards. But the important thing is that, following Plato's lead, it restricts direct divine intervention to the initial moment of creation; what happens thereafter is the result of natural causation, as the elements move and interact in the manner proper to them, and as seeds (the "seminal causes" of Stoic philosophy, borrowed through Augustine) implanted in created things undergo a natural process of development. Even the appearance of Adam and Eve, and of subsequent humans, did not call for miraculous intervention.

This naturalism is one of the most salient features of twelfth-century natural philosophy. It is found in commentaries on the days of creation

Fig. 9.7. God as architect of the universe. Vienna, Österreichische Nationalbibliothek, MS 2554, fol. Iv (13th c.).

(perhaps the best place for a natural philosopher to display his naturalistic inclinations), but also in more general treatises on natural philosophy by scholars such as William of Conches, Adelard of Bath, Honorius of Autun, Bernard Sylvester, and Clarembald of Arras (most of whom were connected with the schools of northern France). These men differed on matters of cosmological and physical detail, of course, but they shared a new conception of nature as an autonomous, rational entity, which proceeds without interference according to its own principles. There was a growing awareness of natural order or natural law and a determination to see how far natural principles of causation would go in providing a satisfactory explanation of the world.[17]

An outspoken advocate of the new naturalism was William of Conches (d. after 1154), who studied and taught at Chartres or Paris or both, before joining the household of Geoffrey Plantagenet, where he tutored the future King Henry II of England. William developed an elaborate cosmology and physics based on Platonic principles (with important additions from some newly translated sources). In his *Philosophy of the World,* William lashed out at those who appeal too readily to direct divine causation:

> Because they are themselves ignorant of nature's forces and
> wish to have all men as companions in their ignorance, they
> are unwilling for anybody to investigate them, but prefer that
> we believe like peasants and not inquire into the [natural]
> causes [of things]. However, we say that the cause of every-
> thing is to be sought. . . . But these people, . . . if they know
> of anybody so investigating, proclaim him a heretic.

William's purpose, as he made clear elsewhere, was not to deny divine agency, but to declare that God customarily works through natural powers and that the philosopher's task is to push these powers to their explanatory limit. Adelard of Bath (fl. 1116–42) made the same point about the same time, urging that only when natural explanations "utterly fail should there be recourse to God." And a little later Andrew of St. Victor, discussing the interpretation of biblical events, advised that "in expounding Scripture, when the event described admits of no natural explanation, then and then only should we have recourse to miracles." [18]

This may seem a sensible position, but it was also a dangerous one. How could a strong commitment to the search for natural causes (the position of the twelfth-century natural philosophers) avoid slipping into outright denial of the miraculous (a totally unacceptable outcome for the Christian scholar)? Would scholars be capable of maintaining the delicate balance between belief and unbelief called for by this position? William of Conches addressed the problem directly, pointing to the difference between acknowledging that it was within God's power to perform some act and maintaining that God actually had performed it; surely God did not do everything of which he was capable. He added that his philosophical position (and that of his fellow "naturalists") did not detract from divine power and majesty, since whatever comes to pass is ultimately of divine origin: "I take nothing away from God; all things that are in the world were made by God, except evil; but he made other things through the operation of nature, which is the instrument of divine operation." Indeed, study of the physical world enables us to appreciate "divine power, wisdom, and goodness." [19] Searching for secondary causes is not a denial, but an affirmation, of the existence and majesty of the first cause.

Several other philosophical maneuvers could also help to relieve the tension. It was possible to reconcile the reality of miracles with the fixity of nature by acknowledging that miracles represent genuine suspensions of the usual laws of nature, while maintaining that these suspensions were planned by God from the first creation and built into the cosmic machinery, so that they remain perfectly natural in the larger sense. Moreover, one could speak about a fixed natural order without infringing on divine omnipotence and freedom by arguing (a) that God had unlimited freedom to create any kind of world he wished, but (b) that in fact he chose to make this world and, having completed his creative activity, was not going to meddle with the product. This latter distinction was to become crucial to developing thought on the subject in the thirteenth and fourteenth centuries.[20]

Some modern readers will be tempted to view all of this as unacceptable intrusion of theology into the scientific realm. However, if we wish to understand the twelfth century, it is important for us to realize that twelfth-century bystanders would have been apt to see these developments in precisely the opposite light—as a possibly dangerous intrusion of philosophy into the theological realm. What was new and threatening was not theological presence in philosophical precincts, where it had always made itself quite at home, but the flexing of philosophical muscle in territory where theology had hitherto reigned without challenge. To critics of the twelfth-century naturalists, it appeared that philosophy might be about to throw off her handmaiden status.

Let us briefly summarize several other aspects of twelfth-century natural philosophy. The *Timaeus* and supporting sources not only promoted the idea of a fixed natural order; they also made humans a part of that order, governed by the same laws and principles, so that an exploration of human nature was understood to be continuous with exploration of the universe more generally. Frequently the point was made even more forcefully, through the macrocosm-microcosm analogy: humans not only belong to the cosmos but are actually miniatures of it. It follows that the cosmos and the individual person are linked by structural and functional similarities, which bind them into a tight unity. For example, just as the cosmos consists of the four elements, animated by the world soul (the exact nature of which gave rise to considerable debate in the twelfth century), so the human is a composite of body (the four elements) and soul.

Having made humankind part of the natural order, twelfth-century scholars increasingly took an interest in the "natural man" and his capacities—that is, humans as they are independently of divine grace. (Thus historians sometimes write of twelfth-century "humanism.") In this connection, there was a strong tendency to affirm the value of human reason; as part of the natural order, and therefore sympathetic to its rhythms and harmonies, reason was regarded as a particularly suitable instrument for the exploration of the cosmos.[21]

Closely associated with the macrocosm-microcosm analogy was the science of astrology. Astrology had fallen into disrepute during the early Middle Ages, owing to the opposition of the church fathers. Augustine attacked it as a form of idolatry (since it had traditionally been associated with worship of the planetary deities) and for its tendency to lead to fatalism and the denial of free will. But under the influence of twelfth-century Platonism, as well as an influx of Arabic astronomical and astrological literature in translation, astrology was restored to at least quasi-respectability. In the *Timaeus,* the Demiurge is credited with making the planets or celestial gods, but then delegating to them the responsibility of bringing forth subsequent forms of life in the lower regions. This suggestive account, coupled with the idea of cosmic unity, the macrocosm-microcosm analogy, and certain long-known correlations between celestial and terrestrial phenomena (the seasons and the tides), and augmented by newly translated Arabic astrological works, led to a resurgence of interest and belief in astrology. This is not the place to enter into a detailed analysis of astrological theory or practice (treated in chap. 11, below). What is important for our purposes is to notice that twelfth-century astrology had nothing to do with the supernatural; quite the contrary, it flourished among the *naturalists* of the twelfth century precisely because it entailed the exploration of the *natural* forces that link heaven and earth.[22]

Finally, did the mathematical tendencies of Platonic philosophy influence twelfth-century thought, as we might expect? Yes, but in a form that may surprise modern readers. Mathematics was employed in the first half of the twelfth century, not to quantify natural laws or to provide a geometrical representation of natural phenomena, but to answer questions that we would regard as metaphysical or theological. This is an exceedingly abstruse subject, which we cannot go into deeply, but one example may help to point the way. Following Boethius, twelfth-century scholars saw the theory of numbers (specifically, the relationship of the number 1 to the remaining numbers) as a vehicle for understanding the relationship between divine unity and the multiplicity of created things. It is the latter point that Thierry of Chartres was addressing when he wrote that "the creation of number is the creation of things." Mathematics also served in the twelfth century as a model of the axiomatic method of demonstration. A broader conception of the scientific uses of mathematics would have to await the translation and assimilation of Greek and Arabic mathematical science later in the century.[23]

THE TRANSLATION MOVEMENT

The revival of learning began as an attempt to master and exploit traditional Latin sources. However, before the end of the twelfth century it was transformed by the infusion of new books, containing new ideas, freshly translated from Greek and Arabic originals. This new material, first a trickle and eventually a flood, radically altered the intellectual life of the West. Up to this point, Western Europe had been struggling to reduce its intellectual losses; hereafter, it would face the altogether different problem of assimilating a torrent of new ideas.[24]

The separation between East and West had never been total, of course. There were always travelers and traders, and near the borders bilingual (or multilingual) people would be numerous. There were also diplomatic contacts between Byzantine, Muslim, and Latin courts: an early and significant case was the exchange of ambassadors (both of them scholars) between the courts of Otto the Great in Frankfurt and 'Abd al-Raḥmān in Cordoba, which occurred about 950. Another kind of contact is illustrated by Gerbert's pilgrimage to northern Spain in the 960s to study Arabic mathematical science. Considered individually, such events may seem of small significance; but added together, they gradually created in Western minds an image of Islam and (to a lesser extent) Byzantium as repositories of great intellectual riches. It became clear to Western scholars wishing to enlarge the body of knowledge in Latin Christendom that they could do no better than to make contact with these intellectually superior cultures.

The earliest translations from the Arabic—several treatises on mathematics and the astrolabe—were made late in the tenth century in Spain. A century later a North African become Benedictine monk, named Constantine (fl. 1065–85), made his way to the monastery of Monte Cassino in southern Italy. There he began to translate medical treatises from Arabic to Latin, including the works of Galen and Hippocrates, which would supply the foundation of medical literature on which the West would build for several centuries.[25]

These early translations whetted the European appetite for more. Beginning in the first half of the twelfth century, translation became a major scholarly activity, with Spain as the geographical focus. (Contact with the Middle East as a result of the Crusades had a minimal impact on translations.) Spain had the advantage of a brilliant Arabic culture, an ample supply of Arabic books, and communities of Christians (known as Mozarabs) who had been allowed to practice their religion under Muslim rule and who could now help to mediate between the two cultures. As a result of the Christian reconquest of Spain, centers of Arabic culture and libraries of Arabic books fell into Christian hands; Toledo, the most important center, fell in 1085, and in the course of the twelfth century the riches of its library began to be seriously exploited, thanks in part to generous patronage from the local bishops.

Some of the translators were native Spaniards, fluent in Arabic from childhood: such a man was John of Seville (fl. 1133–42), probably a Mozarab, who translated a large number of astrological works; another was Hugh of Santalla (fl. 1145), from one of the Christian states in the north of Spain, who translated texts on astrology and divination; yet another, and one of the ablest, was Mark of Toledo (fl. 1191–1216), who translated several Galenic texts. But others came from abroad: Robert of Chester (fl. 1141–50) came from Wales; Hermann the Dalmatian (fl. 1138–43) was a Slav; and Plato of Tivoli (fl. 1132–46) was an Italian. These men came to Spain, presumably without prior knowledge of Arabic. Once there, they found a teacher, learned Arabic, and began to translate. Occasionally they joined forces with a bilingual native (perhaps a Mozarab or a Jew who knew Arabic and the vernacular language) and proceeded to translate cooperatively.

The greatest of the translators from Arabic to Latin was undoubtedly Gerard of Cremona (ca. 1114–87).[26] Gerard, from northern Italy, came to Spain in the late 1130s or early 1140s in search of Ptolemy's *Almagest,* which he had been unable to locate elsewhere. He found a copy in Toledo, remained to learn Arabic, and eventually rendered it into Latin.

But he also discovered texts on all manner of other subjects, and over the next thirty-five or forty years (perhaps with the help of a crew of assistants)[27] produced translations of many of these. His output is absolutely astonishing: at least a dozen astronomical texts, including the *Almagest;* seventeen works on mathematics and optics, including Euclid's *Elements* and al-Khwārizmī's *Algebra;* fourteen works on logic and natural philosophy, including Aristotle's *Physics, On the Heavens, Meteorology,* and *On Generation and Corruption;* and twenty-four medical works, including Avicenna's great *Canon of Medicine* and nine Galenic treatises. The total comes to seventy or eighty books, all translated carefully and literally by a man who possessed a good command of the languages as well as the subject matter.

Translation from the Greek had never entirely ceased: recall Boethius in the sixth century and Eriugena in the ninth. But translation from the Greek resumed and dramatically accelerated in the twelfth century. Italy was the principal location, especially the south (including Sicily), where there had always been Greek-speaking communities and libraries containing Greek books. Italy also benefited from ongoing contact with the Byzantine Empire. One of the important early translators was James of Venice (fl. 1136–48), a legal scholar in touch with Byzantine philosophers, who translated a collection of Aristotle's works. A series of important works on mathematics and mathematical science appeared in Greco-Latin translations about the middle of the century: Ptolemy's *Almagest* (whether before or after Gerard's translation from the Arabic cannot be determined) and Euclid's *Elements, Optics,* and *Catoptrics.*

Greco-Latin translating activity continued in the thirteenth century, most notably in the work of William of Moerbeke (fl. 1260–86). Moerbeke set out to provide Latin Christendom with a complete and reliable version of the Aristotelian corpus, revising existing translations where he could, producing new translations from the Greek where that was required. Moerbeke also translated some of the major Aristotelian commentators, a variety of Neoplatonic authors, and some mathematical works by Archimedes.[28]

Finally, a word about the motivation for the translations and the selection of materials to be translated. The aim was clearly utility, broadly defined. Medicine and astronomy led the way in the tenth and eleventh centuries; early in the twelfth century the emphasis seems to have been on astrological works, along with the mathematical treatises required for the successful practice of astronomy and astrology. Medicine and astrology both rested on philosophical foundations; and it was at least partly to recover and asses those foundations that attention was directed, beginning in the second half of the twelfth century and continuing through the thirteenth, toward the physical and metaphysical works of Aristotle and his commentators (including the Muslims Avicenna and Averroes). Once the full scope of Aristotle's works became known, of course, it became clear that his philosophical system was applicable to an enormous range of scholarly issues treated in the schools.[29]

By the end of the twelfth century, Latin Christendom had recovered major portions of the Greek and Arabic philosophical and scientific achievement; in the course of the thirteenth century many of the remaining gaps would be filled. These books spread quickly to the great educational centers, where they contributed to the educational revolution. In the next chapter we will examine some of the struggles provoked by the newly translated materials.

THE RISE OF UNIVERSITIES

The typical urban school in the year 1100 was small, consisting of a single master or teacher and perhaps ten or twenty pupils. By the year 1200 the schools had grown dramatically in number and size. We have almost no quantitative data, but in leading educational centers like Paris, Bologna, and Oxford, students undoubtedly numbered in the hundreds. Some idea of the explosion in the school population can be gained from the fact that more than seventy masters taught in Oxford between 1190 and 1209.[30] An educational revolution was in progress, driven by European affluence, ample career opportunities for the educated, and the intellectual excitement generated by teachers like Peter Abelard. Out of the revolution emerged a new institution, the European university, which would play a vital role in promoting the natural sciences. Let us briefly examine the process.

The absence of documentary evidence makes it impossible for us to trace in detail the steps by which universities came into existence. But what is clear is that the great expansion of educational opportunity at the elementary level (where instruction was offered in Latin grammar, the art of chanting, and basic arithmetic) led to demand among the intellectually ambitious for higher studies. Certain cities, such as Bologna, Paris, and Oxford, acquired a reputation for advanced studies in the liberal arts, medicine, theology, or law, and to these cities teachers and students gravitated in large numbers. Once there, a teacher would set up shop under the auspices of an existing school or as an independent, free-lance teacher—ad-

Map 6. Medieval universities

vertising for students and teaching them individually or in groups for a fee (rather like a modern teacher of music or dance). Instruction would generally take place in quarters provided by the teacher.

With numerical expansion came the need for organization—to secure rights, privileges, and legal protection (since many of the teachers and students were foreigners, without the rights of the local citizenry), to gain control of the educational enterprise, and generally to promote their mutual well-being. Fortunately, there was an organizational model ready at hand in the guild structure that was developing at the same time within various trades and crafts; it was therefore natural for teachers and students to organize themselves similarly into voluntary associations or guilds. Such a guild was called a "university [*universitas*]"—a term that originally had no scholarly or educational connotations but simply denoted an association of people pursuing common ends. It is important to note, then, that a university was not a piece of land or a collection of buildings or even a charter, but an association or corporation of teachers (called "masters") or students. The fact that a university owned no real estate (to begin with) made it extremely mobile, and the early universities were thus able to use the threat of packing up and moving to another city as leverage to secure concessions from the local municipal authorities.

It is impossible to assign a precise date to the founding of any of the early universities for the simple reason that they were not founded, but emerged gradually out of preexisting schools—their charters coming after the fact. It is customary, however, to see the masters of Bologna as having achieved university status by 1150, those of Paris by about 1200, and those of Oxford by 1220. Later universities were generally modeled on one or another of these three.[31]

Among the aims of these corporations were self-government and monopoly—which amount to control of the teaching enterprise. Gradually the universities secured varying degrees of freedom from outside interference and thus the right to establish standards and procedures, to fix the curriculum, to set fees and award degrees, and to determine who would be permitted to study or teach. They managed this by virtue of high-level patronage from popes, emperors, and kings, who offered protection, guaranteed privileges, granted immunity from local jurisdiction and taxation, and generally took the side of the universities in a variety of power struggles. The universities were considered vital assets, which needed to be carefully nurtured and (if circumstances dictated) judiciously disciplined. The extraordinary thing is how effective the nurturing proved to be, and how rarely and benevolently the disciplinary function was exercised. There were certain episodes, as we shall see, in which the church intervened decisively; but for the most part the universities managed the rare and remarkable feat of securing patronage and protection without interference.[32]

As the universities grew in size, internal organization was required. There were variations, of course, but Paris (the preeminent university in northern Europe) will serve to illustrate. At Paris there came to be four faculties or guilds: an undergraduate faculty of liberal arts (by far the largest of the four) and three graduate facilities—law, medicine, and theology. The liberal arts were considered preparatory for work in the graduate faculties, admission to which ordinarily depended on completion of the course of study in the arts. Because the masters in the arts faculty were more numerous than the teachers in the other faculties, they came to control the university.

A boy came to the university at about age fourteen, having previously learned Latin in a grammar school. In northern Europe, matriculation in the university generally conferred clerical status; this does not mean that students were priests or monks, but simply that they were under the authority and protection of the church and had certain ecclesiastical privileges. The student enrolled under a particular master (the apprenticeship model should be kept in mind), whose lectures he followed for three or four years before presenting himself to be examined for the bachelor's (young man's) degree. If he passed this, he became a bachelor of arts, with the status of a journeyman apprentice, and was permitted to give certain types of lectures under the direction of a master (rather like the modern teaching assistant), while continuing his studies. At about age twenty-one, having heard lectures on all of the required subjects, he could take the examination for the M.A. (master of arts) degree. Passage of this examination brought the student full membership in the arts faculty, with the right to teach anything in the arts curriculum.

The universities were enormously large by comparison with Greek, Roman, or early medieval schools, but they fell far short of the mammoth public universities of the present. There were wide variations, of course, but a typical medieval university was comparable in size to a small American liberal arts college—with a student population falling somewhere between about 200 and 800. The major universities were considerably larger: Oxford probably had between 1,000 and 1,500 students in the fourteenth century; Bologna was of similar size; and Paris may have peaked at 2,500 to 2,700 students.[33] It is evident from these figures that university-educated people were but a minuscule fraction of the European population, but their cumulative influence over time should not be underestimated; that German culture, for example, was profoundly shaped by the more than 200,000 students who passed through the German universities between 1377 and 1520 seems indisputable.[34]

It would be a mistake to suppose that most of these students emerged from their university experience with a degree; the great majority dropped out after a year or two, having acquired sufficient education to meet their needs, having run out of money, or having discovered themselves unsuited

Fig. 9.8. Mob Quad, Merton College, Oxford. Dating from the fourteenth century, this is the oldest complete quadrangle in Oxford.

SCHOLA ASTRONOMIAE ET RHETORICAE

Fig. 9.9. Doorway to one of the late-medieval schools, presently part of the Bodleian Library, University of Oxford.

to the academic life. Substantial numbers died before completing their studies—a reminder of the high mortality rates of the Middle Ages.[35] The student who did earn the M.A. was often required to teach for two years (because of a chronic shortage of teachers in the arts faculty); he might simultaneously embark on a degree in one of the graduate faculties, which promised to lead to far more lucrative employment. Few masters of arts actually made a career of teaching in the faculty of arts. The program of studies in medicine (leading to the master's degree or doctorate—there was no difference) required five or six years beyond the M.A. degree; in law, about seven or eight additional years; and in theology, somewhere between eight and sixteen years of further study. This was a long and demanding program, and those who completed the master's degree in any of the graduate faculties belonged to a small scholarly elite.

We come at last to the curriculum. This evolved as the Middle Ages progressed, of course, but certain generalizations are possible.[36] First, it came to be understood that the seven liberal arts no longer provided an adequate framework within which to conceive the mission of the schools. Grammar declined in significance, yielding its place in the curriculum to a greatly expanded emphasis on logic. The mathematical subjects of the quadrivium, never prominent in the medieval schools, retained their low profile (with some exceptions, to be dealt with below). The arts curriculum was rounded out by the three philosophies: moral philosophy, natural philosophy, and metaphysics. And, of course, medicine, law, and theology came to be viewed as advanced subjects, covered in graduate faculties and requiring study in the arts as a prerequisite.

Second, where does this leave the subjects that we think of as scientific? We will deal with the content of the various sciences in subsequent chapters; here the question concerns their place in the curriculum. The quadrivial arts were generally taught, but rarely stressed. Arithmetic and geometry, between them, occupied perhaps eight to ten weeks in the curriculum of the typical medieval undergraduate; but those who wanted more could frequently obtain it, at least in the larger universities. Astronomy was more highly cultivated, either as the art of timekeeping and the establishment of the religious calendar (especially determination of the movable date of Easter) or as the theoretical substructure for the practice of astrology (frequently in connection with medicine). The teaching texts were Greek and Arabic books in translation (including, on occasion, Ptolemy's *Almagest*) or new ones written expressly for the purpose. The average level of astronomical knowledge must have been quite low, but there were times and places when the subject was taught with skill and sophistication; and there can be no doubt that the universities produced a few highly proficient astronomers (see chap. 11).

If the mathematical sciences remained generally inconspicuous, Aristotelian natural philosophy became central to the curriculum. From modest beginnings late in the twelfth century, Aristotle's influence grew until, by the second half of the thirteenth century his works on metaphysics, cosmology, physics, meteorology, psychology, and natural history became compulsory objects of study. No student emerged from a university education without a thorough grounding in Aristotelian natural philosophy. And finally, we must note that medicine had the good fortune to be cultivated within its own faculty.[37]

Third, one of the most remarkable features of this curriculum was the high degree of uniformity from one university to another. Up to this point, different schools generally represented different schools of thought. In ancient Athens, for example, the Academy, the Lyceum, the Stoa, and the Garden of Epicurus were committed to the propagation of rival and (to some extent) incompatible philosophies. But the medieval universities, while differing somewhat in emphasis and specialty, developed a common curriculum consisting of the same subjects taught from the same texts.[38] This was partly a response to the sudden influx of Greek and Arabic learning through the translations of the twelfth century, which supplied European scholars with a standard collection of sources and a common set of problems. It was also connected, as both cause and effect, with the high level of mobility of medieval students and professors. Professorial mobility was facilitated by the *ius ubique docendi* (right of teaching anywhere) conferred on the master by virtue of completing his course of study. Thus a scholar who earned his degree at Paris could teach at Oxford without interference and, perhaps more importantly, without acquiring a case of intellectual indigestion; this was possible only because subjects taught at the one did not differ markedly in form or content from those same subjects as taught at the other. For the first time in history, there was an educational effort of international scope, undertaken by scholars conscious of their intellectual and professional unity, offering standardized higher education to an entire generation of students.

Fourth, this standardized education communicated a methodology and a worldview based substantially on the intellectual traditions traced in the early chapters of this book. Methodologically, the universities were committed to the critical examination of knowledge claims through the use of Aristotelian logic. And the system of belief that emerged from the application of this method integrated the content of Greek and Arabic learning with the claims of Christian theology. We will deal below (especially in chap. 10) with struggles over the reception of the new learning and the form and content of the resulting synthesis; at present it is sufficient to note that these struggles were won by the liberal party, which wished to enlarge the store of European learning by assimilating the fruits of Greek and Arabic scholarship. Thus in the medieval universities, Greek and Arabic science (almost in their entirety) at last found a secure institutional home.

Finally, it must be emphatically stated that within this educational system the medieval master had a great deal of freedom. The stereotype of the Middle Age pictures the professor as spineless and subservient, a slavish follower of Aristotle and the church fathers (exactly how one could be a slavish follower of both, the stereotype does not explain), fearful of departing one iota from the demands of authority. There *were* broad theological limits, of course, but within those limits the medieval master had remarkable freedom of thought and expression; there was almost no doctrine, philosophical or theological, that was not submitted to minute scrutiny and criticism by scholars in the medieval university. Certainly the medieval master, particularly the master who specialized in the natural sciences, would not have thought of himself as restricted or oppressed by either ancient or religious authority.

TEN

The Recovery and Assimilation of Greek and Islamic Science

THE NEW LEARNING

The educational revival of the eleventh and twelfth centuries was broadened and transformed in the course of the twelfth century by the acquisition of new sources. In 1100, the revival could still be construed as an attempt to recover and master the Latin classics: Roman and early medieval authors, including the Latin church fathers, and a few Greek sources (Plato's *Timaeus* and parts of Aristotle's logic, for example) that existed in early Latin translations. A trickle of new translations from both the Greek and the Arabic had begun to flow, but their impact was still modest. A hundred years later, this trickle had become a torrent, and scholars found themselves struggling valiantly to organize and assimilate a body of new learning overwhelming in scope and magnitude.

The existence of this new learning was the central feature of intellectual life in the thirteenth century, setting an agenda that would preoccupy the best scholars of the century. The task was to come to terms with the contents of the newly translated texts—to master the new knowledge, organize it, assess its significance, discover its ramifications, work out its internal contradictions, and apply it (wherever possible) to existing intellectual concerns. The new texts were enormously attractive because of their breadth, their intellectual power, and their utility. But they were also of pagan origin; and, as scholars gradually discovered, they contained material that was theologically dubious. It was thus a sobering intellectual challenge that thirteenth-century scholars confronted; their approach to the new material and their skill in dealing with it would contribute permanently to the shape of Western thought.

Most of the translated works were benign. The very fact that a text was translated tells us that somebody thought its usefulness outweighed any potential dangers. Technical treatises on all manner of subjects (mathematics, astronomy, statics, optics, meteorology, and medicine) were, in fact, received with unqualified enthusiasm: they were obviously superior to anything previously available on their respective subjects; in many cases they filled an intellectual void; and they contained no unpleasant philosophical or theological surprises. Thus Euclid's *Elements,* Ptolemy's *Almagest,* al-Khwārizmī's *Algebra,* Ibn al-Haytham's *Optics,* and Avicenna's *Canon of Medicine* were peacefully added to the corpus of Western knowledge. The process by which these and other technical treatises were mastered and assimilated will be treated in subsequent chapters.

Insofar as there was trouble, it appeared in broader subject areas that impinged on worldview or theology—subjects such as cosmology, physics, metaphysics, epistemology, and psychology. Central to these subjects were the works of Aristotle and his commentators, which successfully addressed a multitude of critical philosophical problems, while promising untold future benefits from the proper employment of their methodology. The explanatory power of the Aristotelian system was obvious, and it proved exceedingly attractive to Western scholars. But it bestowed its benefits at a certain price, for Aristotelian philosophy inevitably touched on many issues already addressed by the blend of Platonic philosophy and Christian theology that had gradually become entrenched over the previous millennium. Unlike treatises on narrower, more technical subjects, Aristotelian philosophy did not fill an intellectual vacuum, but invaded occupied territory. This led to a variety of skirmishes, which ended (as we shall see) in a negotiated settlement. Let us examine the steps by which this occurred.

ARISTOTLE IN THE UNIVERSITY CURRICULUM

Most of Aristotle's works and some of the commentaries on them (especially those of the eleventh-century Muslim Avicenna) were available in translation by 1200. We know very little about their early circulation or their use in the schools, but they seem to have made an appearance at both Oxford and Paris during the first decade of the thirteenth century. At Oxford, no obstacles arose in the next few decades to the slow, but steady, growth of Aristotelian influence.[1] At Paris, however, Aristotle ran into early trouble: allegations were made that pantheism (roughly, the identification of God with the universe) was being taught by masters of arts under Aristotelian inspiration. The outcome of these charges was a decree, issued by a council of bishops meeting in Paris in 1210 and reflecting conservative opinion in the faculty of theology, forbidding instruction on Aristotle's natural philosophy within the faculty of arts. This decree was renewed in 1215 by the papal legate Robert de Courçon, but it was still applicable only to Paris.[2]

Pope Gregory IX became directly involved in 1231, in the course of promulgating regulations governing the University of Paris. Gregory acknowledged the legitimacy of the ban of 1210, and renewed it, specifying that Aristotle's books on natural philosophy were not to be read in the faculty of arts until they had been "examined and purged of all suspected error." Gregory explained himself ten days later, in a letter appointing a commission to act on the matter: "Since the other sciences should serve the wisdom of holy Scripture, they are to be appropriated by the faithful insofar as they are known to conform to the good pleasure of the Giver." However, it had come to Gregory's attention that "the books on natural philosophy that were prohibited in a provincial council at Paris . . . contain both useful and useless matter." Therefore, "in order that the useful not be contaminated by the useless," Gregory admonished his newly appointed commission to "eliminate all that is erroneous or that might cause scandal or give offense to readers, so that when the dubious matter has been removed, the remainder may be studied without delay and without offense."[3]

What is noteworthy is that Gregory acknowledged both the utility and the dangers of Aristotelian natural philosophy. Aristotle remained under the ban until purged of error; but once error had been eliminated, scholars were encouraged to put him to use. It is also important to note that the commission appointed by Gregory seems never to have met, perhaps because one of its leading members, the theologian William of Auxerre, died within the year, and no purged version of Aristotle has ever been discovered. The subsequent acceptance of Aristotle was based on a complete, uncensored version of his works.

Various documents address the fate of Aristotle's works during the next twenty-five years. They reveal that the bans of 1210, 1215, and 1231 were partially successful for a time, but that they began to lose their effectiveness around 1240. One reason for this may have been Gregory IX's death in 1241, after which his regulations of a decade earlier may have lost some of their coercive power; another may have been a growing awareness among the Parisian masters of arts that they were losing ground (and reputation) to their counterparts at Oxford and other universities. We should also reckon with the possibility that the teaching of Aristotle's logic (not covered by the bans), ready availability of Aristotle's works on natural philosophy (despite the ban on teaching them), and the recovery of new

Aristotelian commentators (especially Averroes) elevated Aristotle's reputation to the point where the juggernaut of Aristotelian philosophy was un-
stoppable. And, of course, we need to remember that it had always been
legitimate for theologians to make such use of Aristotle as they saw fit.

Whatever the causes, Aristotle's works on natural philosophy seem to
have become the subject of lectures in the faculty of arts in the 1240s or
shortly before; one of the earliest to teach them was Roger Bacon.[4] About
the same time, more liberal attitudes toward the use of Aristotle were pen-
etrating the theological faculty at Paris, as we see in a growing tendency
to allow Aristotelian philosophy to shape theological speculation and
thought. By 1255 the tables had turned completely, for in that year the faculty of arts passed new statutes making mandatory what was apparently already the practice—namely, lectures on all known works of Aristotle.
Aristotle's natural philosophy had not merely created a place for itself in
the arts curriculum; it had become one of the principal ingredients.

POINTS OF CONFLICT

It is time to pinpoint the features of Aristotelian philosophy that were
worth worrying or fighting about. But first we must note that the content of
Aristotle's philosophy, as understood by his Western readers, was in a state
of flux. Because Aristotle was exceedingly difficult to follow, readers naturally availed themselves of whatever explanatory aids they could acquire;
fortunately, commentators in both late antiquity and medieval Islam had
paraphrased Aristotle or explained difficult points in the various Aristotelian texts, and the works of these commentators were progressively
translated along with Aristotle's own works and used wherever Aristotle
was seriously studied. In the closing decades of the twelfth century and the
early decades of the thirteenth, the principal commentator was the Muslim
Avicenna (Ibn Sīnā, 980–1037), who presented a Platonized version of Aristotelian philosophy.[5] The charge of pantheistic teaching at Paris made in
1210 undoubtedly reflects the inroads of Avicenna's Neoplatonic reading of
Aristotle. However, beginning about 1230, the commentaries of Avicenna
began to be displaced by those of the Spanish Muslim Averroes (Ibn
Rushd, 1126–98).[6] There is no question that Averroes too was capable of
extending or distorting Aristotle's meaning, and that he did so on occasion,
but in general the switch from Avicenna's guidance to that of Averroes
meant a return to a more authentic, less Platonized, version of Aristotelian
philosophy. So influential did Averroes become in the West that he came to
be known simply as "*the* Commentator."

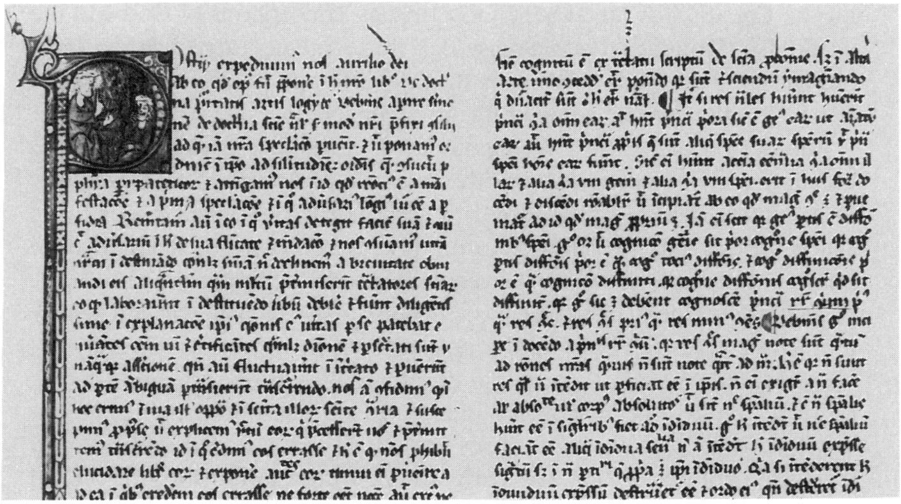

Fig. 10.1. The beginning of Avicenna's *Physics* (*Sufficientia,* pt. II), Graz, Universitätsbiblio-thek, MS II.482, fol. 111r (13th c.).

What was there in the Averroistic (or more authentic) reading of Aris-totle that caused trouble? Some specific claims appeared (with varying de-grees of clarity) to violate orthodox Christian doctrine; and beneath these claims there lay a general outlook, rationalistic and naturalistic in tone, which struck some observers as antithetical to traditional Christian thought. The simplest way of discussing these issues will be to begin with the specific claims.

A prominent feature of the Aristotelian cosmos was its eternity, de-fended by a variety of arguments in a variety of Aristotle's works. Bearing, as it did, on the doctrine of creation, this was a claim that could hardly be overlooked by Aristotle's Christian readers. Aristotle's position was that the cosmos did not come to be and cannot cease to be. The elements, he ar-gued, have always behaved according to their natures; consequently, there cannot have been a moment when the universe as we know it came into being, and no moment will come when it ceases to be; it follows that the universe is eternal. Aristotle thus repudiated the evolutionary cosmology of the pre-Socratic philosophers.[7]

From a Christian standpoint, however, this is an intolerable conclusion. Not only does the Bible contain an account of creation in the opening chapters of Genesis, but the absolute dependence of the created universe on the Creator was fundamental to Christian conceptions of God and the world. Consequently, among Aristotle's thirteenth-century Christian commentators, we find an unbroken string of attempts to resolve this problem.[8] We will consider some of the arguments below.

Another problem, also bearing on the relationship between Creator and creation, is that of determinism. The question of deterministic tendencies in Aristotle's natural philosophy is a very thorny one. What needs to be stated here is that the universe as he described it contains unchangeable natures, which are the basis of a regular cause-and-effect sequence. This is particularly evident in the heavens, where that which is will always be. Moreover, Aristotle regarded the deity, the Prime Mover, as eternally unchanging and therefore incapable of intervening in the operation of the cosmos; the cosmic machinery thus runs inevitably and unchangeably onward, initiating a chain of causes and effects that descend into and pervade the sublunar realm. The danger here is that within the Aristotelian framework no room can be found for miracles.[9] And finally, attached to Aristotelian philosophy were astrological theories, which threatened the freedom of human choice (essential to Christian teaching on sin and salvation) if celestial influences could be shown to act on the will.

All of these deterministic tendencies or elements were viewed in the thirteenth century as a challenge to Christian doctrine—particularly divine freedom and omnipotence, divine providence, and miracles. Aristotle's Prime Mover, who does not even know of the existence of individual humans and certainly does not intervene on their behalf, is a far cry from the Christian God, who knows when the sparrow falls and numbers the very hairs on our head.[10]

As a final example of troublesome Aristotelian ideas, we turn to the nature of the soul. Aristotle argued that the soul is the form or organizing principle of the body—the full actualization of the potentialities inherent in the matter of the individual. It follows that the soul cannot have independent existence, since form, even if it can be distinguished from matter, cannot exist independently of matter. To suppose that the soul might be separated from the body would be as foolish as to suppose that the sharpness of an axe could be separated from the matter of the axe. At death, therefore, when the individual dissolves, its form or soul simply ceases to be.[11] Such a conclusion is clearly incompatible with Christian teaching on the immortality of the soul.

The immortality of the individual soul was also called into question by another psychological doctrine developed by Averroes, as he attempted to work out certain difficulties in Aristotle's epistemology. The full Averroistic theory, known as "monopsychism," is extremely intricate. What is important for us is his claim that the immaterial and immortal part of the human soul, the "intellective soul," is not individual or personal but a unitary intellect shared by all humans. It would seem to follow that what survives death is not personal but collective; immortality is preserved, but not personal immortality. The violation of Christian teaching is again clear.[12]

Such claims as these were not isolated pieces of philosophy, but manifestations of basic attitudes toward reason and its proper relationship to faith and theology; they entered Western Europe as specific instances of an outlook and a methodology. The champions of the new Aristotelianism were inclined to enlarge the scope of rational activity, naturalistic explanation, and Aristotelian demonstration; philosophy was their game, and they wished to display its virtues in every intellectual arena. When philosophy entered the theological faculty and began to influence theological method, coming to rival biblical studies as the focus of theological education, traditionalists understandably reacted in anger and frustration. Charges of intellectual arrogance and vain curiosity became commonplace. Were the articles of the faith to be tested by the content and methods of pagan philosophy? Were the teachings of Christ, the Apostle Paul, and the church fathers subordinate to those of Aristotle?

A particularly acute example of this outlook in the realm of natural philosophy was the tendency to restrict analysis to causal principles discoverable through the exercise of human observation and reason, without regard for the teachings of biblical revelation or church tradition. Divine or supernatural causation was never denied, but it was placed (by the more aggressive proponents of the new methodology) outside the province of natural philosophy. This naturalism, the seeds of which are visible in such twelfth-century thinkers as William of Conches (see chap. 9, above), blossomed under the stimulus of Aristotle and his commentators. Perhaps the most threatening manifestation of these naturalistic inclinations was the growing tendency of some philosophers to make a distinction between "speaking philosophically" and "speaking theologically" and, what is much worse, to acknowledge that philosophical and theological methods might lead to incompatible conclusions.

The advocates of the new methods no doubt saw the introduction of philosophical rigor into theological debate as a great step forward. But to the traditionalists, this seemed to be a serious case of insubordination and a violation of traditional distinctions between the enterprises of philosophy and theology. Viewed in the worst light, it appeared that Jerusalem was being asked to yield to the authority of Athens.

Fig. 10.2. The Basilica of St. Francis, Assisi. Begun within a few years of Francis's death in 1226, to contain his tomb, this church became the "head and mother" of the Franciscan order and an important pilgrimage site. Courtesy of Christopher Kleinhenz.

Before considering thirteenth-century attempts to resolve these difficulties, we must briefly examine the institutional framework within which the attempts took place. The debates over the new Aristotle were scholarly in nature, and all of the participants were products of the universities. Many were active teachers; others were university alumni who had ascended to positions of leadership and authority in the church. It will help us to understand the persistent medieval tendency to mingle philosophy and theology if we grasp the career patterns of the university scholar: virtually all theologians had studied philosophy in the arts faculty before embarking on theological studies; moreover, theological students frequently found themselves teaching simultaneously in the faculty of arts, as a means of support.

Consequently, some of the most influential philosophical treatises of the Middle Ages were written by scholars who were teaching philosophy while studying theology.[13]

Some of the leading figures, by the middle of the century, were Franciscans or Dominicans—members, that is, of the mendicant orders founded early in the thirteenth century. The mendicants were "regular clergy," because they lived under a *regula* or rule (which included a vow of poverty), unlike "secular clergy" (such as parish priests) who did not. In contrast to the monastic orders, which stressed withdrawal from the world in the pursuit of personal holiness, the mendicants were committed to an active ministry within an urban setting; this eventually propelled them into the educational arena, including the universities, where they became actively involved in all of the great philosophical and theological controversies.

These institutional details contributed in subtle ways to the intellectual developments with which we are concerned. The struggles surrounding the new learning were not purely ideological, but were complicated by disciplinary and institutional affiliations and rivalries. Philosophers and theologians were united by the educational experience of the arts faculty; but this did not prevent them from skirmishing periodically over disciplinary boundaries. Within theology, the mendicants were locked for a time in a power struggle with secular theologians at the University of Paris over the right to hold professorial chairs. And within the mendicant orders, Franciscans and Dominicans developed somewhat different philosophical loyalties and characteristic approaches to the problem of faith and reason. If we wish to gain a nuanced understanding of the course of events, we need to be sensitive to these disciplinary and institutional undercurrents.

RESOLUTION: SCIENCE AS HANDMAIDEN

Despite the dangers that we have recounted, Aristotelian philosophy proved too attractive to ignore or suppress permanently. Since the translations of Boethius early in the sixth century, Aristotle's name had been synonymous with logic, and that logic had insinuated itself deeply into scholarship on almost every subject; now an enlarged corpus of Aristotelian logic was available and ready to go to work. Aspects of Aristotelian metaphysics had also filtered through the literature of the early Middle Ages, and now, with access to the full Aristotelian text, Western scholars had in their hands a powerful engine for understanding and analyzing their universe. Form, matter, and substance, actuality and potentiality, the four causes, the four elements, contraries, nature, change, purpose, quantity, quality, time, and space—Aristotle's discussion of all of these topics, and more, furnished a persuasive conceptual framework through which to experience and talk about the world. In his various psychological works Aristotle treated the soul and its faculties, including sense perception, memory, imagination, and cognition. He also offered a cosmology, in which the universe was convincingly mapped and its operation explained, from the outermost heaven to the earth at the center. Aristotle gave an account of motion, of what we would call matter theory, and of meteorological phenomena that went far beyond anything previously available. And finally, he offered a biological corpus unmatched for size and for descriptive and explanatory detail. It was inconceivable that these intellectual treasures would be simply repudiated; and there was never a serious movement with that as its aim. The problem was not how to eradicate Aristotelian influence, but how to domesticate it—how to deal with points of conflict and to negotiate boundaries, so that Aristotelian philosophy could be put to work on behalf of Christendom.

The process of reconciliation began as soon as the works of Aristotle and his commentators became available. An early attempt was made by Robert Grosseteste (ca. 1168–1253), a formidable Oxford scholar and first chancellor of the university; though not himself a Franciscan, Grosseteste was the first lecturer in the Franciscan school at Oxford, thereby exercising a formative influence on the intellectual life of the Order. Grosseteste's commentary on Aristotle's *Posterior Analytics,* written probably in the 1220s, was one of the earliest efforts to deal seriously with Aristotle's scientific method.[14] Grosseteste was also acquainted with Aristotle's *Physics, Metaphysics, Meteorology,* and biological works; he reveals their influence in his commentary on the *Physics* and in a series of short treatises on various physical subjects. However, Grosseteste's intellectual formation was strongly shaped by Platonic and Neoplatonic influences, and also by some of the newly translated works on mathematical science, and in his physical works we find a rather uneasy juxtaposition of Aristotelian and non-Aristotelian elements. Grosseteste's cosmogony (his account of the origin of the cosmos), for example, while set within a broadly Aristotelian framework, should be seen primarily as an attempt to reconcile Neoplatonic emanationism (the idea that the created universe emanated from God, as light emanates from the sun) with the biblical account of creation *ex nihilo.*[15]

Important aspects of Grosseteste's program were continued by a younger Englishman, Roger Bacon (ca. 1220–ca. 1292). An admirer of Grosseteste (but probably never his student), Bacon was inspired by

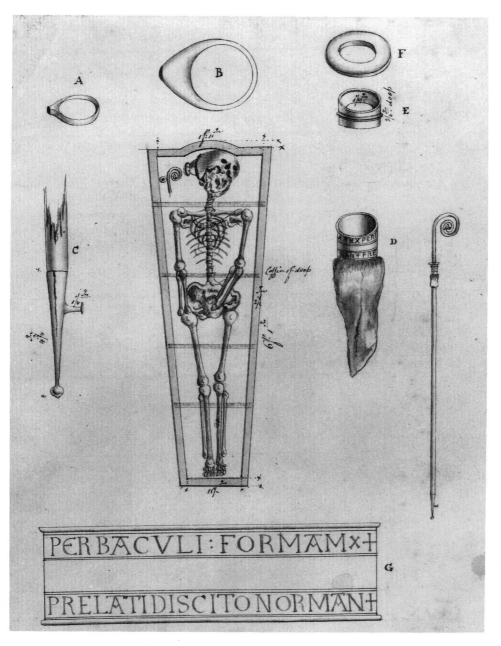

Fig. 10.3. The skeleton of Robert Grosseteste. Sketched when Grosseteste's tomb in Lincoln Cathedral was opened in 1782, this is one of a very few drawings of a medieval scholar prepared, so to speak, "from life." Shown with the skeleton are the other items found in the coffin, including the episcopal ring and the remains of the pastoral staff. For a fuller description, see D. A. Callus, ed., *Robert Grosseteste, Scholar and Bishop,* pp. 246–50. By permission of The Natural History Museum, London.

Grosseteste's scholarly example, especially his mastery of the mathematical sciences. The details of Bacon's education are obscure, but it is clear that he studied at both Oxford and Paris. He began to teach in the faculty of arts at Paris in the 1240s, where he was one of the first to lecture on Aristotle's books on natural philosophy—specifically, the *Metaphysics, Physics, On Sense and the Sensible;* probably *On Generation and Corruption* (which deals with theory of matter), *On the Soul,* and *On Animals;* and perhaps *On the Heavens.*[16] Later he joined the Franciscan order and spent the remainder of his life in study and writing.

We will deal with various aspects of Bacon's scientific thought in subsequent chapters; what is important here is his campaign to save the new learning from its critics. Bacon's major scientific writings were not "pure" pieces of philosophy or science, but passionate attempts to persuade the church hierarchy (these works were addressed to the pope) of the utility of the new learning—not just Aristotelian philosophy, but the totality of new literature on natural philosophy, mathematical science, and medicine. Bacon argued that the new philosophy is a divine gift, capable of proving the articles of the faith and persuading the unconverted, that scientific knowledge contributes vitally to the interpretation of Scripture, that astronomy is essential for establishing the religious calendar, that astrology enables us to predict the future, that "experimental science" teaches us how to prolong life, and that optics enables us to create devices that will terrorize unbelievers and lead to their conversion. The object of Bacon's campaign was to take the handmaiden formula of Augustine and apply it to new circumstances, in which the quantity of purported knowledge waiting to be enlisted as handmaiden was vastly larger and more complicated.[17] The natural sciences were thus justified by their religious utility. There is "one perfect wisdom," Bacon argued in his *Opus maius,*

> and this is contained in holy Scripture, in which all truth is
> rooted. I say, therefore, that one discipline is mistress of the
> others—namely, theology, for which the others are integral
> necessities and which cannot achieve its ends without them.
> And it lays claim to their virtues and subordinates them to its
> nod and command.[18]

Theology does not oppress the sciences, in Bacon's view, but puts them to work, directing them to their proper end.

As for the points of alleged conflict with Christian belief, Bacon dismissed them as problems arising from faulty translation or ignorant interpretation; if philosophy is truly God-given, there can be no genuine conflict between it and the articles of the faith. To reinforce this point, Bacon marshaled the authority of Augustine and other patristic writers who urged Christians to reclaim philosophy from its pagan possessors. And just in case these arguments failed, he shouted down the critics with a blast of rhetoric about the wonders of science.

Despite Bacon's enthusiasm, a cautious attitude toward the new philosophy, especially the new Aristotle, became typical of the Franciscan order about mid-century. One of the people most instrumental in shaping this attitude was the Italian Franciscan Bonaventure (ca. 1217–74). Bonaventure studied both the liberal arts and theology at the University of Paris, then remained to teach theology from 1254 to 1257, resigning to become minister general of the Franciscan order. There can be no doubt that Bonaventure respected Aristotelian philosophy, drawing his logic and much of his metaphysics from it; but, like Grosseteste and Bacon, he was heavily influenced by Augustine and the Neoplatonic tradition, and in his thought we find a rich synthesis of Aristotelian and non-Aristotelian elements.

Bonaventure certainly agreed with Bacon on the validity and applicability of Augustine's handmaiden formula: pagan philosophy was an instrument, to be used for the benefit of theology and religion. But he was much more cautious than Bacon about the utility of philosophy and more sharply aware of the risks of promoting it. He was pessimistic about the capacity of reason alone, without the assistance of divine illumination, to discover truth; and as a result, he was apt to keep philosophy on a short leash and quick to abandon Aristotle or his commentators on any issue where they departed from the teachings of revelation. Thus he flatly rejected any possibility of an eternal world; he defended the immortality of the individual soul, dismissing monopsychism and arguing that each soul is itself a substance (a composite of spiritual form and spiritual matter) that survives the dissolution of the body; and he vigorously fought any suggestion of astrological determinism. Finally, in opposition to Aristotelian naturalism, Bonaventure stressed God's providential participation in every case of cause and effect.[19]

In the careers of Grosseteste, Bacon, and Bonaventure we can see several important tendencies of the early and middle years of the thirteenth century: growing knowledge of the Aristotelian corpus, a mixture of admiration and suspicion about its contents, and a tendency to read various Augustinian or Platonic ideas into the Aristotelian text. Two Dominicans active in the middle and later years of the century, Albert the Great and Thomas Aquinas, contributed to a much fuller mastery of Aristotelian philosophy and a more open attitude toward its claims.

Fig. 10.4. Albert the Great. Fresco by Tommaso da Modena (1352), located in the Monastery of San Niccolò, Treviso. Alinari/Art Resource N.Y.

German by birth, Albert the Great (ca. 1200–1280) was educated at Padua and the Dominican school in Cologne. In the early 1240s he was sent to Paris to study theology, becoming master of theology in 1245. For the next three years he occupied one of the two Dominican professorships at Paris. Thomas Aquinas studied under him during this period, and when Albert was called back to Cologne in 1248 to reorganize the Dominican school there, Thomas accompanied him. Most of Albert's Aristotelian commentaries were composed after his departure from Paris; they are not (except for his commentary on Aristotle's *Ethics*) the products of Albert's teaching, but extracurricular writings intended for the benefit of Dominican friars.[20]

Albert was the first to offer a comprehensive interpretation of Aristotle's philosophy in Western Christendom; on these grounds, he is often regarded as the effective founder of Christian Aristotelianism. This should not be taken to mean that Albert achieved philosophical purity; some of his early commentaries were devoted to Neoplatonic authors, and to the end of his life he retained allegiance to portions of Platonic philosophy; moreover, he was always ready to correct or discard Aristotelian doctrines that he considered false and to introduce pieces of truth found elsewhere. Nonetheless, Albert perceived the profound significance of Aristotelian philosophy and set out to interpret the whole of it for his fellow Dominicans. In the prologue to his commentary on Aristotle's *Physics,* he explained:

> Our purpose . . . is to satisfy as far as we can those brethren
> of our order who for many years now have begged us to com-
> pose for them a book on physics in which they might find a
> complete exposition of natural science and from which also
> they might be able to understand correctly the books of
> Aristotle.[21]

Albert responded not only with a *Physics* commentary but with commentaries on, or paraphrases of, all available Aristotelian books—an output that occupies twelve fat volumes (more than 8,000 pages) in the nineteenth-century edition of Albert's works. Included in these commentaries are long digressions, in which Albert lays out the results of his own investigations and reflections. Nobody before Albert had given the Aristotelian corpus this kind of painstaking attention, and few have done it since.

His purpose in doing so was to exhibit and make available the explanatory power of Aristotelian philosophy, which he regarded as the necessary preparation for theological studies. He had no intention of releasing Aristotelian philosophy from handmaiden status, but he did mean to give it substantially larger responsibilities. Among Albert's contemporaries, only Roger Bacon had as grand a vision of the importance of the new learning for the practice of theology; but, apart from his youthful lectures on Aristotle at Paris, Bacon devoted his best efforts to the mathematical sciences (especially optics) and the writing of propaganda on behalf of the new learning in general, while Albert committed himself to the work of mastering and interpreting the Aristotelian corpus. Historians have tended to honor those who *broke* with the Aristotelian tradition; Albert deserves our attention and respect as the man who put Western Christendom in touch with the Aristotelian tradition.

At the same time, Albert recognized his obligation to supplement the Aristotelian text on subjects that Aristotle had overlooked or explored superficially, and to correct Aristotle wherever he was wrong; though mightily impressed by Aristotle's achievement, Albert was never tempted to become his slave. To this end, Albert read everything he could lay his hands on: he was heavily dependent on Avicenna; he knew the works of Plato, Euclid, Galen (to a limited degree), al-Kindī, Averroes, Constantine the African, and a host of other Greek, Arabic, and Latin authors. And he brought these other sources to bear, whenever they were relevant, on the problems that he confronted as he interpreted the Aristotelian text.[22]

Albert was also a remarkably acute firsthand observer of plant and animal life: for example, he corrected Avicenna regarding the mating of partridges on the basis of personal observation and reported that he had visited a certain eagle's nest six years in a row; and he was perhaps the best field botanist of the entire Middle Ages.[23] His intellectual energy was boundless, and his nontheological writings (less than half the total) include works on physics, astronomy, astrology, alchemy, mineralogy, physiology, psychology, medicine, natural history, logic, and mathematics. The authority with which he could address any subject explains why Albert was referred to as "the great" already during his lifetime; it also helps to explain why Roger Bacon (who was intolerant of intellectual rivals) viewed him with such hostility.

What did Albert have to say about the sensitive Aristotelian doctrines that had led to the banning of Aristotle early in the century and still threatened the acceptance of his works? On the critical problem of the eternity of the world, Albert never wavered in his commitment to the Christian doctrine of creation. His early view was that philosophy is incapable of addressing this issue definitively, so that one's obligation is simply to accept the teaching of revelation. Later he became convinced that the idea of an eternal universe is philosophically absurd, so that philosophy could settle the matter without theological assistance. In neither case did philosophy (properly practiced) and theology find themselves at odds.

Albert devoted substantially more attention to the nature of the human soul and its faculties. The trick was to account for the soul as a separate immortal substance, independent of the body and able to survive its death, while also accounting for the unification of the soul with the body, as the agent of perception and vitality. Albert could see no way of defending the immortality of the soul without denying Aristotle's claim that the soul is the form of the body; in its place, he substituted the opinion of Plato and Avicenna, that the soul is a spiritual and immortal substance, separable from the body. It was not necessary, however, to repudiate Aristotle totally: Albert argued that although soul is not actually the form of the body, it performs the functions of form.[24]

Finally, how did Albert respond to the "rationalism" of Aristotelian philosophy—the commitment, that is, to the application of philosophical method in all areas of human enterprise? In setting himself the task of showing his colleagues how to look at the world through Aristotelian eyes, Albert was committing himself to a fairly strong form of the rationalist program. He proposed to distinguish between philosophy and theology on methodological grounds and to find out what philosophy alone, without any help from theology, could demonstrate about reality. Moreover, Albert did nothing to diminish or conceal the "naturalistic" tendencies of the Aristotelian tradition. He acknowledged (with every other medieval thinker) that God is ultimately the cause of everything, but he argued that God customarily works through natural causes and that the natural philosopher's obligation was to take the latter to their limit. What is remarkable is Albert's willingness to adhere to this methodological prescription even in his discussion of a biblical miracle—Noah's flood. Noting that some people wish to confine the discussion of floods (including Noah's) to a statement of divine will, Albert pointed out that God employs natural causes to accomplish his purposes; and the philosopher's task is not to investigate the causes of God's will, but to inquire into the natural causes by which God's will produces its effect. To introduce divine causality into a philosophical discussion of Noah's flood would be a violation of the proper boundaries between philosophy and theology.[25]

Albert's program of understanding and disseminating Aristotelian philosophy, while respecting its utility for theology and religion, was continued by his pupil Thomas Aquinas (ca. 1224–74). Thomas, who was born into the minor nobility in south-central Italy, received his first education at the ancient Benedictine Abbey of Monte Cassino (founded by Benedict of Nursia in the sixth century); he continued his studies in the faculty of arts at the University of Naples, where he was introduced to Aristotelian philosophy. After joining the Dominican order, Thomas was sent to Paris, earning the theological doctorate there in 1256. He devoted the remainder of his life to teaching and writing, including two stints teaching theology at Paris, 1257–59 and 1269–72.

Like Albert, Thomas hoped to solve the problem of faith and reason by defining the proper relationship between pagan learning and Christian theology.[26] Against those who would dismiss philosophy as contrary to the faith, he argued that

> even though the natural light of the human mind [i.e., philosophy] is inadequate to make known what is revealed by faith, nevertheless what is divinely taught to us by faith cannot be contrary to what we are endowed with by nature. One or the other would have to be false, and since we have both of them from God, he would be the cause of our error, which is impossible.[27]

Aristotelian philosophy and Christian theology, though methodologically distinct, are compatible roads to truth. Philosophy employs the natural human faculties of sense and reason to arrive at such truths as it can. Theology offers access to truths given by revelation that go beyond our natural capacities to discover and understand. The two roads may sometimes lead to different truths, but they never lead to contradictory truths.

Does this mean that philosophy and theology are equals? Certainly not. Thomas points out that theology is to philosophy as the complete to the incomplete, the perfect to the imperfect. If this is so, why go to the trouble of doing philosophy? Because it renders vital services to the faith. In the first place, it can demonstrate what Thomas calls "the preambles of the faith"—certain propositions, such as God's existence or his unity, which faith takes for granted as its starting points. Second, philosophy can elucidate the truths of the faith by the use of analogies drawn from the natural world; Thomas refers to the doctrine of the trinity as a case in point. And third, philosophy can disprove objections to the faith.[28]

This may seem like a simple reassertion of the Augustinian handmaiden formula, but in fact Thomas has subtly but significantly altered its content. The handmaiden named "philosophy" is still subordinate to the theological enterprise, and therefore still a handmaiden; but in Thomas's view, she has amply demonstrated her usefulness and her reliability, and he therefore offers her enlarged responsibilities and elevated status. Thomas also thinks she will do her job better if relieved of overly close theological supervision. Philosophy and theology both have their spheres of competence, he argues, and each can be trusted within its proper sphere: if we wish to know the details or causes of planetary motion, for example, we must look to the philosophers; on the other hand, if we wish to understand the divine attributes or the plan of salvation, we must be prepared to enter the precincts of theology. In Thomas there is a respect for the philosophical enterprise, and a determination to employ it whenever possible, that takes him beyond the Augustinian position and places him in the forefront of the liberal or progressive wing of theologians in the second half of the thirteenth century.

Despite the methodological differences between philosophy and theology, there are regions where they overlap. For example, the existence of the Creator is known both by reason and by revelation; it can be proved by the philosopher, but it is also given to us by Scripture, as expounded by the theologian. What rules govern the relations of theology and philosophy in such cases? The fundamental principle is that there can be no true conflict between theology and philosophy, since both revelation and our rational capacities are God-given. Any conflict must, therefore, be apparent rather than real—the result of bad philosophy or bad theology. The remedy in such cases is to reconsider both the philosophical and the theological argument.

How did this prescription work itself out in Thomas's practice? In particular, how successfully was it applied to the troublesome Aristotelian doctrines enumerated in the preceding section of this chapter? The short answer is that Thomas confronted all of the problems raised by Aristotelian philosophy, and did so with extraordinary rigor. He directly addressed the Aristotelian controversies in two books: *On the Eternity of the World* and *On the Unicity of the Intellect, against the Averroists* (concerned with monopsychism and the nature of the soul). His position on the eternity of the world was that we know, thanks to revelation, that the world was created at a point in time, but that philosophy cannot settle the matter one way or the other. Those (like Bonaventure) who argued that the eternity of the world was philosophically absurd were wrong, for it is not contradictory to maintain that the universe is created (that is, dependent on divine creative power for its existence) and yet that it has existed eternally. On the nature of the soul, Thomas agreed with Aristotle that the soul is the substantial form of the body (that which combines with the matter of the body to produce the individual human being); but he argued that this form is a special kind of form, capable of existing independently of the body and therefore imperishable. He also claimed that this solution was compatible with Aristotle's own thought.[29]

This, then, is Thomas's solution to the problem of faith and reason. He has made room for both, subtly merging Christian theology and Aristotelian philosophy into what we may call "Christian Aristotelianism." In the process it was necessary for Thomas to Christianize Aristotle, by confronting and wrestling with the Aristotelian doctrines that appeared to conflict with the teachings of revelation and correcting Aristotle where he had fallen into error; at the same time he "Aristotelianized" Christianity, importing major portions of Aristotelian metaphysics and natural philosophy into Christian theology. In the long run (in the nineteenth century), Thomism came to represent the official position of the Catholic church; in the short run, as we shall see, Thomas was viewed by theologians of more conservative persuasion as a dangerous radical.

RADICAL ARISTOTELIANISM AND THE CONDEMNATIONS OF 1270 AND 1277

Albert the Great and Thomas Aquinas were the leaders of a liberal movement, which favored a robust philosophy. But however robust philosophy might become, it would, in their view, forever remain a handmaiden; reason would never be allowed to prevail over revelation. Albert and Thomas pushed philosophy as far as it would go, but they never gave up on a philosophical problem until reason and faith had been harmonized.

But how robust can a handmaiden become before she begins to think about insubordination or insurrection?[30] When biblical miracles are reduced to their natural causes, as in Albert's discussion of Noah's flood, aren't things already out of hand? These were the concerns of conservative theologians, observing developments at Paris. And, as it turns out, their fears were not entirely groundless. Our evidence is fragmentary, but it is apparent that while Albert and Thomas were harmonizing philosophy and theology, certain masters of arts began to teach dangerous philosophical doctrines, without regard for their theological consequences. These were committed philosophers, aggressively practicing their trade and recognizing no need to yield, or even pay attention, to any outside authority. The harmonization of philosophy and theology was not their problem.

The best known of this radical faction, and its leader, was Siger of Brabant (ca. 1240–84). Siger, a brash young master of arts, began his teaching career defending the eternity of the world and Averroistic monopsychism, with its dangerous implications for personal immortality. His aim was to do philosophy without so much as a glance at theological teaching on any of the subjects touched, and he maintained that the conclusions he reached were the necessary and inevitable conclusions of philosophy, properly practiced. After the appearance of Thomas's treatise *On the Unicity of the Intellect,* which was directed specifically at his teaching, Siger modified his position on the nature of the soul, bringing it into conformity with orthodox Christian teaching.[31] Older and wiser after this run-in with the theologians, Siger was thereafter careful to make clear that although his philosophical conclusions were not wrong, but in fact necessary philosophical conclusions, nonetheless they need not be true; when it came to *truth,* he affirmed the articles of the faith. Historians have been divided over whether to take this profession of faith at face value or to conclude that Siger was merely attempting to placate the ecclesiastical power structure. Either way, the dangerous implications of Siger's public position are clear: philosophical inquiry properly conducted can lead to conclusions that contradict those of theology.

The position of the radicals is nicely illustrated by a little treatise, *On the Eternity of the World,* written by Boethius of Dacia (fl. 1270), a member of Siger's circle. One of the most striking characteristics of this work is its rigorous separation of philosophical and theological argument. Boethius systematically assembled and then refuted the philosophical arguments that had been used to defend the Christian doctrine of creation against the Aristotelians. He proceeded to demonstrate that the philosopher, speaking as a philosopher, had no alternative but to defend the eternity of the world. He made clear, however, that according to theology and faith he himself accepted the doctrine of creation, as any Christian must.

Boethius thus yielded in the end to the articles of the faith, but in the meantime he displayed an intensely rationalistic orientation. He argued that there is no question capable of rational investigation that the philosopher is not entitled to investigate and resolve. "It belongs to the philosopher," he wrote,

> to determine every question which can be disputed by reason;
> for every question which can be disputed by rational arguments falls within some part of being. But the philosopher investigates all being—natural, mathematical, and divine.
> Therefore it belongs to the philosopher to determine every question which can be disputed by rational arguments.

Boethius went on to argue that the natural philosopher cannot even consider the possibility of creation, because to do so would introduce supernatural principles that are out of place in the philosophical realm. Likewise the philosopher denies the resurrection of the dead, because according to natural causes (to which the natural philosopher limits himself) such a thing is impossible.[32]

This is an attempt, impressive for its rigor, to follow philosophical argument to its logical conclusion, without regard for the faith, while still acknowledging the ultimate authority of theology. However, nobody should

Fig. 10.5. Cathedral of Notre Dame, Paris, built in the twelfth and thirteenth centuries.

be surprised to learn that the theological faculty and the religious authorities were neither convinced nor pleased, but regarded Siger, Boethius, and their group as a growing menace. If philosophy was consistently going to reach conclusions at odds with the faith, it could no longer be regarded as a faithful handmaiden; rather, it began to appear as a hostile force and a threat requiring decisive action.

That decisive action came in two condemnations issued by the bishop of Paris, Etienne Tempier, in 1270 and 1277. The former condemned thirteen philosophical propositions allegedly taught by Siger and his fellow radicals in the faculty of arts. Coming, as it apparently did, with the encouragement of both Bonaventure and Thomas Aquinas, this condemnation represents a reaction by the theological establishment to the activities of the radical fringe in the faculty of arts. By 1277, the menace seemed broader and more serious: it was clear by this time that the earlier condemnation had not stamped out radical Aristotelianism, and conservatives within the faculty of theology moved with increased vigor to meet what they perceived as a growing threat. Indeed, there was a tendency among the conservatives to view everybody significantly more liberal than themselves as dangerous; the result was the issuance (on the third anniversary of Aquinas's death) of a greatly enlarged list of forbidden propositions, 219 in all, the teaching of which was declared grounds for excommunication. Included in this list were fifteen or twenty propositions drawn from the teaching of Aquinas. Let us pause to examine the content of some of the condemned propositions and the import of Tempier's action.[33]

The obviously dangerous elements of Aristotelian philosophy are all represented on Tempier's two lists of forbidden propositions: the eternity of the world, monopsychism, denial of personal immortality, determinism, denial of divine providence, and denial of free will. The rationalistic tendencies of Siger and the radicals are also explicitly targeted: after 1277 it is forbidden, for example, to proclaim the right of philosophers to resolve all disputes on subjects to which rational methods are applicable; or to maintain that reliance on authority never yields certainty. The naturalism of the Aristotelian tradition also figured prominently in the condemnation of 1277: Tempier condemned the opinion that secondary causes are autonomous, so that they would continue to act even if the first cause (God) were to cease to participate; also the claim that God could not have created a man (an obvious reference to Adam) except through the agency of another man; and the methodological principle that natural philosophers, because they restrict their attention to natural causes, are entitled to deny the creation of the world.

This is the sort of list of condemned propositions we might have expected. But in addition the condemnation of 1277 included a heterogeneous assortment of other propositions that impinge in various ways on natural philosophy. Several astrological propositions were condemned: that the heavens influence the soul as well as the body and that events will repeat themselves in 36,000 years, when the celestial bodies return to their present configuration. Also forbidden was the claim that the celestial spheres are moved by souls. A particularly important set of condemned propositions—important because of their implications for fourteenth-century debates—dealt with things that God allegedly could not do, because Aristotelian philosophy had demonstrated their impossibility. It was apparently being argued by philosophers that God could not have created additional universes (Aristotle had argued that multiple universes are impossible); that God could not move the outermost heaven of this universe in a straight line (because a vacuum, which Aristotelian philosophy ruled out, would thus be left behind in the vacated space);[34] and that God could not create an accident without a subject (for example, redness without something to be red). All of these propositions were condemned in 1277 on the grounds that they fly in the face of divine freedom and omnipotence. The position of Tempier, or of those who formulated the list of propositions on his behalf, was that Aristotle and the philosophers must not be allowed to place a lid on God's freedom or power to act; God can do anything that involves no logical contradiction, including create multiple universes or accidents without subjects.

What can we learn from these events? The condemnations have been much discussed and their importance often inflated or misunderstood. Pierre Duhem, writing early in the twentieth century, saw the condemnation of 1277 as an attack on entrenched Aristotelianism, especially Aristotelian physics, and therefore as the birth certificate of modern science. This is a clever interpretation, and it is not entirely wrong: there can be no doubt (as we shall see below) that the condemnations encouraged scholars to explore non-Aristotelian physical and cosmological alternatives.[35] But to place the emphasis here is to miss the primary significance of the condemnations. Duhem viewed the condemnations as the key event in the shattering of Aristotelian orthodoxy, but in 1277 no such orthodoxy existed; the boundaries and the power relationship between Aristotelian philosophy and Christian theology were still being negotiated, and the degree to which Aristotelianism would acquire the status of orthodoxy was not yet clear.

Or to express the same point a little differently, the condemnations are significant not so much for the effect they had on the future course of natural philosophy as for what they tell us about what had already occurred. Coming at the end of nearly a century of struggle over the new learning, the condemnations represent a conservative backlash against liberal and radical efforts to extend the reach and secure the autonomy of philosophy, especially Aristotle's philosophy. They reveal the extent of that reach and the power of the opposition—the fact that a sizable and influential group of traditionalists was not yet ready to accept the brave new world proposed by the liberal, and especially the radical, Aristotelians. Thus, to put the event in its proper light, the condemnations represent a victory not for modern science but for conservative thirteenth-century theology. The condemnations were a ringing declaration of the subordination of philosophy to theology.

They were also an attack on Aristotelian determinism and a declaration of divine freedom and omnipotence. We have noted that a number of the propositions condemned in 1277 dealt with things that God could not do—for example, endow the heaven with rectilinear motion (on the grounds that a vacuum, which Aristotelian philosophy forbids, would thereby be created in the vacated space). In condemning this proposition, Tempier certainly did not mean to debate a point of natural philosophy with Aristotle but to announce that whatever the natural state of things might be (and we may presume that he accepted Aristotle's account of this), God has the power to intervene, should he wish; a vacuum may not exist naturally, but it can surely exist supernaturally; it may not exist in this universe, but a free and omnipotent God could have created a different universe.[36] Aristotle had attempted to describe the world not simply as it is, but as it must be. In 1277 Tempier declared, in opposition to Aristotle, that the world is whatever its omnipotent Creator chose to make it.[37]

What were the implications of these theological points for the practice of natural philosophy? In the first place, certain articles in the condemnations raised new and pressing questions, which required further analysis. For example, the claim that God could supernaturally create accidents without subjects (important because it impinged on the doctrine of transubstantiation)[38] provoked serious debate about a fundamental point in Aristotelian metaphysics—the nature and relationship of accidents and their subjects. The anti-astrological article condemning the idea that history will repeat itself every 36,000 years, when the planetary bodies return to their original configuration, provoked Nicole Oresme (ca. 1320–82) to write an entire mathematical treatise in which he explored questions of commensurability and incommensurability and demonstrated the unlikelihood of all the planetary bodies returning to their original configuration within a finite period of time. Articles about the celestial movers gave rise to animated debates about this important feature of cosmic operation. And the articles that stressed God's unlimited creative power gave license to all manner of speculations about possible worlds and imaginary states of affairs that it was evidently within God's power to create. This led to an avalanche of speculative or hypothetical natural philosophy in the fourteenth century, in the course of which various principles of Aristotelian natural philosophy were clarified, criticized, or rejected.[39]

Second, many of the articles in the condemnations were motivated by concern over the element of necessity that Aristotle had attached to his natural philosophy—the claim that things cannot be otherwise than they are. When Aristotelian necessity was forced to yield before claims of divine omnipotence, other Aristotelian principles immediately became vulnerable. For example, the mere possibility that God could create other universes outside our own requires a conception of space outside our universe compatible with that possibility. Consequently, in the aftermath of the condemnations, many scholars came to agree that there must be a void space, perhaps even an infinite void space, outside the cosmos suited to receive these possible universes. Likewise, if it is supernaturally possible for the outermost heaven or perhaps the entire cosmos to be moved in a straight line, it follows that motion must be the kind of thing that can be meaningfully applied to the outermost heaven or the cosmos as a whole. But Aristotle had defined motion in terms of surrounding bodies, and there is nothing outside the outermost heaven to surround it. It is evident, therefore, that Aristotle's definition of motion requires revision or correction.[40]

THE RELATIONS OF PHILOSOPHY AND THEOLOGY AFTER 1277

The condemnations are important as benchmarks in the gradual assimilation of Aristotelian philosophy by medieval Christendom. They reveal the strength of conservative sentiment in the 1270s and signal a provisional conservative victory. But it may be well to pause and consider precisely what had been won.

In the first place, even the most conservative of those involved in promulgating the condemnations were not aiming for the elimination of Aristotelian philosophy. Their purpose was merely to administer a healthy dose of discipline, which would unforgettably remind philosophy of her handmaiden status, while at the same time settling certain points of dispute. Second, although this was, strictly speaking, a local victory (since Tempier's decree was formally binding only in Paris), its influence was, in fact, substantially wider. For one thing, Paris was the premier European university for theological studies (the only one on the Continent at the time), and a decree such as this would inevitably have reverberations throughout Christendom. For another, the pope was known to be in touch with Parisian developments, concerned about the dangers of radical Aristotelianism, and possibly willing to intervene on behalf of the conservatives. Moreover, eleven days after Tempier promulgated his 1277 decree, the archbishop of Canterbury, Robert Kilwardby, issued a smaller, but in many respects similar, condemnation applicable to all of England. And in 1284 Kilwardby's decree was renewed by his successor as archbishop of Canterbury, the Franciscan John Pecham, an old adversary of Aquinas and one of the leaders among the traditionalists.

We have no exact knowledge about the force of the condemnations late in the thirteenth century or early in the fourteenth; we can assume that their power to compel obedience and to shape philosophical thought varied widely. By 1323 Thomas Aquinas's reputation had recovered to the point where Pope John XXII could elevate him to sainthood; and in 1325 the bishop of Paris revoked all articles of the condemnation of 1277 applicable to Thomas's teaching. Nonetheless, we can still detect the shadow of the condemnations a century after their promulgation. John Buridan, a Parisian master of arts and twice rector of the university, who flourished around the middle of the fourteenth century, was one of many who continued to wrestle with the difficulties posed by the condemnations. Indeed, on several occasions Buridan revealed a sharp awareness of the threat of theological censure (particularly acute for masters of arts) when his scholarly labors carried him into theological territory. In his *Questions on Aristotle's Physics,* where he found it necessary to comment on the movers of the celestial spheres, he concluded by declaring his willingness to yield to theological authority: "this I do not say assertively but [tentatively], so that I might seek from the theological masters what they might teach me in these matters." And in 1377, a full century after the condemnation, the distinguished Parisian theologian Nicole Oresme defended his opinion that the cosmos is surrounded by an infinite void space by advising potential critics that "to say the contrary is to maintain an article condemned in Paris."[41]

Meanwhile, Aristotelian philosophy had come to stay. It became firmly entrenched in the arts curriculum and came more and more to dominate undergraduate education; in 1341 new masters of arts at Paris were required to swear that they would teach "the system of Aristotle and his commentator Averroes, and of the other ancient commentators and expositors of the said Aristotle, except in those cases that are contrary to the faith." At the same time Aristotelian philosophy was becoming an indispensable tool for practitioners of the advanced disciplines of medicine, law, and theology, and increasingly served as the foundation of serious intellectual effort on any subject.[42]

It does not follow, however, that an enduring solution to the problem of faith and reason had been found. There has been no adequate historical analysis of fourteenth-century developments, and it is not possible at this time even to draw an adequate sketch. However, a few modest generalizations are possible.

First, there was a rapid growth of epistemological sophistication and a broad retreat from the ambitious claims made on behalf of philosophy (by liberal and radical Aristotelians) in the thirteenth century. The ability of philosophy to measure up to traditional Aristotelian standards of certainty or successfully address certain subject matters was increasingly called into question, as skeptical tendencies asserted themselves. In particular, the ability of philosophy to address theological doctrine was drastically curtailed. For example, John Duns Scotus (ca. 1266–1308) and William of Ockham (ca. 1285–1347), while not seeking a total separation between philosophy and theology, diminished their area of overlap by questioning the ability of philosophy to address articles of the faith with demonstrative certainty. Deprived of its ability to achieve certainty, philosophy no longer threatened theology, at least to the same degree; the articles of the faith were not open to philosophical demonstration, but were to be accepted by faith alone. In short, a workable peace was produced by compelling philosophy and theology to disengage—to acknowledge their methodological differences and, on that basis, to accept different spheres of influence. In the case of natural philosophy, this was clearly a smaller sphere.[43]

Second, theologians and natural philosophers in the fourteenth century became heavily preoccupied with the theme of divine omnipotence—a traditional theme within Christian theology, but one whose importance had been reemphasized by the condemnations. If God is absolutely free and omnipotent, it follows that the physical world is contingent rather than necessary; that is, there is no necessity that it should be what it is, for it is dependent solely on the will of God for its form, its mode of operation, and its very existence. The observed order of cause and effect is not necessary, but freely imposed by divine will. A fire has the power to warm, for example, not because fire and warmth are necessarily connected, but because God chose to connect them, endowing fire with this power and choosing continually to concur as it performs its warming function. God is free, however, to introduce exceptions: when Shadrach, Meshach, and Abednego were cast into the fiery furnace without harm, as the book of Daniel (chap. 3) recounts, this miracle represented a perfectly permissible decision on God's part to suspend the usual order.[44]

This much is widely accepted by historians; but two divergent lines of argument have developed from it. According to one, if nature does not have its own permanently assigned powers but owes its behavior at every moment to the (possibly capricious) divine will, the idea of a fixed natural order is severely compromised, and serious natural philosophy becomes impossible. According to the other, acknowledgment that God could have created any world he wished led fourteenth-century natural philosophers to perceive that the only way to discover which one he did create is to go out and look—that is, to develop an empirical natural philosophy, which helped to usher in modern science. Both arguments require brief comment.

The former, which sees the doctrine of divine omnipotence as a destructive influence on natural philosophy, exaggerates the level of divine interference envisioned by medieval natural philosophers—none of whom believed that God meddled frequently or arbitrarily with the created universe. A formula that was regularly invoked distinguished between God's absolute and ordained power. When we consider God's power absolutely or in the abstract, we acknowledge that God is omnipotent and can do as he wishes; at the moment of creation there were no factors other than the law of noncontradiction limiting the kind of world he might create. But in fact we recognize that God chose from among the infinity of possibilities open to him and created *this* world; and, because he is a consistent God, we can be confident that he will (but for a rare exception)[45] abide by the order thus established, and we need not worry about perpetual divine tinkering. In short, the infinite scope of God's activity guaranteed by the doctrine of divine omnipotence (God's absolute power) was, for practical purposes, restricted to the initial act of creation; thereafter, what was at issue was God's activity within the existing order (his ordained power). This formula was attractive precisely because it safeguarded absolute divine omnipotence *without* sacrificing the kind of regularity required for serious natural philosophy.[46]

The latter argument, which finds the origins of experimental science in the doctrine of divine omnipotence, is plausible enough. We might expect medieval natural philosophers to have recognized that the behavior of a contingent world cannot be inferred with certainty from any known set of first principles and, therefore, to have set out to develop empirical methodologies. The only trouble with this conclusion is that the historical record does not appear to bear it out. The ringing proclamation of divine omnipotence and nature's contingency, whether in the condemnations or in the writings of philosophers and theologians, was not accompanied or quickly followed by a dramatic increase in the frequency of observation and experiment. Natural philosophers and theologians continued to believe that both the world and the proper method for exploring it were more or less as Aristotle had described them—though they were prepared, just as they had always been, to read Aristotle critically and to question this or that *detail* of Aristotelian natural philosophy or methodology. Modern experimental science was still centuries away; when it eventually emerged, it doubtless owed something to the theological doctrine of divine omnipotence, but to claim a simple causal connection would be extremely reckless.[47] This is a problem that requires further analysis; that analysis, if it is to be useful, must respond to the subtlety and complexity of historical reality.

ELEVEN

The Medieval Cosmos

In previous chapters we have examined the reception of the new learning and the struggles surrounding its assimilation in the thirteenth and fourteenth centuries. In this and the next two chapters, we must undertake a more systematic survey of the natural philosophy that emerged from these struggles. As a means of organizing this material, we will work our way down from the top, from the outermost reaches of the cosmos to the earth at its center. We will also employ the distinction (familiar to Aristotle and his medieval followers) between the organic and inorganic realms. In this chapter, we begin with the basic architecture of the cosmos, emphasizing the heavens but touching also on the structure of the terrestrial region. In the next chapter, we will deal with the behavior of inanimate things in the sublunar realm. And in the following chapter, we will turn to the domain of living creatures.[1]

THE STRUCTURE OF THE COSMOS

We have already touched on early medieval and twelfth-century cosmologies.[2] We saw that encyclopedic writers of the early Middle Ages communicated a modest assortment of basic cosmological information, drawn from a variety of ancient sources, especially Platonic and Stoic. These writers proclaimed the sphericity of the earth, discussed its circumference, and defined its climatic zones and division into continents. They described the celestial sphere and the circles used to map it; many revealed at least an elementary understanding of the solar, lunar, and other planetary motions. They discussed the nature and size of the sun and moon, the cause of eclipses, and a variety of meteorological phenomena.

This picture was enriched in the twelfth century by renewed attention to the content of Plato's *Timaeus* (and Calcidius's commentary on it) and by early contact with Greek and Arabic books in translation. One of the result-ing changes was increased emphasis (surpassing that of the early church fathers) on reconciling Platonic cosmology and the biblical account of creation. Another novelty was the frequent argument by twelfth-century authors that God limited his creative activity to the moment of creation; thereafter, they held, the natural causes that he had created directed the course of things. Twelfth-century cosmologists stressed the unified, or-ganic character of the cosmos, ruled by a world soul and bound together by astrological forces and the macrocosm-microcosm relationship. In an important continuation of early medieval thought, twelfth-century scholars described a cosmos that was fundamentally homogeneous, composed of the same elements from top to bottom: Aristotle's quintessence or aether and his radical dichotomy between the celestial and terrestrial regions had not yet made their presence felt.[3]

I introduced Thierry of Chartres in chapter 9 to illustrate some of these features of twelfth-century cosmology. Another representative of the same tradition, more useful to us because he wrote more voluminously, is Robert Grosseteste (ca. 1168–1253), one of the most celebrated figures of medieval science.[4] Grosseteste is also important as an illustration of the continuation of Platonic currents into the thirteenth century; for although he was educated late in the twelfth century, his major writings date from the first half of the thirteenth.

Central to Grosseteste's cosmology was light: the cosmos came into exis-tence when God created a dimensionless point of matter and its form, a dimensionless point of light.[5] The point of light instantaneously diffused itself into a great sphere, drawing matter with it and giving rise to the cor-poreal cosmos. Subsequent radiation (from the outermost limit of the cosmos back toward the center) and differentiation gave rise to celestial spheres and the characteristic features of the sublunar region. In his early writings Grosseteste seems to have accepted the idea of a world soul—an idea from which he later retreated. The theme of microcosm and macro-cosm is fundamental to Grosseteste's works: humanity represents the pin-nacle of God's creative activity, simultaneously mirroring the divine nature and the structural principles of the created cosmos. Finally, Grosseteste shared the early medieval and twelfth-century belief in a homogeneous cosmos: the heavens in his cosmology are made of finer (specifically, more rarefied) stuff than are terrestrial substances, but the difference is quan-titative rather than qualitative.[6]

Cosmology, like so many other subjects, was transformed by the whole-sale translation of Greek and Arabic sources in the twelfth and thirteenth centuries. Specifically, the Aristotelian tradition gained center stage in the thirteenth century and gradually substituted its conception of the cosmos for that of Plato and the early Middle Ages. This is not to suggest that Aristotle and Plato disagreed on all of the important issues; on many of the basics they were in full accord. Aristotelians, like Platonists, conceived the cosmos to be a great (but unquestionably finite) sphere, with the heavens above and the earth at the center. All agreed that it had a beginning in time—although, as we have seen, some Aristotelians of the thirteenth century were prepared to argue that this could not be established by philosophical arguments. And nobody representing either school of thought doubted that the cosmos was unique: although nearly everybody acknowledged that God could have created multiple worlds, nobody seriously believed that he had done so.

But where Aristotle and Plato disagreed, the Aristotelian world picture gradually displaced the Platonic. One of the major differences concerned the issue of homogeneity. Aristotle divided the cosmic sphere into two distinct regions, made of different stuff and operating according to different principles. Below the moon is the terrestrial region, formed out of the four elements. This region is the scene of generation and corruption, of birth and death, and of transient (typically rectilinear) motions. Above the moon are the celestial spheres, to which the fixed stars, the sun, and the remaining planets are attached. This celestial region, composed of aether or the quintessence (the fifth element), is characterized by unchanging perfection and uniform circular motion. Other Aristotelian contributions to the cosmological picture were his elaborate system of planetary spheres and the principles of causation by which celestial motions produced generation and corruption in the terrestrial realm.

A variety of Aristotelian features, then, merged with traditional cosmological beliefs to define the essentials of late medieval cosmology—a cosmology that became the shared intellectual property of educated Europeans in the course of the thirteenth century. Universal agreement of such magnitude emerged not because the educated felt compelled to yield to the authority of Aristotle, but because his cosmological picture offered a persuasive and satisfying account of the world as they perceived it. Nonetheless, certain elements of Aristotelian cosmology quickly became the objects of criticism and debate, and it is here, in the attempt to flesh out and fine-tune Aristotelian cosmology and bring it into harmony with the opinions of other authorities and with biblical teaching, that medieval scholars made their cosmological contribution. It would be impossible in a single chapter, or even a single book, to deal fully with medieval cosmological thought (Pierre Duhem devoted ten volumes to the subject), and we must limit ourselves to the most important and most hotly debated questions.[7]

THE HEAVENS

Before entering the cosmos, let us pause just outside it: what, if anything, exists there? All agreed that no material substance is found outside the cosmos; if the cosmos is held to contain all the material substance that God created, this conclusion is unavoidable. But what about space devoid of corporeal substance? Aristotle had explicitly denied the possibility of place, space, or vacuum outside the world, and this conclusion was generally accepted until a reevaluation of the issue was provoked by the condemnation of 1277. Two articles in the condemnation bore directly on the problem. In one of them it was declared that God has the power to create multiple worlds, in the other that God is capable of endowing the outermost heaven with rectilinear motion. Now if an additional cosmos could be deposited outside ours, then it must be possible for space to exist there, capable of receiving it; likewise, a heavenly sphere in rectilinear motion would inevitably vacate one space and move into another. Most writers were satisfied to acknowledge the possibility that God *could* create a void space outside the cosmos; a few, like Thomas Bradwardine (d. 1349) and Nicole Oresme (ca. 1320–82), argued that he had actually done so. Bradwardine identified this void space with God's omnipresence and argued that since God is infinite, extracosmic void space must likewise be infinite.

Christian considerations seem to have been paramount in this modification of Aristotelian cosmology, but Stoic influence is also apparent. The notion of extracosmic void came to the West with a Stoic pedigree. Western scholars even borrowed specific Stoic arguments, such as the often-repeated thought experiment about what would occur if somebody situated at the *very periphery* of the material universe, at the outermost limit of all material substance, were to thrust an arm beyond that periphery. It seemed obvious that the arm must be received by a space that had hitherto been empty. Through a combination of Christian and Stoic influence, then, an important modification was imposed on Aristotelian cosmology—a modification that was to figure prominently in cosmological speculations through the end of the seventeenth century and beyond.[8]

As we enter the cosmos, we immediately encounter celestial spheres. How many of these exist, what is their nature, and what are their functions? There were seven known planets—moon, Mercury, Venus, sun, Mars, Jupiter, and Saturn, generally thought to be arranged in that order. In the simplified version of the cosmos preferred by medieval writers on cos-

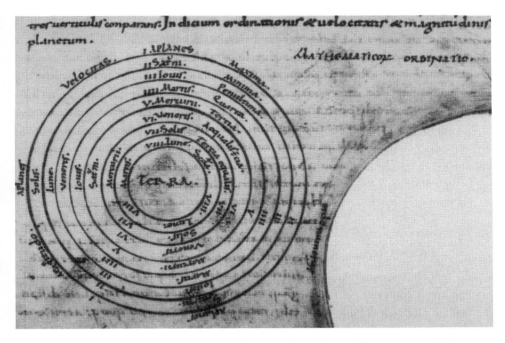

Fig. 11.1. The simplified Aristotelian cosmology popular in the Middle Ages. Paris, Biblio-thèque Nationale, MS Lat. 6280, fol. 20r (12th c.).

mology, which ignored most of the astronomical details, each planet required a single sphere to account for its motion. In addition, according to Aristotle, outside the planetary spheres, defining the outer limit of the cosmos, is the sphere of the fixed stars or *primum mobile*. Several problems arose as medieval scholars thought about this outermost sphere.

One of them was to define its place. The place of a thing, according to Aristotle, is determined by the body or bodies that contain it. But if the sphere of fixed stars is itself the outermost body, there is nothing outside it to serve as container. The natural conclusion of this line of argument—that the *primum mobile* is not in a place—was too paradoxical to be accepted by all but a few of the toughest minds. Various solutions were therefore proposed, including an attempt to redefine place so as to allow it to be determined by the contained, rather than the containing, body.[9]

Another problem for Aristotle's outermost sphere grew out of the account of creation in the book of Genesis, where a distinction was made between the "heaven [*caelum*]" created on the first day and the "firmament [*firmamentum*]" created on the second—obviously two different things, since created on different days. Moreover, the biblical text states that the firmament separates waters beneath it from waters above it; the waters beneath the firmament could be equated with the sphere of water in the terrestrial region, but the waters above the firmament apparently constituted yet another celestial sphere. Discussion of this problem led some Christian commentators to postulate three spheres beyond the seven planetary spheres: the outermost of these, the invisible and motionless *empyreum,* served as the abode of the angels; next came the aqueous or crystalline heaven, perfectly transparent, consisting of water (possibly in a hard or crystallized form but more likely fluid, and possibly water only in a figurative sense); and then the firmament, bearing the fixed stars. The total number of heavenly spheres, for those who accepted this line of argument, came to ten. In time, all three outer spheres were assigned cosmological and astronomical functions; some scholars, wishing to account for an additional stellar motion, postulated an eleventh sphere as well. It is important to note the mutual interaction between cosmology and theology in these discussions: Aristotelian cosmology was adjusted to meet the demands of biblical interpretation; at the same time, the biblical account absorbed the fundamentals of Aristotelian cosmology, with its medieval modifications, and took substantial portions of its meaning from contemporary cosmological theory.[10]

Medieval cosmologists were, of course, interested in the substance or material cause of the celestial region. Many writers of the early Middle Ages, drawing on the Stoic tradition, supposed the heavens to consist of a fiery substance. After the recovery of Aristotle's works, some version of Aristotle's opinion that the heavens were formed out of the quintessence or aether (a perfect, transparent substance not subject to change) was generally accepted. There were debates about the nature of this aether—for example, whether it was a composite of form and matter. Among those who admitted the existence of form and matter in the heavens, some argued that the matter of the heavens was similar in kind to terrestrial matter, while others maintained that the two matters must be totally different. Whatever the nature of the aether might be, everybody agreed that it was divided into distinct spheres, in perfect contact (for otherwise there would be void space), all rotating frictionlessly in their proper directions and with their characteristic speeds. Individual spheres were assumed to be continuous—that is, without interstices or gaps. Seldom did a writer inquire whether they were fluid or hard; both alternatives found support among the few who addressed this issue. The planets were judged to be small spherical regions of greater density or lucidity in the transparent, lucid aether.[11]

A much more hotly debated question was the nature of the celestial movers. Aristotle had identified a set of Unmoved Movers as the causes of celestial motion—the objects of desire of the planetary spheres, which do their best to imitate the changeless perfection of the Unmoved Movers by rotating with eternal, uniform circular motion. The Unmoved Movers are thus final, rather than efficient, causes. Now the Unmoved Mover of the uppermost movable sphere (the "Prime Mover") was customarily identified with the Christian God; but the identity of the additional Unmoved Movers was a more troublesome problem. It would have been easy to identify them with the planetary deities described in Plato's *Timaeus;* but to acknowledge the existence of any deity besides the Creator would have been a clear case of heresy within the Christian tradition, and it was therefore important for Christian scholars to distance themselves from such notions by assigning the Unmoved Movers a status well short of divinity. A common solution was to conceive of them as angels or some other kind of separated intelligences (minds without bodies). There were alternative solutions, however, which dispensed entirely with angels and intelligences. Robert Kilwardby (ca. 1215–79) assigned the celestial spheres an active nature or innate tendency to move spherically. John Buridan (ca. 1295–ca. 1358) argued that there is no need to postulate the existence of celestial intelligences, since they have no scriptural basis; it is possible, therefore, that the cause of celestial motion is an impetus or motive force, analogous to the impressed force that moves a projectile (see the discussion below, chap. 12), which God imposed on each of the celestial spheres at the moment of creation.[12]

The analysis thus far has assumed that the heavens consist of a simple set of tightly nested, concentric spheres. This seems to have been Aristotle's view; it was articulated and vigorously defended by Ibn Rushd (Averroes) in Muslim Spain; and it had a number of important Western adherents. But some medieval cosmological writers modified their cosmology to take into account the eccentric deferents and epicycles of Ptolemaic astronomy—an attempt, obviously, to harmonize cosmology and planetary astronomy. We will deal more fully with these developments later in this chapter; here it will be sufficient to note that the solution was to endow each of Aristotle's planetary spheres with thickness sufficient to contain the Ptolemaic deferent and epicycle for that planet within it (see fig. 11.10, p. 266). The radius of the inside of a given planetary sphere would then equal the minimum distance between the earth and that planet on the Ptolemaic model; the radius of the outside of that planetary sphere would equal the planet's maximum distance from the earth.

The packing principle followed in this system—thick planetary spheres, packed contiguously, without wasted space—made it possible to calculate the size of the various planetary orbits and ultimately the dimensions of the cosmos. To get the calculation started, an estimate of the size of the innermost sphere, that of the moon, was needed. Several Muslim astronomers, including al-Farghānī and Thābit ibn Qurra in the ninth century and al-Battānī in the ninth or tenth, performed the calculation, borrowing the required data from Ptolemy's *Almagest,* with modifications. In the West, Campanus of Novara (d. 1296), gave his version of the calculation, assigning a figure of 107,936 miles to the radius of the inner surface of the moon's sphere (the moon's closest approach to the earth) and 209,198 miles to the radius of the outer surface of the moon's sphere (the moon's farthest retreat from the earth). Similar calculations for Mercury and Venus produced a "theoretical" distance for the sun that accorded roughly with the parallax calculated for the sun by astronomers in antiquity. Continuation of the computation for the superior planets yielded a radius of 73,387,747 miles for the outside of Saturn's sphere and the inside of the stellar sphere. These figures, or figures close to them, prevailed until revised by Copernicus in the sixteenth century.[13]

THE TERRESTRIAL REGION

A detailed analysis of natural operations in the terrestrial realm will be presented in the next chapter. But here we must touch on various macroscopic features of the sublunar region that bear on the larger cosmological issues to which this chapter is devoted.

We enter the terrestrial region by descending beneath the lunar sphere. This is the region of the four elements, which (in the idealized model) are arranged in concentric spheres, each in its proper place: first fire, then air, followed by water, and finally earth at the center. Two of the elements—fire and air—are intrinsically light and naturally ascend; the other two—water and earth—are intrinsically heavy and naturally descend. The elements are continuously transmuted one into another owing to the influence of the sun and other celestial bodies. Thus, for example, water is transformed into air in the process that we know as evaporation; conversely, air can be transformed into water to produce rain.

The fiery and aerial spheres were held to be the scene of various other meteorological phenomena, such as comets, shooting stars, rainbows, lightning, and thunder. Comets were considered to be atmospheric phenomena, the burning of a hot and dry exhalation that has ascended from the earth into the sphere of fire. Rainbows, it was generally held, are produced when sunlight is reflected from the droplets of water in a cloud;

various authors introduced refraction of light into the process; and early in the fourteenth century Theodoric of Freiberg (d. ca. 1310) offered an explanation very close to the modern one, employing the combined reflection and refraction of light in individual droplets (see fig. 11.2).[14]

At the center of everything is the sphere of the earth. Every medieval scholar of the period agreed on its sphericity, and ancient estimates of its circumference (about 252,000 stades) were widely known and accepted.[15] The terrestrial land mass was typically divided into three continents—Europe, Asia, and Africa—surrounded by sea. Sometimes a fourth continent was added. Beyond such basics, knowledge of the surface features of the earth and their spatial relationships varied radically with time, place, and individual circumstances. Let us undertake a brief sketch of medieval geographical knowledge.

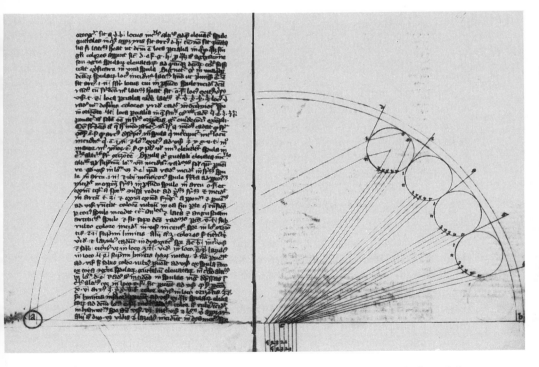

Fig. 11.2. Theodoric of Freiberg's theory of the rainbow. The sun is shown at the lower left, a set of raindrops at the upper right, and the observer is located at the lower center. The drawing aims to demonstrate how two refractions and an internal reflection within individual drops can produce the observed pattern of colors. Basel, Öffentliche Bibliothek der Universität, MS F.IV.30, fols. 33v–34r (14th c.).

Geographical knowledge existed in many forms during the Middle Ages, and we must be careful not to indulge the modern tendency to identify it exclusively with maps or map-like mental images.[16] Of course, medieval people had first-hand, experiential knowledge of their native region. Knowledge of more distant places could come from travelers, of which there were many sorts: merchants, craftsmen, laborers, pilgrims, missionaries, warriors, troubadours, itinerant scholars, civil and ecclesiastical officials, even fugitives and the homeless. For the few fortunate enough to have access to libraries, books such as Pliny's *Natural History* or Isidore of Seville's *Etymologies* offered geographical knowledge of a more exotic sort and on a grander scale in the form of written descriptions. Pliny and Isidore communicated a substantial collection of geographical lore (some of it mythological) through use of the "periplus"—a sequential list of the cities, rivers, mountains, and other topographical features encountered as one navigated a coastline. This information was usually accompanied by interesting historical, cultural, and anthropological detail. Drawing on earlier compilations, Pliny and Isidore led their readers on a swift tour of the periphery of the European and African continents.[17] Towards the end of the Middle Ages, new travel literature began to enrich the store of such knowledge.

Traditional literary sources also dealt with climate, dividing the terrestrial globe into climatic zones or "climes." In a typical scheme, there were five of these: two frigid zones (the arctic and antarctic) around the poles, a temperate zone adjacent to each of these, and a torrid zone straddling the equator and (according to some) divided into two distinct rings by a great equatorial ocean. The torrid zone was considered uninhabitable on account of its heat—though some scholars disputed this claim. Medieval Europeans, of course, found themselves living in the northern temperate zone. On the opposite side of the earth, in the southern temperate zone, are the antipodes. Whether the antipodes are inhabited by antipodeans (people who walk upside down) was a matter of dispute.

There is a natural tendency for those of us familiar with modern maps to organize our geographical knowledge spatially, by the use of map coordinates, thereby reducing geography to geometry. But this was not true of medieval people, most of whom had never seen a map of any kind, let alone a map constructed on geometrical principles. Such maps as medieval people produced were not necessarily intended to portray in exact geometrical terms the spatial relationships of the topographical features indicated on them, and the notion of scale was almost nonexistent. Their function may have been symbolic, metaphorical, historical, decorative, or didactic. For example, the thirteenth-century Ebstorf map employs the world as a symbol of the body of Christ. And a representation of the terrestrial globe in a fifteenth-century manuscript illustrates the division of the world into three continents, each ruled by one of Noah's sons.[18] If, therefore, we wish to avoid misrepresenting medieval aims and achievements, we must be careful not to regard medieval maps as failed attempts at modern mapping.

Among the most numerous, most interesting, and most studied medieval maps are the *mappaemundi,* or world maps. The most common form of *mappamundi* was the T-O map, associated with Isidore of Seville, which gave a schematic representation of the three continents—Europe, Africa, and Asia. In figure 11.3, the "T" inserted within the "O" represents the waterways (the Don and Nile Rivers and the Mediterranean Sea) believed to divide the known land-mass into its major parts: Asia at the top of the map, Europe at the lower left, and Africa at the lower right. Nonschematic

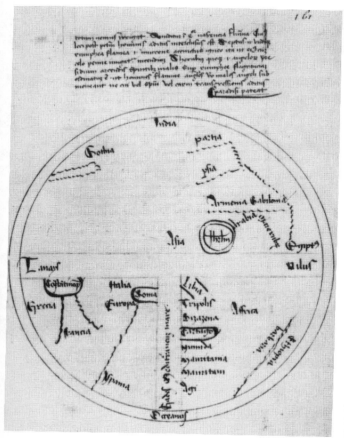

Fig. 11.3. A T-O map. Paris, Bibliothèque Nationale, MS Lat. 7676, fol. 161r (15th c.).

versions of the T-O map, which departed from the rigid T-O diagram in order to incorporate a variety of geographical detail, were also produced (see fig. 11.4). Another common type of map was zonal, featuring the climatic zones as its organizing principle.[19]

Medieval mapping took a mathematical turn (and thereby a turn toward modern cartography) in the form of portolan charts, embodying the prac-

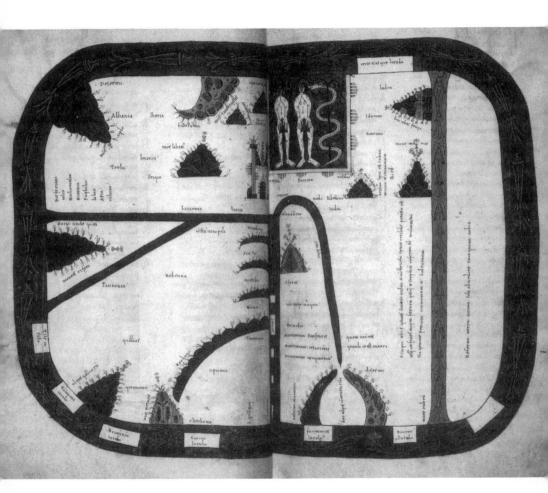

Fig. 11.4. A modified T-O map, the Beatus map (1109 A.D.). A fourth continent is shown at the far right. London, British Library, MS Add. 11695, fols. 39v–40r. By permission of the British Library. For further discussion, see J. B. Harley and David Woodward, eds., *The History of Cartography,* vol. 1, plate 13.

tical knowledge of sailors and designed to facilitate travel by sea. These maps, invented perhaps in the second half of the thirteenth century, offered a "realistic" representation of the coastline and employed a network of "rhumb lines" arranged around a compass rose to convey the distances and directions between any two points (see fig. 11.5). First applied to the Mediterranean Sea, portolan charts were later produced for the Black Sea and the Atlantic coastline of Europe. The use of portolan charts made possible more adventurous voyages of exploration, which in turn greatly expanded European geographical knowledge. Cartography was de-

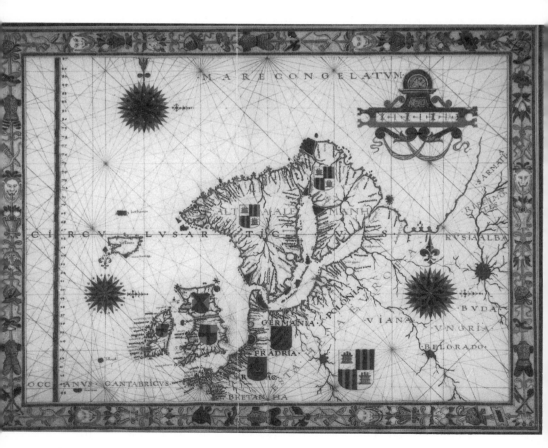

Fig. 11.5. A portolan chart by Fernão Vaz Dourado (ca. 1570). The Huntington Library, HM 41(5).

cisively transformed, finally, by the acquisition of Ptolemy's *Geography,* translated into Latin early in the fifteenth century, which taught Europeans the mathematical techniques by which to represent a spherical body on a two-dimensional surface.[20]

If map-making seems impressive for its practicality, we may do well to restore balance to this section by concluding with an investigation of a question that will appear (at least on the surface) to have no practical application whatsoever—namely, whether the earth rotates on its axis, and what would happen if it were to do so. Aristotle had convincingly presented the grounds for believing that the earth is stationary; and although all medieval scholars agreed with this, several thought the arguments for a rotating earth worth exploring. In looking at this question, they joined good ancient company, for the idea had never entirely disappeared from ancient cosmological and astronomical literature: Aristotle, Ptolemy, and Seneca all discussed it. The most searching explorations of the implications of a rotating earth came in the fourteenth century from John Buridan and Nicole Oresme.

There was no thought here of removing the earth from the center of the cosmos; what Buridan and Oresme had in mind was simply a daily rotation of the earth about its axis. The obvious advantage of postulating such a rotational motion was that doing so would eliminate the necessity of assigning a daily rotation to each of the celestial spheres; it would mean replacing many fast motions with a single slower one, an economy that nearly everybody could appreciate.[21] Buridan pointed out that astronomers observe relative, rather than absolute, motions and that assigning the earth a daily rotation would have no effect on astronomical calculations. Consequently the question of a rotating earth could not be decided on astronomical grounds but must depend on physical arguments. Buridan himself provided such an argument, pointing out that an arrow shot vertically upward (on a windless day) from the surface of a rotating earth would not return to its starting point, because while it was in the air the earth would be moving beneath it; since an arrow shot vertically upward *does* return to its starting point, he argued, we can be certain that the earth is stationary.

A fuller investigation of the problem was produced a few years later by Oresme. One of the most acute natural philosophers of the medieval period, Oresme began by replying to the standard objections to a rotating earth. He argued that all we ever perceive is relative motion and, there-

Fig. 11.6. Nicole Oresme. Paris, Bibliothèque Nationale, MS Fr. 565, fol. 1r (15th c.). The large instrument is an armillary sphere, a teaching aid that offers a physical representation of the ecliptic, celestial equator, and other celestial circles.

fore, that observation cannot settle the issue. He replied to Buridan's arrow argument, maintaining that on a rotating earth, while the arrow is
moving vertically upward and then vertically downward, it would also be
moving horizontally; the arrow would therefore remain above the point
on the earth from which it was shot and return eventually to its starting
point. He reinforced this argument with a shipboard example similar to
the one that Galileo would use in the seventeenth century to defend the
relativity of motion:

> Such a thing appears to be possible in this way, for, if a man
> were in a ship moving very rapidly eastward without his
> being aware of the movement, and if he drew his hand in a
> straight line downward along the ship's mast, it would seem
> to him that his hand was moving with a [downward] rec-
> tilinear motion; so, according to this opinion it seems to us
> that the same thing happens with the arrow which is shot
> straight down or straight up. Inside the boat so moved, there
> can be all kinds of movements—horizontal, criss-cross, up-
> ward, downward, in all directions—and they seem to be
> exactly the same as those when the ship is at rest. Thus, if a
> man in this boat walked toward the west less rapidly than the
> boat was moving toward the east, it would seem to him that
> he was moving west when actually he was moving east; and
> similarly as in the preceding case, all the motions here below
> would seem to be the same as though the earth were at rest.

Oresme proceeded to argue that scriptural passages seeming to teach the
fixity of the earth can be interpreted as an accommodation on the part of
the biblical text, "which conforms to the customary usage of popular
speech."[22] Having thus refuted the objections to a moving earth, he completed the argument by presenting the positive case—a set of arguments
for the economy of moving the earth instead of all the heavens.

This is a powerful and (for Copernicans like ourselves) convincing argument for the rotation of the earth on its axis. Did it convince Oresme's contemporaries? No; and, in fact, it apparently did not even convince Oresme
himself. His argument represented the best philosophical or rational argument for the mobility of the earth that Oresme could construct. But the
doctrine of divine omnipotence guaranteed that it was at best a probable
argument, which could not be allowed to place limits on God's creative
freedom; for all we know, God might prefer an uneconomical world. In
the end, therefore, Oresme accepted the traditional opinion that the earth

is fixed, supporting it with a quotation from Psalm 92:1: "For God hath established the world, which shall not be moved."[23] Apparently this scriptural passage would not yield (as the others had) to the principle that Scripture accommodates itself to popular speech.

Historians have been unsure how to interpret Oresme's apparent turnabout. Many have been tempted to suppose that he saw that he was heading for theological trouble and decided to save himself with a disclaimer. In fact, Oresme took the trouble to explain what he was doing, and we should certainly take his own account seriously. His purpose, he revealed, was to offer an object-lesson for those who would impugn the faith by rational argument. His success at formulating persuasive philosophical arguments for an idea as "opposed to natural reason" as the rotation of the earth demonstrated, he thought, the unreliability of rational argument and, therefore, the caution to be used where rational argument touches the faith, as this one did. His purpose was both cosmological and theological from the beginning.[24]

THE GREEK AND ISLAMIC BACKGROUND TO WESTERN ASTRONOMY

We have considered the overall structure of the cosmos and some of the principles by which it was believed to operate. We now turn to the endeavor to make exact planetary observations and to develop models that account quantitatively for the planetary data. We must begin by disposing of a certain widely influential interpretive scheme. Pierre Duhem built this interpretation on the distinction between two possible ways of viewing astronomical models. On the "realist" view, astronomical models are expected to represent physical reality and answer to the physical criteria of the physicist or natural philosopher. On the "instrumentalist" view, astronomical models are nothing more than convenient fictions—useful mathematical instruments for predicting planetary positions but without any physical truth-value.

According to Duhem, ancient astronomy was overwhelmingly an instrumentalist enterprise. Astronomy and physics came to be defined, Duhem thought, as mutually exclusive endeavors. The task of the physicist (or natural philosopher) was conceived to be the investigation of the structure and nature of things as they really exist, while the astronomer's job was to develop mathematical models that yield quantitative predictions. The astronomer who allowed physical considerations to interfere with his mathematical task was violating disciplinary boundaries. Duhem employed the same conceptual framework and the same categories to understand medieval developments, tracing the fortunes of realist and instrumentalist assumptions.[25]

There are reasons for believing that Duhem seriously exaggerated the instrumentalist tendencies in ancient astronomical thought. In only one or two sources from late antiquity are the realist and instrumentalist alternatives defined, and rarely if ever did anybody defend instrumentalism as a way of doing astronomy.[26] This is not to deny that the Greeks distinguished between a physical and mathematical approach, that Ptolemy (for example) was committed to a program that was primarily mathematical, or that the quest for mathematical success could induce Ptolemy and other astronomers to ride roughshod over physical concerns. It is simply to insist that distinguishing between physics and mathematics or finding them sometimes in conflict is not the same as calling for their divorce; it is also to maintain that the long-term goal of mathematical astronomers, even if they could not always achieve it in practice, was to create a mathematical astronomy that paid attention to, and was consistent with, the received principles of natural philosophy. We must remember that Ptolemy wrote both the highly mathematical *Almagest* and the more physical *Planetary Hypotheses;* moreover, even when most narrowly focused on mathematical goals in the *Almagest,* he did not entirely overlook the physical realities.

As we look at medieval astronomy, we will find that it remained predominantly a mathematical enterprise; from Roman times it belonged to the mathematical quadrivium, and it never lost its association with mathematics. But we must be careful not to assume that its mathematical goals were an expression of mathematical instrumentalism. Medieval mathematical astronomers, like their ancient predecessors, were interested in geometrical models and even quantitative predictions, but none of them concluded from this that astronomy ought to be divorced from physical reality. It follows that during the Middle Ages astronomy and cosmology were not glaring at each other across a methodological chasm, but rubbing shoulders along a methodological continuum.

If we cannot sharply distinguish between astronomy and cosmology on methodological grounds, is there any justification for treating them as distinct enterprises or disciplines? Yes. One of the best ways of distinguishing medieval disciplines is to forget about their formal definitions and to examine them as textual traditions. The cosmological questions that occupied us in the opening sections of this chapter tended to appear in commentaries on certain texts—Aristotle's physical works (especially *On the Heavens* and the *Metaphysics*), John of Sacrobosco's *Sphere,* Peter Lombard's *Sentences,* and the creation account in Genesis.[27] The mathematical analysis of the heavens belonged to a different textual tradition, springing from Ptolemy's *Almagest* and other works of mathematical astronomy produced during the Hellenistic period. That esoteric skills had to be acquired by anybody wishing to practice (or even understand) mathematical astronomy certainly helped to discourage any attempt to merge these two traditions into a general "celestial science."

Islamic astronomy has already made an appearance in this book. In order to prepare ourselves to understand Western astronomical developments, we must now add some detail. The very earliest influences on Islamic astronomy were Indian and Persian versions of Greek astronomy. However, during the ninth century Muslim astronomers gained direct access to Greek sources, the most important of which was Ptolemy's *Almagest,* translated several times in the course of the ninth century—the final and best version being that of Isḥāq ibn Ḥunayn produced at the House of Wisdom in Baghdad. A vigorous tradition of Islamic astronomy, based largely on Ptolemaic principles, developed over the next few centuries. Problems of chronology, time-keeping, and the calendar were major motivating factors behind this astronomical effort: the need to work out the relationship between the lunar calendar and the solar year, to predict the beginning of the lunar month, and to determine times of prayer were all pressing problems, which required astronomical know-how for their solution. Another motivating factor was undoubtedly the close connection between astronomy and the practice of astrology—the latter a heavily patronized activity at Islamic courts.[28]

It is impossible to capture the richness of the Islamic astronomical achievement in a brief summary account. We may make some headway, however, by taking note of the categories under which this astronomical achievement fell. First, a great deal of effort was devoted to mastery, improvement, and dissemination of Ptolemaic astronomical theory. The astronomical textbooks of al-Farghānī and al-Battānī (both of which were subsequently translated into Latin) are good representatives of this achievement. Second, the calculational aspect of Ptolemaic astronomy was improved by developments in spherical trigonometry, including the use of all six modern trigonometric functions (as opposed to the one function, the "chord," employed by Ptolemy).[29]

Third, important progress was made in astronomical observation and instrumentation. Many observatories or observation posts were established on Islamic soil—some relatively permanent, others fairly ephemeral—in order to improve and supplement Ptolemaic data. Tables of numerical data, with instructions for their use, were prepared and widely disseminated. Instruments were constructed, including large stationary quadrants or sextants for measuring the altitudes of stars and planets: the quadrant at the Maragha Observatory, built in the second half of the thirteenth century, had a radius of more than four meters; the gigantic meridian arc at Samarkand, built in the fifteenth century by Ulugh Beg and used primarily for solar observations, had a radius of more than forty meters.[30]

The most impressive and useful of the astronomical instruments from a mathematical standpoint was the astrolabe, invented during the Hellenistic period but perfected in Islam. The astrolabe was a hand-held instrument consisting of a graduated circle and a sighting rule (the alidade), which pivoted about a pin and allowed observations of the altitude of a star or planet, and a set of circular brass plates that fit into a brass body or "mother" and made the astrolabe into an astronomical computer (see figs. 11.7 and 11.8). The mathematical principle that helped convert the astrolabe into a computer was stereographic projection, by which the spherical heavens could be projected (for convenience) onto a set of flat plates (see fig. 11.9). The uppermost plate (the "rete"), designed to represent the rotating heavens, contained a star map (limited to a few of the most prominent stars) and an eccentric circle representing the ecliptic (figs. 11.7 and 11.8); much of this plate was cut away, so that the user could see through it to a plate fixed beneath, called the "climate." The "climate" bore the projection of a fixed coordinate system defined for the latitude of the user, consisting of a horizon line, circles of equal altitude, and lines of equal azimuth, as well as the celestial equator, tropic of cancer, and tropic of capricorn (fig. 11.8). The rete could then be rotated over the climate to simulate the rotation of the heavens with respect to a terrestrial observer; the position of the sun on the ecliptic could be marked, and a variety of useful calculations then became possible.[31]

Fourth, Islam saw substantial criticism of Ptolemaic astronomical theory and attempts to improve or correct it. One of the early critics was Ibn al-Haytham (d. ca. 1040, known in the West as Alhazen), who objected to Ptolemy's use of the equant on the grounds that it violated the principle of uniform motion. Ibn al-Haytham also attempted a physical interpretation of Ptolemaic eccentrics and epicycles along lines developed by Ptolemy himself in his *Planetary Hypotheses*. This was an attempt to unite the mathematical and the physical approaches to astronomical phenomena— to integrate the mathematical techniques of the *Almagest* with the physical scheme of the *Planetary Hypotheses* more fully and successfully than Ptolemy himself had done. The basic idea was to thicken each of the planetary spheres to the point where it could contain within itself an eccentric channel or ring, through which the epicycle would pass.

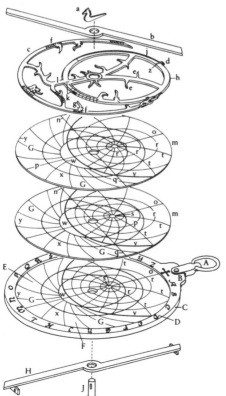

Fig. 11.8. An "exploded" view of the astrolabe. Courtesy of J. D. North. Originally published in J. D. North, *Chaucer's Universe,* p. 41.

a Horse (a wedge ending in a horse's head)
b Rule
c Rete
e Star pointers
h Ecliptic circle
k Lines demarcating the signs of the zodiac
m Climates
r Almucantars (circles of equal altitude)
s Zenith
t Lines of equal azimuth
v Horizon line
w Tropic of Cancer
x Equator
y Tropic of Capricorn
C Mother
G Hour angle lines
H Alidade (with sighting holes)
J Pin

Fig. 11.9. Sterographic projection of the almucantars. The circles of equal altitude (top) are projected onto a horizontal plane passing through the equator of the celestial sphere, as they would be seen by an observer situated at the south celestial pole. These circles of equal altitude or almucantars, along with lines of equal azimuth, were primary features of the "climate" of the astrolabe. Courtesy of J. D. North. Originally published in North's *Chaucer's Universe,* p. 53.

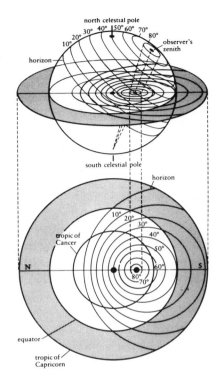

Fig. 11.10. Ibn al-Haytham's solid-sphere model of the Ptolemaic deferent and epicycle.

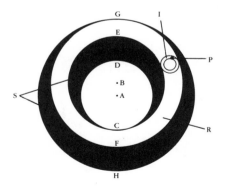

In figure 11.10, the thickened space, S, is bounded by spherical surfaces CD and GH. A is the center of the universe, where the earth is situated. Cutting through the sphere and eccentric to it is a ring R, centered on B and bounded by surfaces CE and FG. Situated within the ring is the epicycle I, bearing the planet P. The entire sphere rotates about its center A on a daily basis, carrying the ring with it; meanwhile, the epicycle "rolls" through the ring in the sidereal period of the planet (the time required for the planet to complete one circuit of the ecliptic), and through it all the planet is carried around the rotating epicycle. Similar thickened spheres are required for each of the remaining planets. If all of these spheres are tightly nested, one inside another, we have a physical model of the planetary system that incorporates the fundamentals of Ptolemaic planetary astronomy, while offering a tolerable rendition of Aristotle's system of concentric spheres.[32]

The attack on Ptolemy was especially vigorous in twelfth-century Spain, where a series of scholars, including Ibn Bājja (Avempace), Ibn Ṭufayl, Ibn Rushd (Averroes), and al-Biṭrūjī (Alpetragius), criticized Ptolemaic planetary models on the grounds that they were physically impossible; these scholars proceeded to call for an astronomy consistent with Aristotelian physics. Ibn Rushd (1126–98) attacked the use of eccentrics, epicycles, and especially equants, maintaining that they tell us nothing about physical reality; in their place he urged a return to the concentric spheres of Aristotle. Al-Biṭrūjī (fl. 1190) went considerably beyond Ibn Rushd in the effort (ultimately unsuccessful) to show how a simple system of concentric spheres might yield predictions comparable to those of Ptolemaic astronomy. In al-Biṭrūjī's scheme, there is a set of simple concentric spheres, one for each planet; all rotate uniformly from east to west with a motion propagated inward from the *primum mobile,* diminishing as it goes (thereby eliminating the need for both east-to-west and west-to-east motions that many natural philosophers found disagreeable in Aristotelian cosmology). To account for the observed irregularity of planetary motions, al-Biṭrūjī allowed each planet to crawl around on the surface of its sphere (with a motion governed by what have been described as a deferent and epicycle drawn on the surface of the sphere).[33]

ASTRONOMY IN THE WEST

During the early Middle Ages, the West was without access to the Greek sources of mathematical astronomy—the works of Hipparchus, Ptolemy, and others. Astronomy was certainly understood to be a mathematical art, a member of the mathematical quadrivium, but the amount of mathematical astronomy actually known to early medieval scholars was minimal. Authors such as Pliny, Martianus Capella, and Isidore of Seville offered an elementary description of the celestial sphere and its major circles; of the seven planets and their west-to-east motion through the band of the zodiac, including retrograde motion; and of the sun-linked motion of Mercury and Venus. The ability to deal with problems of chronology and the calendar was also a well-developed art. But knowledge of Ptolemaic models or any other scheme for the practice of serious mathematical astronomy was nonexistent.[34]

The state of Western astronomical knowledge was radically altered in the tenth and eleventh centuries by contact with Islam, principally through Spain. It is certain that Gerbert of Aurillac (ca. 945–1003) had something to do with this; it is possible that he returned from his studies in northern Spain bearing astronomical treatises. Whatever the exact details, these early contacts brought Christendom a versatile astronomical instrument, the astrolabe, along with the mathematical knowledge required to put it to use. Several treatises on the construction and use of the astrolabe, translated from Arabic to Latin, circulated in the eleventh century. The astrolabe, in turn, was responsible for a reorientation of Western astronomy, away from qualitative and toward quantitative concerns.[35]

Serious quantitative astronomy, however, required substantial bodies of observational data. By the early twelfth century, we know, Western scholars had begun to gather such data firsthand. But much larger and more useful bodies of data were obtained through translation of Arabic sources. The astronomical tables of al-Khwārizmī (d. after 847), along with instructions (*canons*) for their use, were translated by Adelard of Bath in 1126. The *Toledan Tables* (compiled in Toledo by al-Zarqālī during the eleventh century) were translated a little later.[36] These translated tables were treasuries of quantitative astronomical information, but they had been constructed for earlier eras and locations other than the ones where they were to be used. Consequently they required adaptation—work carried out by a number of twelfth-century scholars, including Raymond of Marseilles and Robert of Chester. In their work we have the beginnings of a genuine Western tradition of mathematical astronomy.

Although astronomical instruments and tables of astronomical data were necessary for the practice of mathematical astronomy, they were not sufficient. A third requirement was astronomical theory. The instructions accompanying a set of astronomical tables might offer a glimpse of its theoretical underpinnings, but this was limited in quantity and confusing. Treatises of

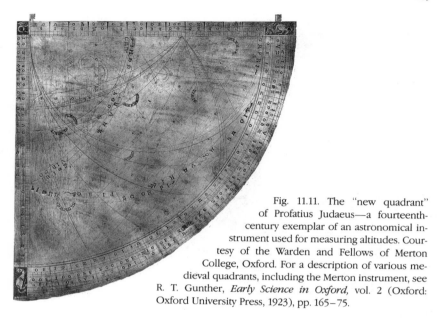

Fig. 11.11. The "new quadrant" of Profatius Judaeus—a fourteenth-century exemplar of an astronomical instrument used for measuring altitudes. Courtesy of the Warden and Fellows of Merton College, Oxford. For a description of various medieval quadrants, including the Merton instrument, see R. T. Gunther, *Early Science in Oxford,* vol. 2 (Oxford: Oxford University Press, 1923), pp. 165–75.

theoretical astronomy, presenting the mathematical models that lay behind the data and the calculations, were needed; and again they were supplied through translation, in this case from both the Arabic and the Greek. Al-Farghānī's elementary handbook of Ptolemaic astronomy was translated in 1137 by John of Seville as *The Rudiments of Astronomy.* In the second half of the twelfth century the more technical astronomical works of Thābit ibn Qurra, Ptolemy, and others became available: Ptolemy's *Almagest* was rendered into Latin twice, once from the Greek and subsequently (by Gerard of Cremona) from the Arabic. Astrological texts that appeared about the same time contributed to the interest in astronomical theory and calculations. Indeed, the astrologer's need for astronomical calculations, along with the growing connection between astrology and medicine, helps to explain the growth of astronomical studies.

By the end of the twelfth century, the most important astronomical texts were available in Latin. The history of Western astronomy from this point onward is a story of growing mastery and increasing dissemination of astronomical knowledge, primarily within the universities. One of the necessities in the universities was for textbooks that would bring the complexities of Ptolemaic astronomy within the reach of students. An introductory treatise such as al-Farghānī's *Rudiments of Astronomy* could, of course, be put to use; but teachers in the universities soon produced books of their

11.12. An astronomer observing with an astrolabe. Paris, Bibliothèque de l'Arsenal, MS 1186, fol. 1v (13th c.).

own. One of the earliest and most popular was *The Sphere* of Johannes de Sacrobosco (John of Holywood), written at Paris about the middle of the thirteenth century. This work, which continued to be commented upon and used as a university textbook as late as the seventeenth century, contained an elementary account of spherical astronomy and a few brief remarks on planetary motions. For example, Sacrobosco described the west-to-east motion of the sun around the ecliptic at the rate of about 1°/day; he noted that each of the planets except the sun is carried around on an epicycle, which in turn is carried around on a deferent circle, and explained how the epicycle-on-deferent model accounts for retrograde motion; and he attributed lunar and solar eclipses, respectively, to the shadows cast by the earth and moon. Beyond this his planetary astronomy did not go.[37]

Sacrobosco's *Sphere* was obviously meant to convey only the most elementary astronomical knowledge, perhaps for the benefit of students interested in chronology, time-keeping, and calendar-construction ("computus"). Another treatise, the *Theorica planetarum* (*Theory of the Planets*), composed a little later by an anonymous author, possibly also a Parisian teacher, raised the discussion of planetary astronomy to a substantially higher level. The *Theorica* sketched the basic Ptolemaic theory for each of the planets, supplementing the description with geometrical diagrams. For example, the motion of the sun around the ecliptic was explained as the result of uniform west-to-east motion about an eccentric deferent circle at the rate of 59′8″ (just short of 1°) per day; meanwhile that eccentric is carried uniformly east-to-west at the rate of one full rotation per day by the "universe" or stellar sphere. In the model for the superior planets—Mars, Jupiter, and Saturn—the planet P (fig. 11.13) moves uniformly around the epicycle from west to east, while the center of the epicycle moves in the same sense around the deferent. The motion of the epicycle around the deferent is uniform with respect to equant point Q; the center of the deferent is halfway between the equant point and the center of the earth.[38] The *Theorica* seems quickly to have become the standard textbook of astronomical theory, firmly establishing the Ptolemaic models against any possible rivals and fixing astronomical terminology for several centuries.

A serious problem posed by the establishment of Ptolemaic theory was how to bring it into harmony with Aristotelian cosmology. It appeared to scholars that the eccentric and epicyclic circles of Ptolemaic astronomy were not easily reconciled with Aristotelian concentric spheres or the principles of Aristotelian natural philosophy; anybody who needed help perceiving the magnitude of the problem could obtain it from Averroes' attack on the paraphernalia of Ptolemaic astronomy. The only system to achieve success from a quantitative standpoint seemed questionable from

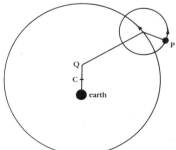

Fig. 11.13. The model for one of the superior planets, according to the *Theorica planetarum*.

a physical or philosophical standpoint. There was considerable thrashing about on this issue in the thirteenth and fourteenth centuries, as scholars explored the status of theoretical claims or looked for compromise positions. The astronomers, who inevitably demanded quantitative results, had no choice but to retain the Ptolemaic models. For some of the more philosophically inclined, the idea of creating a quantitatively exact astronomy on Aristotelian principles remained an elusive dream.[39]

The solid-sphere version of the Ptolemaic system developed by Ibn al-Haytham made an appearance in the thirteenth century. Roger Bacon, writing in the 1260s, appears to have been the first Western scholar to give this scheme a thorough discussion; and after him there seems to have been a brief flurry of interest in it among the Franciscans, for Bernard of Verdun and Guido de Marchia (both Franciscans) offered descriptions of it. The idea seems also to be implicit in certain astronomical works of the period, such as that of Campanus of Novara, but it did not again become the object of serious cosmological debate until picked up by Georg Peuerbach, one of the Viennese scholars who contributed to the revival of astronomy in the fifteenth century.[40]

One thing the *Theorica* did not do was convey the quantitative content of Ptolemaic astronomy or the means of making actual astronomical calculations. This function came to be served by the *Toledan Tables* and then, after about 1275, the *Alfonsine Tables* (prepared at the court of Alfonso X of Castile), which frequently accompanied the *Theorica*. The *Alfonsine Tables* (fig. 11.14) served as the standard guide to the practice of mathematical astronomy until confronted by new competitors in the sixteenth century.[41]

While a modest amount of elementary astronomical knowledge may have become quite common among those with a university education, advanced knowledge of the sort represented by the *Toledan* or *Alfonsine Tables,* or even the *Theorica,* no doubt remained quite rare. Seldom did the universities require astronomical knowledge for a degree in arts, though some instruction in astronomy (most often in the form of lectures on the *Theorica,* occasionally through lectures on Ptolemy's *Almagest*) was frequently available. However, it does not follow from the low profile of astronomy in the university curriculum that astronomy was going nowhere. In fact, despite the relative paucity of practitioners, knowledge of mathematical astronomy among those who did practice it became steadily more sophisticated. Out of this medieval tradition would emerge, in the fifteenth and sixteenth centuries, astronomers of the stature of Johannes Regiomontanus and Nicolaus Copernicus.[42]

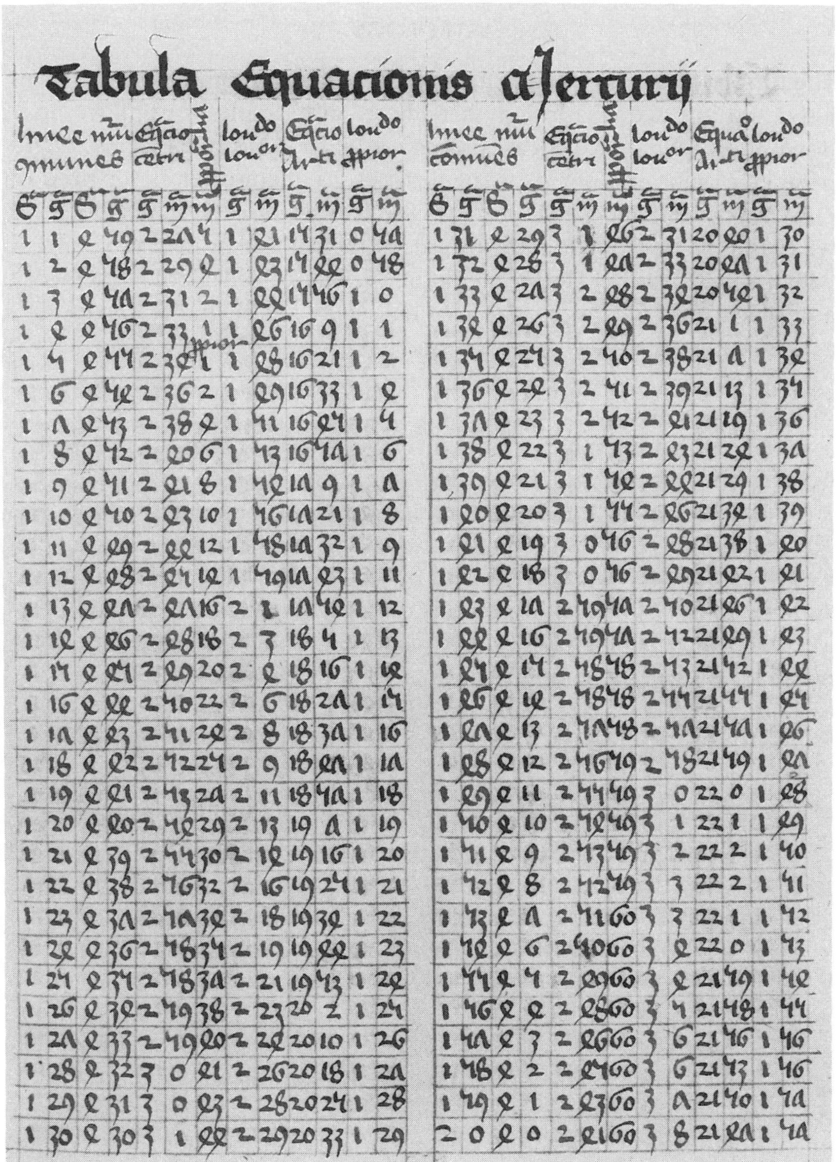

Fig. 11.14. The Alfonsine Tables. A page from the table for Mercury. Houghton Library, Harvard University, fMS Typ 43, fol. 46r (ca. 1425). By permission of the Houghton Library.

ASTROLOGY

The history of astrology has suffered from a tendency among historians to judge the practice of astrology harshly, as an example of primitive, irrational, or superstitious ideas, promoted by fools and charlatans. There *were* charlatans, of course, as medieval critics themselves never tired of pointing out. But medieval astrology also had a serious scholarly side, and we must not allow our attitude toward it to be colored by the low regard in which astrology is held today. Medieval scholars judged astrological theory and practice by *medieval* criteria of rationality and by the *contemporary* evidence to which they had access; and it is only as we do the same that we can hope to understand the importance and the changing fortunes of astrology during the Middle Ages.[43]

It will help if we begin by distinguishing between (1) astrology as a set of beliefs about physical influence within the cosmos and (2) astrology as the art of casting horoscopes, determining propitious moments, and the like. The former was a respectable branch of natural philosophy, the conclusions of which were rarely called into question. The latter, by contrast, was vulnerable to a variety of objections (empirical, philosophical, and theological) and remained a subject of contention throughout the Middle Ages. Although we will touch upon astrology in the second sense, it is astrology as an aspect of cosmic physics that will primarily concern us.

There were compelling reasons for believing that the heavens and the earth were physically connected. First, there were observational data that made the connection obvious: nobody could doubt that the heavens were the major source of light and heat in the terrestrial region; the seasons were plainly connected with solar motion around the ecliptic; the tides were apparently connected with lunar motion; and it seemed clear enough, once the compass made its appearance (late in the twelfth century), that the poles of the celestial sphere exercised a magnetic influence on certain minerals.

Observational arguments of this sort were reinforced by traditional religious beliefs. The association of the heavens with divinity and the understanding that divinity exercised influence in the terrestrial realm were prominent features of ancient religions. The belief that stellar and planetary events were omens (signs rather than causes) of terrestrial events was widespread in ancient Mesopotamia, where reading the omens became a specialized art, demanding a measure of astronomical knowledge. Such beliefs were gradually enriched and transformed by the addition of new elements, including the notion that the celestial configuration at the time of a person's conception or birth could be used as a means of predicting certain details of that person's life (see above, chap. 1).[44]

Within Greek culture, astrological ideas obtained support from a variety of philosophical systems. In the *Timaeus,* Plato's Demiurge explicitly delegated to the planets or planetary deities the task of bringing into existence things in the sublunar realm; and this suggested the possibility of an ongoing relationship. Plato also stressed the unity of the cosmos, including parallels between the cosmos as a whole and individual humans (the macrocosm-microcosm analogy). In Aristotle's cosmos the Unmoved Mover was the source not simply of the motions of the celestial spheres, but also of motion and change in the sublunar realm. In his discussion of meteorological phenomena, Aristotle argued that the terrestrial region "has a certain continuity with the upper [celestial] motions; consequently all its power is derived from them." Elsewhere he attributed seasonal changes, as well as all generation and corruption in the terrestrial realm, to the motion of the sun around the ecliptic. Finally Stoics, with their vision of an active, organic cosmos characterized by unity and continuity, seem to have embraced and defended the science of astrology. It should be clear, then, that astrology in its physical or cosmological form was the empirical and rational investigation of the causal connections between the heavens and the earth. Almost any ancient philosopher would have considered it extraordinarily foolish to deny the existence of such connections.[45]

Ptolemy is an excellent case in point—excellent not only because he addressed the question fully and clearly, but also because he exercised a powerful influence on both the Islamic and Western astrological traditions. In his astrological handbook, the *Tetrabiblos,* Ptolemy acknowledged that astrological prognostications cannot match the certitude of astronomical demonstrations; nonetheless, he affirmed the existence of celestial forces and the validity of astrological prognostications of a general sort. It is apparent to everybody, he argued,

> that a certain power emanating from the eternal ethereal substance . . . permeates the whole region about the earth. . . . For the sun . . . is always in some way affecting everything on the earth, not only by the changes that accompany the seasons of the year to bring about the generation of animals, the productiveness of plants, the flowing of waters, and the changes of bodies, but also by its daily revolutions furnishing heat, moisture, dryness, and cold in regular order and in correspondence with its positions relative to the zenith. The

moon, too, . . . bestows her effluence most abundantly upon mundane things, for most of them, animate or inanimate, are sympathetic to her and change in company with her. Moreover, the passages of the fixed stars and the planets through the sky often signify hot, windy, and snowy conditions of the air, and mundane things are affected accordingly.

The practitioner who understands these influences and who has also mastered the celestial motions and configurations ought to be able to predict a wide variety of natural phenomena:

If, then, a man knows accurately the movements of all the stars, the sun, and the moon, . . . and if he has distinguished in general their natures as the result of previous continued study . . . ; and if he is capable of determining in view of all these data, both scientifically and by successful conjecture, the distinctive mark of quality resulting from the combination of all the factors, what is to prevent him from being able to tell on each given occasion the characteristics of the air from the relations of the phenomena at the time, for instance, that it will be warmer or wetter? Why can he not, too, with respect to an individual man, perceive the general quality of his temperament from the atmosphere at the time of his birth, as for instance that he is such and such in body and such and such in soul, and predict occasional events, by use of the fact that such and such an atmosphere is attuned to such and such a temperament and is favourable to prosperity, while another is not so attuned and conduces to injury?[46]

A certain amount of anti-astrological sentiment surfaced within Hellenistic philosophy and subsequently within both the Islamic and Christian traditions. The object of attack, however, was not belief in the reality of celestial influence, but the threat of determinism and (among the church fathers) the assignment of divinity to the stars and planets. The most influential voice within Christendom was that of Augustine (354–430). Augustine attacked vulgar astrology as a fraudulent enterprise, practiced by impostors; but his greatest concern was for what he regarded as the tendency toward fatalism or determinism within astrological theory. At all costs, the freedom of the will must be protected, for otherwise there would be no human responsibility. Augustine appealed frequently to the "twins problem" (not original with him), pointing out that twins, conceived at the same instant and born almost simultaneously, often experience dramatically different fates. But Augustine opened the door to the possibility of physical influence, as long as it was held to affect only the body, when he wrote:

> It is not entirely absurd to say, with reference only to physical differences, that there are certain sidereal [i.e., stellar] influences. We see that the seasons of the year change with the approach and the receding of the sun. And with the waxing and waning of the moon we see certain kinds of things grow and shrink, such as sea-urchins and oysters, and the marvelous tides of the ocean. But the choices of the will are not subject to the positions of the stars.[47]

The anti-astrological polemics of Augustine and other church fathers helped to create a climate of opinion hostile to astrology during the early Middle Ages. In early medieval literature we find regular condemnation of the practice of horoscopic astrology—often accompanied, however, by admission of the reality of celestial forces and their influence on a variety of terrestrial phenomena.[48]

The flowering of Platonic philosophy and the recovery of Greek and Arabic astrological writings in the twelfth century led to a resurgence of interest in astrology and a more favorable attitude toward its doctrines. Any suggestion of astrological determinism remained anathema, of course, but assertions about the reality of stellar and planetary influence and the possibility of successful astrological prognostications now became commonplace. For example, in his influential *Didascalicon* (written in the late 1120s) Hugh of St. Victor (d. 1141) expressed approval of the "natural" part of astrology, which deals with the "temper or 'complexion' of physical things, like health, illness, storm, calm, productivity, and unproductivity, which vary with the mutual alignments of the astral bodies." An anonymous author writing near the end of the twelfth century or early in the thirteenth noted that "we do not believe in the deity of either the stars or the planets, nor do we worship them, but we believe in and worship their Creator, the omnipotent God. However, we do believe that the omnipotent God endowed the planets with the power that the ancients supposed came from the stars themselves." Another author from the twelfth century, addressing the issue of determinism, wrote that "the stars . . . can produce an aptitude for having wealth, never the fact of having it."[49]

The translation of astrological treatises from Greek and Arabic was of critical importance in shaping these new attitudes. The major works were

Fig. 11.15. The Arabic astrologer Albumasar or Abū Maʿshar, presumably holding his *Introduction to the Science of Astrology*. Paris, Bibliothèque Nationale, MS Lat. 7330, fol. 41v (14th c.).

Ptolemy's *Tetrabiblos,* translated in the 1130s, and Albumasar's *Introduction to the Science of Astrology,* translated twice in the 1130s and 1140s; these were accompanied by various smaller astrological tracts and joined eventually by works of Aristotle that addressed the question of celestial influence. The *Tetrabiblos* offered a defense of astrological belief and introduced its readers to some of the technical principles of the art. For example, it identified the various planets with specific terrestrial effects: the sun heats and dries, the moon chiefly humidifies, Saturn principally cools but also dries, and Jupiter heats and humidifies in moderation; the influence of certain planets is favorable, that of other planets unfavorable; certain planets are masculine, others feminine. The *Tetrabiblos* also explained how the powers of the planets are strengthened or weakened according to their geometrical relationship to the sun (their "aspect"). It assigned specific qualities to the signs of the zodiac. And it explained the general traits of people dwelling in different regions of the terrestrial globe by "familiarity" or sympathy between those regions and the planets and signs of the zodiac that govern them.

The contribution of Albumasar's *Introduction* was to elaborate on the astrological principles found in Ptolemy's *Tetrabiblos* and other astrological literature (including Persian and Indian sources), but more especially to establish astrology on proper philosophical foundations by integrating traditional astrological lore with Aristotelian natural philosophy. In practical terms, what this meant was the adoption of Aristotle's metaphysics of matter, form, and substance, as well as Aristotle's claim that the celestial bodies are the source of all motion in the terrestrial region and the agents of generation and corruption. Through planetary influence, forms are imposed on the four elements to produce the physical substances of daily experience; changes in the planetary configuration bring about a perpetual cycle of transmutations, birth and death, coming-to-be and passing-away. Aristotle's account of generation and corruption had concentrated on the motion of the sun around the ecliptic; while attaching priority to the sun, Albumasar (following long astrological tradition) brought the remaining planets, as well as their geometrical relationship to the sun and to the signs of the zodiac, into the causal picture.[50]

The Aristotelianizing of astrology was furthered, of course, by the acquisition of Aristotle's own works in the course of the twelfth century. During the thirteenth, astrological belief took root and became a standard part of the medieval world view. Astrology also became closely associated with the practice of medicine: no reputable physician of the later Middle Ages would have imagined that medicine could be successfully practiced without it.[51]

Philosophers and theologians continued to worry about astrological determinism—a subject that surfaced in the condemnation of 1277—and astrological practitioners were regularly denounced as charlatans. But even the
most vigorous opponents of astrology were ready to acknowledge the reality of celestial influence. Nicole Oresme, who wrote whole books attacking
astrology, admitted that the part of astrology that deals with large-scale
events, such as "plagues, mortalities, famine, floods, great wars, the rise
and fall of kingdoms, the appearance of prophets, new religions, and similar changes, . . . can be and is sufficiently well known but only in general
terms. Especially we cannot know in what country, in what month, through
what persons, or under what conditions, such things will happen." As for
the influence of the heavens on health and disease, "we can know a certain
amount as regards the effects which ensue from the course of the sun and
moon but beyond this little or nothing."[52] Astrology as an aspect of natural
philosophy would flourish until the seventeenth century and beyond.

TWELVE

The Physics of the Sublunar Region

The decision to employ the term "physics" in the title of this chapter is not without risk. The risk is that readers will equate medieval physics with modern physics on the basis of the shared name. It will then be natural for these readers to conclude that medieval physicists were trying to be modern physicists, but with limited success, and that medieval physics was a primitive or failed version of modern physics. And as long as we view medieval physics as failed modern physics, we close the door on the possibility of grasping its own distinctive aims and notable achievements.

The fact is that medieval physics was a remarkably coherent theoretical system in its own right, which achieved considerable success in answering the questions to which it was addressed. And those questions were, on the whole, broader questions than the ones that concern a modern physicist. The breadth of medieval physics becomes clear when we examine the relevant medieval terminology. The Latin nouns *physica* and *physicus* (meaning "physics" and "physicist," respectively) derive from the Greek word *physis,* usually translated "nature." For Aristotle (whose influence here was paramount), the *physis* or nature of a thing was the inner source of its character or behavior, responsible for all natural changes that occur in it. Nature in the collective sense included all things that possess such a nature. And the physicist was the person who investigates natural things and the natural changes that occur in them—simply stated, the student or philosopher of nature in all of its manifestations.[1]

This is not to say that there are no significant continuities between medieval and modern physics. Some of the questions that occupied medieval scholars continued, with little or no change, to occupy their successors in the sixteenth and seventeenth centuries and beyond, and the Middle Ages supplied significant pieces of the scientific vocabulary and the conceptual

framework of early modern physics. Surely these continuities are a legitimate and important object of historical investigation, and we will not entirely ignore them in this chapter; but they cannot be our main business if
our goal is to understand the aims and achievements of *medieval* thought
about nature.[2] We must never succumb to the temptation of supposing,
when we have identified the pieces of medieval physics appropriated by
later ages, that we have thereby figured out what medieval physicists themselves regarded as the essential features of their discipline.

MATTER, FORM, AND SUBSTANCE

What were the fundamental explanatory principles of medieval physics or
natural philosophy? After the reception and assimilation of Aristotle's philosophy in the twelfth and thirteenth centuries, the principles in question
were broadly Aristotelian—though obscurity, incompleteness, and incon-
sistency in the various Aristotelian texts where these principles were set
out left plenty of room for further articulation of the theory and for discussion and debate about the fine points. Let us begin with a brief review of
some of the basics of Aristotelian natural philosophy.[3]

According to Aristotle, all objects in the terrestrial realm ("substances"
he called them) are composites of form and matter. Form, the active principle or agent, bearer of the properties of the individual thing, combines
inseparably with matter, the passive recipient of the form, to produce a
concrete corporeal object. If the object in question is a "natural" object (as
opposed to one produced artificially, by a craftsman), it also has a nature
(determined primarily by its form but secondarily by its matter), which
disposes it to certain kinds of behavior. Thus fire naturally communicates
warmth, rocks naturally fall (if lifted out of their natural place), babies naturally grow and mature, and acorns naturally develop into oak trees. These
natures we discern through long and persistent observation: whatever cannot be the product of chance (because of the regularity of its occurrence)
or of artifice (because no artificer had anything to do with it) must be the
result of nature. Because natures are the determining factors in all cases of
natural change, they are necessarily of great interest to the physicist or
natural philosopher.

Aristotle's medieval followers, contemplating this scheme, identified
two kinds of form—one of them associated with essential properties, the
other with incidental properties. The defining characteristics of a thing,
which make it what it is, are conveyed by what came to be called its "substantial form." Substantial form combines with absolutely propertyless first

matter to give being or existence to a substance and to endow it with those properties that make it the kind of thing it is. However, besides essential properties, every substance also has properties of an incidental or accidental sort, associated with "accidental form." Thus the family dog may be short-haired or long-haired, lean or fat, friendly or ferocious, housebroken or not, and yet it retains the characteristics (supplied by its substantial form) that enable us to identify it unmistakably as a dog.

Aristotle's theory of form, matter, and substance is nicely exemplified in his theory of the elements. Aristotle accepted the position of his predecessors, Plato and the pre-Socratics, to the effect that the familiar materials or substances of everyday experience are complex rather than simple. That is, sensible things in the sublunar world are compounds or mixtures, reducible to a small set of fundamental roots or principles, called "elements." Aristotle adopted Empedocles' and Plato's list of four elements—earth, water, air, and fire—and argued that these combine in various proportions to produce all of the common substances. Aristotle agreed with Plato that the four elements are not fixed and immutable, but undergo transmutations; and the scheme that explained how this was possible was his theory of form and matter.

Each of the elements, he argued, is a composite of form and matter; since the matter in question is capable of assuming a succession of forms, the elements can be transformed into one another. The forms instrumental in producing the elements are those associated with the four primary or "elemental" qualities: hot, cold, wet, and dry. Primary matter informed by coldness and dryness yields the element earth; primary matter informed by coldness and wetness yields water; and so forth. But this primary matter has the capacity to receive any of the four elemental qualities. Therefore, if the quality of dryness in a piece of the element earth yields to wetness through the action of a suitable agent, that piece of earth will cease to exist, and an appropriate amount of the element water will take its place. Aristotle argued that such transformations are occurring constantly, and the elements are therefore constantly being transmuted one into another. Changes of this kind proved capable of accounting for many of the familiar phenomena that we associate today with the disciplines of chemistry and meteorology.[4]

The basic form-matter theory was easily understood, but its application to the real world posed a variety of problems. The world seemed to contain a hierarchy of forms and matters, and the Aristotelian definitions outlined above worked better at some levels than at others. Aristotle's definition of matter as the totally unqualified recipient of forms applies nicely to the constitution of the elements: the matter that receives the elemental forms of the primary qualities (hot, cold, wet, dry) is totally without properties of its own, apart from the ability to receive the elemental forms. Of itself, it is imperceptible, unknowable, and without actual existence. Aristotle referred to this as "primary matter." But accidental forms are imposed on matter that already has independent, substantial existence: the marble out of which a statue is to be made exists as a concrete thing, with a variety of properties (size, shape, color, density, and hardness), before the sculptor endows it with the accidental forms that make it into a specific statue. In the same way, the hair that turns gray (thus serving as the matter for the accidental form of grayness) was already a substantial thing, with specific, identifiable characteristics, before it changed color. Reflection on problems such as this caused Aristotle's ancient and medieval followers to sharpen his definitions and to clarify the distinction between the insubstantial primary matter of the elements and the substantial secondary matter encountered in cases of accidental change.[5]

The matter-form theory was elaborated in Islam by Avicenna (Ibn Sīnā, 980–1037) and Averroes (Ibn Rushd, 1126–98) in ways that would prove influential in the West. The two Muslim commentators thought it impossible to derive the elements from the imposition of the elemental forms directly on primary matter. An intermediate step was required, which would first invest the primary matter with three-dimensionality. To this end, they developed the notion of "corporeal form," which must first be imposed on primary matter to yield a three-dimensional body. The elements emerge, then, when this three-dimensional body (a kind of secondary matter) receives the elemental forms. The idea of corporeal form was transmitted to Christendom, where it proved both influential and controversial. We have seen its adoption by Robert Grosseteste, who identified corporeal form with light.[6]

Aristotle had placed form and matter on essentially equal footing—neither was subordinate to the other, and each had its function—but this balance proved difficult to maintain. Within the Neoplatonic tradition (Avicenna is a good example) there was a tendency to demote matter, to see it as virtual nothingness, while elevating form to a position of quasi-autonomy. Avicenna's younger contemporary Avicebron (d. 1058) veered in the other direction, elevating matter at the expense of form. Avicebron's influence may help to explain the willingness of Western scholars (especially Franciscans, such as Richard of Middleton and Duns Scotus) to argue that God can create matter without form.[7]

COMBINATION AND MIXTURE

One very important class of phenomena to which the theory of matter, form, and substance was applicable was that associated with what we would today call "chemical combination." The centrality of this class of phenomena is apparent when we recall that, according to Aristotle, all substances encountered in the real world, including organic tissue, are compounds of the four elements. It comes as no surprise, therefore, that Aristotle should have inquired into the nature of chemical combination and the status of the original ingredients in a compound. He distinguished between a mechanical aggregate, in which the small particles of two substances are situated side by side without loss of individual identity, and a true blending of the ingredients into a homogeneous compound in which the original natures disappear; he called the latter a "mixt" or "mixture" (we will employ the Latin terms *mixtio* for the process and *mixtum* [plural, *mixta*] for the product, in order to preserve the technical meaning that Aristotle had in mind), and it is this kind of combination that he considered applicable to the mixing of the elements.

In a *mixtum,* according to Aristotle, the individual natures of the ingredients are replaced by a new nature that permeates the compound down to its smallest parts. The properties of the *mixtum* represent an averaging of the properties of the ingredients. If, for example, we combine a wet and a dry element (say, water and earth), the wetness or dryness of the resulting compound will fall on the scale that runs between the extremes of wetness and dryness, at a point determined by the relative abundance of those two qualities. Although the original elements no longer have actual existence in the *mixtum,* Aristotle made remarks that suggested that they maintain a virtual or potential presence that permits them to exercise some kind of continuing influence.[8]

Aristotle's discussion left a number of problems for his commentators. One was to recast the theory of combination or *mixtio* in the language and conceptual framework of matter and form, for those terms do not appear in Aristotle's account. In the course of that effort it was necessary to inquire how the new substantial form of the *mixtum* emerges from the forms of the constituent elements. Another problem of critical importance was to determine in what sense the forms of the original elements continue to exist in the *mixtum;* since it was acknowledged that when the *mixtum* is destroyed the elements out of which it was formed reappear, it seemed evident that they survive in some way within the *mixtum.* Debates on these matters became extremely intricate, and we must limit ourselves to a few introductory remarks.

Everybody agreed that the substantial forms of the constituent elements are replaced by a new substantial form of the *mixtum*. But how does this come about? It was generally agreed that the way was paved for the emergence of the new substantial form by the mingling of the elements, the interaction of their respective qualities, and possibly the corruption of their substantial forms. However, there were good reasons (drawn from Aristotle) for believing that the new substantial form could not be generated out of these antecedent substantial forms or out of the qualities of the original elements; outside intervention seemed to be required. The usual solution was to invoke higher powers—celestial forces or celestial intelligences, possibly even God himself—assigning to them the responsibility for infusing the new substantial form into the primary matter when the preconditions had been met.

As for survival of the elements in the *mixtum,* everybody saw that it was necessary to find some way of allowing the elements to lurk in the *mixtum* potentially or virtually, awaiting a suitable opportunity to reveal themselves. Avicenna argued that the forms of the elements survive intact, while their qualities are weakened to the point of insensibility. Averroes maintained that both the forms of the elements and their qualities are reduced in strength or intensity and maintain a potential existence within the *mixtum*. Since, according to Aristotle, substantial forms do not admit of degrees—that is, cannot be strengthened or weakened (after all, a given four-legged mammal is either a dog or not a dog; in this context talk of more and less makes no sense)—Averroes concluded that the forms of the original elements must not be substantial forms but have a status between that of substantial and accidental form. Thomas Aquinas (ca. 1124–74) argued that the forms of the elements are extinguished in the process of *mixtio,* but that their qualities retain some kind of virtual influence in the *mixtum*. These and other positions became the basis of lively debate among late medieval natural philosophers.[9]

A final question with which we must deal has to do with the physical divisibility of corporeal substances—say, wood or stone or organic tissue. Is there a limit to the process of division, and what are the properties of the smallest pieces? Are they anything like atoms? Aristotle had alluded to the smallest pieces of the ingredients of a *mixtum,* which mingle and interact, and on these remarks subsequent commentators based a theory of what came to be called *minima* or *minima naturalia* (smallest natural parts). The theory acknowledged that in principle divisibility should be endless; however small the piece before you, there is no physical reason why you cannot divide it again. But it was argued that there is nonetheless a smallest quantity of each substance, below which it will no longer be that substance because the form of the substance cannot be preserved in a smaller quantity.

There were attempts in the Middle Ages to construe the theory of *minima* as a variant of atomism. It is true that both theories acknowledged the particulate structure of matter, but otherwise they were far apart. The particles of the atomists were unbreakable least parts; the *minima* of the Middle Ages were divisible, though if divided they would lose their identity. All atoms were of identical stuff, differing only in size and shape; *minima* were as different as the substances to which they belonged. In the atomist vision, properties in the macroscopic world did not, in general, have exact counterparts in the microscopic world: atomists did not explain the redness of a flower, for example, by the redness of its constituent particles. Rather, the atomist program was to reduce the qualitative richness of the world of sense experience to austere, qualitatively bare atoms (characterized only by size, shape, motion, and possibly weight). Minimists, by contrast, continued the Aristotelian program, assigning to the least parts precisely the properties of the whole to which they integrally belonged: *minima* of wood are still wood.[10]

ALCHEMY

Closely associated with medieval theories of corporeal substance, combination, and mixture was the art or science of alchemy. This is one of the least studied and most poorly understood of all aspects of medieval science, and here we can do no more than offer the barest sketch of its aims, achievements, and theoretical foundations.[11]

Alchemy was both an empirical art, which sought to transmute base metals into gold (or other precious metals), and the theoretical science that explained and guided this effort. Of the reality of the transmutation of substances into one another there could not be the slightest doubt. Consider the case of a plant or tree, where water and soil nutrients are transformed into a delicate blossom or succulent fruit; or the even more extraordinary case of a lamb, which seems to have the ability to convert water and grass into wool and meat. Now this is possible, according to alchemical theory, because of the fundamental unity of all corporeal substance. Aristotle's natural philosophy offered an explanation of this unity, portraying the four elements as products of prime matter and pairs of the four elemental qualities: hot, cold, wet, dry. Alter the qualities, and you transmute the elements one into another. Alter the proportions of the elements in a *mixtum,* and you transform the *mixtum* into a different substance.

But alchemists were interested primarily in the metals. According to a widely held theory deriving from Aristotle, all metals are compounds or *mixta* of sulphur and mercury.[12] The *mixtio* of sulphur and mercury was conceived as a process of development or maturation that takes place naturally in the earth, under the influence of heat. The particular metal that emerges depends on all of the factors that go into the maturation process, including the purity and homogeneity of the sulphur and the mercury, their proportions in the *mixtum,* and the degree of heat. Now it was the aim of the alchemist to short-cut and accelerate the process of maturation—to reproduce in a short time, by artifice, what nature took perhaps a thousand years to accomplish in the womb of the earth. The goal and endpoint of the process, if perfectly carried out, was gold; imperfection or shortfall gave rise to one of the other metals.

In practical terms, the alchemist aimed to reduce a base metal to prime matter by stripping off its substantial and accidental forms; then to add forms by following the appropriate alchemical recipes, in such a way as to reconstitute the metal as one of the precious metals. Alternatively, alchemists sought to discover the recipe for the "elixir" or "philosopher's stone," a substance believed to have the power to permeate base metals and transform them into gold. In the course of these efforts, alchemists developed many chemical processes, including solution, calcination, fusion, distillation, putrefaction, fermentation, and sublimation. They also produced the required apparatus, including a great variety of furnaces for heating and melting, the alembic for distillation, and various flasks, receivers, and other vessels for the melting, mixing, pulverizing, and collecting of alchemical substances.[13]

Alchemy seems to have had Greek origins, perhaps in Hellenistic Egypt. Greek texts were subsequently translated into Arabic and gave rise to a flourishing and varied Islamic alchemical tradition. Among the outstanding Arabic alchemical writings were the corpus attributed to Geber (Jābir ibn Ḥayyān, fl. 9th-10th c.) and the *Book of the Secret of Secrets* by Muḥammad ibn Zakariyyā al-Rāzī (d. ca. 925). Beginning about the middle of the twelfth century, this body of alchemical writings was translated into Latin, initiating a vigorous Latin alchemical tradition. Belief in the truth of alchemical theory and the validity of alchemical aims was widespread but far from universal; from Avicenna onward, a strong critical tradition devel-

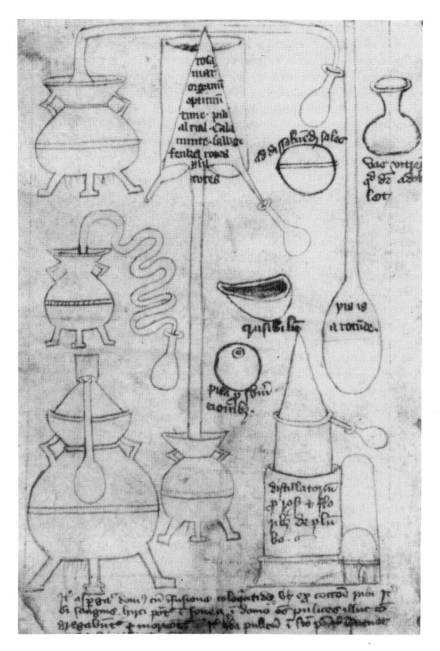

Fig. 12.1. Alchemical apparatus, including furnaces and stills. British Library, MS Sloane 3548, fol. 25r (15th c.). By permission of the British Library.

oped, and as much ink was devoted to polemics about the possibility of alchemy as to its theory and practice. In the course of its long history, alchemy became affiliated with many other technical arts (metallurgy and dye-making, for example) and systems of thought. It acquired theological, magical, and allegorical overtones and was gradually transformed into an all-embracing mystical philosophy; by the end of the Middle Ages, for example, alchemical transformation was frequently linked to the spiritual transformation of the alchemical experimenter, and it was believed by some that the elixir not only transformed base metals into gold but also conferred immortality.[14]

CHANGE AND MOTION

Historians frequently contrast the static character of the Aristotelian universe with the dynamism of the atomic philosophy. It is easy to see what they have in mind. In Aristotle's sublunar realm, natural motion ceases when the moving object reaches its natural place, and violent motion comes to an end when the external force no longer acts. If we put everything in its natural place and get rid of external movers, Aristotle's world will screech to a halt. By contrast, the world of the atomists is in a permanent state of motion—atoms moving, colliding, and forming temporary clusters in an eternal maelstrom.

However, the impression that Aristotle's cosmos is static comes from restricting our attention to one kind of change—change of place or "local motion." Look beneath the surface, not at the location of an object, but at the nature of the object, and the true dynamism of the Aristotelian cosmos becomes apparent. For Aristotle, natural things are always in a state of flux; it is part of their essential nature to be in transition from potentiality to actuality. This is no doubt most obvious in the biological realm, where growth and development are inescapable, but Aristotle's biological studies powerfully shaped his entire philosophy of nature. His definition of nature, as the inner source of change found in all natural bodies, may well have had biological origins, but it was applicable to both the organic and the inorganic realms. The central object of study in Aristotle's natural philosophy, then, was change in all of its forms and manifestations. Aristotle stated bluntly in his *Physics* (book 3) that if we are ignorant of change, we are ignorant of nature.[15] If the gross objects that fill the Aristotelian cosmos seem to prefer rest over motion, beneath the surface they are seething with change.

Aristotle and his medieval followers identified four kinds of change:

(1) generation and corruption, (2) alteration, (3) augmentation and diminution, and (4) local motion. Generation and corruption occur when individual things (that is, substances) come into existence and go out of existence. Alteration is change of quality, as when the cold object becomes warm. Augmentation and diminution refer to quantitative change—that is, change of size, as in rarefaction and condensation. And local motion is change of place—the kind of change that seventeenth-century scientists elevated to a place of centrality that it did not have within Aristotelian physics.

When we examine Aristotle's theory of motion, therefore, we are looking at one aspect of his theory of change. It was change in general that interested Aristotle and his commentators, and local motion was but one of several varieties and by no means the most fundamental. It will save us from a great deal of confusion if we keep this in mind. Features of Aristotelian and medieval theories of motion that seem strange and idiosyncratic when examined from the standpoint of modern dynamics frequently take on quite a different appearance when we judge them in the light of the questions they were meant to answer.

This brings us face-to-face with an important and difficult methodological issue. The customary way of approaching medieval theories of motion is to carry the conceptual framework of modern dynamics back to the Middle Ages and use it as a grid through which to view medieval developments. This procedure has the enormous advantage of keeping us on familiar intellectual ground; it has the disadvantage of bringing into focus only those medieval developments that resemble some piece of modern theory. The alternative is to adopt a medieval perspective—an approach that has the obvious advantage of fidelity to the system of ideas that we are endeavoring to understand, but one that may be nearly impossible to carry out in practice. The intellectual framework of medieval theories of motion is a conceptual jungle, suitable only for hardened veterans and certainly no place for day-trips from the twentieth century. Faced with a choice between viewing this jungle from the safe distance of the seventeenth or twentieth century and not viewing it at all, most historians of medieval science have understandably chosen the former alternative. My own view is that we must make some pragmatic compromises in the effort to find a middle way. In the pages that follow, we will take several short excursions into regions of the medieval jungle judged safe for tourists, in order to give some sense of the lay of the land. We will also examine certain medieval developments important for their later influence, endeavoring to describe those developments in ways that will help the reader to grasp the medieval framework out of which they grew.

THE NATURE OF MOTION

When an ancient or medieval natural philosopher turned his attention to any area of inquiry, the first thing he wanted to know was: what things (relevant to the inquiry) exist? This is a question about the entities that populate the universe. Once he had settled this question, he could move on to others, such as: What is the nature of the things that exist? What kind of existence do they have? How do they change? How do they interact? And how do we know about them? If the object of study was motion, the first task would be to figure out whether motion exists and, if so, what sort of thing it is.

Aristotle had addressed this question with enough ambiguity to give his commentators plenty to chew on. In Islam, the two great Aristotelian commentators, Avicenna and Averroes, both joined the fray. And in the West the problem was reopened by Albert the Great. We cannot probe the fine points of this extremely technical debate, but we can reveal the broad outlines by calling attention to two prominent alternatives that emerged by the end of the thirteenth century and a few of the arguments employed to adjudicate between them. According to one opinion, which came to be designated by the phrase *forma fluens* (flowing form), motion is not a thing separate or distinguishable from the moving body, but simply the moving body and its successive places. When Achilles runs a race, the existing things are Achilles and the objects that define the places successively occupied; no additional entity is present, and the word "motion" denotes not an existing *thing,* but merely the *process* by which Achilles comes to occupy successive places. This view was developed by Averroes and Albert the Great. The alternative opinion, known by the name *fluxus formae* (flow of a form), maintained that in addition to the moving body and the places it successively occupies, there is some *thing* inherent in the moving body, which we may call "motion."[16]

We can perhaps begin to perceive the rationale behind this debate by examining a pair of famous arguments, one for each of the alternatives. William of Ockham (ca. 1285–1347) defended the *forma fluens* opinion with characteristic logical rigor. In Ockham's view, "motion" is an abstract, fictional term—a noun that corresponds to no really existing entity. This was not an attempt on Ockham's part to deny that things move, but simply a declaration that *motion* is not a *thing.* The way to get clear on this, Ockham argued, is to consider a sentence such as the following: "Every motion is produced by a mover." A naive reader might suppose that the noun "motion" stands for a real thing (a substance or a quality), for nouns often serve that function. However, we can replace this sentence with another that has identical dynamic content, but different implications for the *nature* of motion: "Each thing that is moved is moved by a mover." Here the noun "motion" has disappeared, and with it the implication that motion might be a real thing. But how are we to choose between the two sentences and the alternative worlds they describe? On the basis of economy. Although the two sentences make the same dynamic claim (things move only if moved by movers), the world in which motion is not an existing thing is a more economical world, because there are fewer things in it; consequently, we should regard it as the real world unless there are convincing arguments to the contrary.[17]

An altogether different set of considerations led John Buridan (ca. 1295–ca. 1358) to defend the *fluxus formae* view. In his commentary on Aristotle's *Physics,* Buridan answered the now familiar question—whether local motion is a thing distinct from the moved object and the places it successively occupies—by reference to theological doctrine. The theological starting point of Buridan's argument was the assumption that God, in his absolute power, could have endowed the cosmos as a whole with a rotational motion. Buridan knew this by virtue of the principle that God can do anything that involves no self-contradiction; moreover, one of the articles of the condemnation of 1277 (according to Buridan's reading) explicitly affirmed God's power to accomplish the analogous feat of moving the entire cosmos in a straight line. But if we adopt the *forma fluens* view that motion is nothing more than the moving object and the places it successively occupies, a serious problem arises. Aristotle had defined place in terms of surrounding bodies. Since the cosmos is not surrounded by anything (for any container would have to be considered part of the cosmos), it seems to have no place. If the cosmos has no place, it obviously cannot change places; and if it does not change places, it cannot be said to move. But this conclusion is incompatible with the starting point of the argument—the unquestionable assumption that God is capable of giving the cosmos a rotational motion. The solution, Buridan thought, was to adopt the broader, *fluxus formae* conception of motion. If motion is not simply the moving body and its successive places, but an additional attribute of the moving body analogous to a quality, then the cosmos could possess this attribute even in the absence of place, and the difficulty would be at least partially overcome. The implication of this theory—that motion is a quality, or something that could be treated as a quality—became quite common among natural philosophers in the second half of the fourteenth century.[18]

THE MATHEMATICAL DESCRIPTION OF MOTION

Nowadays the application of mathematics to motion needs no defense. Theoretical mechanics, the parent discipline of theories of motion, is mathematical by definition, and to anybody with a grasp of modern physics the mathematical way would seem to be the only way. But perhaps it is only by hindsight and from a modern perspective that this conclusion is obvious; it would not have seemed plausible to Aristotle or many who worked within the Aristotelian tradition. We must remember that Aristotle and his medieval followers regarded motion as one of four kinds of change and that they expected the analysis of motion to imitate (in large measure) the analysis of change in general. We also need to see that there is nothing intrinsically mathematical about most cases of change. When we observe sickness yielding to health, virtue replacing vice, and warmth su-perseding cold, no numbers or geometrical magnitudes leap out at us. The generation or corruption of a substance and the alteration of a quality are not obviously mathematical processes, and it is only by heroic efforts over the centuries that scholars have found ways of placing a mathematical handle on a few kinds of change, including local motion. Let us examine the early stages of this process in the later Middle Ages.

The mathematization of nature, of course, had ancient proponents, in-cluding the Pythagoreans, Plato, and Archimedes; and early success was achieved in the sciences of astronomy, optics, and the balance (see chap. 5, above). It was inevitable that the success of these efforts would provide encouragement for those interested in mathematizing other subjects. We find primitive beginnings of the mathematical analysis of motion in Aris-totle's *Physics,* where distance and time, both quantifiable, were employed as measures of motion. Aristotle argued that the quicker of two moving objects covers a greater distance in the same time or the same distance in less time, while two objects moving with equal quickness traverse equal distances in equal times. A generation after Aristotle, the mathematician Autolycus of Pitane (fl. 300 B.C.), took a further step, defining a uniform motion as one in which equal distances are traversed in equal times. It is important to note that in these ancient discussions distance and time were taken, as the critical measures of motion, to which a numerical value might be assigned, while "quickness" or speed never achieved that status, re-maining a vague, unquantifiable conception.[19]

The first impact of this mathematical analysis in medieval Christendom can be seen in the work of Gerard of Brussels, a mathematician who may have taught at the University of Paris in the first half of the thirteenth century. For our purposes the most important characteristic of Gerard's brief *Book on Motion* is the restriction of its contents to what we now call "kinematics." To understand what this means we must briefly examine the distinction between kinematics and dynamics—a distinction that can then serve as one of the organizing principles for the remainder of our discussion of medieval theories of motion. If we wish to investigate the motion of a body, there are basically two ways of going about it. We can concentrate on the causes of the motion, giving an account of the agents or forces that produced it and correlating them, perhaps, with the amount or speed of motion produced; or we can describe the motion without any reference to causation. The former enterprise, which focuses on causation, is known as "dynamics"; the latter, limited to description (usually mathematical description), is known as "kinematics." Gerard is important, then, as a harbinger of the kinematic tradition that was to develop in the Latin West.[20]

That tradition flowered in the fourteenth century among a group of distinguished fourteenth-century logicians and mathematicians associated with Merton College, Oxford, between about 1325 and 1350. This group included Thomas Bradwardine (d. 1349), subsequently appointed archbishop of Canterbury; William Heytesbury (fl. 1335); John of Dumbleton (d. ca. 1349); and Richard Swineshead (fl. 1340–55). To begin with, members of the Merton group made explicit the distinction between kinematics and dynamics that was implicitly present in Gerard's *Book on Motion,* noting that motion can be examined from the standpoint either of cause or of effect. The Merton scholars proceeded to develop a conceptual framework and a technical vocabulary for dealing with motion kinematically. Included in this conceptual framework and its vocabulary were the ideas of velocity and instantaneous velocity, both treated as scientific concepts to which magnitude could be assigned.[21] The Mertonians distinguished between uniform motion (motion at constant velocity) and nonuniform (or accelerated) motion. They also devised a precise definition of uniformly accelerated motion identical to our own: a motion is uniformly accelerated if its velocity is increased by equal increments in equal units of time. Finally, the Merton scholars developed a variety of kinematic theorems, several of which we will examine below.[22]

Before we do that, we must consider the philosophical underpinnings of this kinematic achievement. The emergence of velocity as a new measure of motion, to go along with the ancient measures (distance and time), is a development that needs to be explained. Velocity, after all, is quite an abstract conception, which did not force itself on the observer of moving bodies but had to be invented by natural philosophers and imposed on the phenomena. How did this come about? The answer is found in the philosophical analysis of qualities and their strength or intensity.

The fundamental idea was that qualities or forms can exist in various degrees or intensities: there is not just a single degree of warmth or cold, but a range of intensities or degrees running from very cold to very hot. Moreover, it was acknowledged that forms or qualities can vary within this range; that is, they can be strengthened and weakened, or, to employ the technical medieval terminology, undergo intensification and remission.[23] Now when this general discussion of qualities and their intensification and remission was transferred to the particular case of local motion (motion being conceived as a quality or something closely analogous to a quality), the idea of velocity quickly emerged. The intensity of the quality of motion—that which measured its strength or degree—could be none other than swiftness or (to employ the technical medieval term) velocity. Intensification and remission of the quality of motion must then refer to variations in velocity.

Thinking about qualities, their intensity, and their intensification also gave rise to the distinction between the intensity of a quality and its extension or quantity. An example will help us to understand this distinction. It was obvious enough in the case of heat that one object could be hotter than another; here we have the idea of intensity or degree (more or less the same as our concept of temperature).[24] But it was apparent that there was also something more—namely, distribution of the quality of heat in a subject (a hot object). Suppose that we deposit heat of the same degree or intensity in two bodies, identical except that one has twice the volume of the other; it seems that the larger must contain twice the heat of the smaller. The *intensity* of heat does not vary from one body to the other, but the larger has twice the *quantity* of heat. A consideration of weight yields a comparable distinction between degree or intensity of weight (our density or specific gravity) and the distribution of weight in a body (total quantity of weight). It was assumed that every other quality (including motion) could be looked at in the same way; and thus arose the general distinction between the intensity of a quality and the quantity of that quality.[25]

News of the Merton College achievements in the analysis of qualities was transmitted quickly to other European intellectual centers. In the process, the analysis was enriched and clarified by the addition of a sytem of geometrical representation. The original analysis of qualities at Merton College was carried out verbally, in much the same way as we have been analyzing it. However, the advantages of geometrical analysis were quickly recognized, and fairly elaborate systems of geometrical representation were eventually worked out. One of the first to develop such a system was Giovanni di Casali, a Franciscan from Bologna (who had also spent time in Cambridge), writing about 1351; a far more elaborate geometrical analysis was formulated by Nicole Oresme at the University of Paris later in the same decade. An examination of Oresme's scheme may prove as illuminating for us as it no doubt did for his medieval readers.

The first step was to represent the intensity of a quality by means of a line segment—a relatively easy step for medieval scholars brought up on Aristotle (who employed lines to represent time) and Euclid (who used lines to represent numerical magnitudes). If line segment AB (fig. 12.2) represents a given intensity of some quality, then line segment AC represents twice that intensity. This is fine, but it has not yet gotten us very far. The critical next step was to employ this line to represent the intensity of the quality at any point of the subject. Take a rod AE (fig. 12.3), heated differentially, so that one end is hotter than the other. At point A and at every other point on the rod erect a vertical line representing the intensity of heat at that point. If the temperature increases uniformly from A to E, then the figure will reveal a uniform lengthening of the vertical lines. Now Oresme made the system a good deal more abstract by substituting a horizontal line for the drawing of the rod. This has the effect of creating a generalized system of representation (see fig. 12.4) in which the horizontal

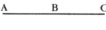

A B C

Fig. 12.2. The use of a line segment to represent the intensity of a quality.

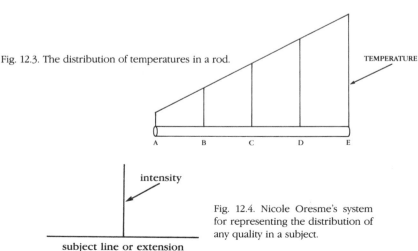

Fig. 12.3. The distribution of temperatures in a rod.

TEMPERATURE

A B C D E

intensity

Fig. 12.4. Nicole Oresme's system for representing the distribution of any quality in a subject.

subject line or extension

line (called the "subject line" or the "extension") represents the subject, whatever it might be, while vertical lines represent the intensity of any quality we choose at the points of the subject where they are erected.

What Oresme has produced is a form of geometrical representation—a forerunner of modern graphing techniques—in which the shape of the figure (as in fig. 12.3) informs us about variations in the intensity of a quality over its subject. But how do we make the transition from qualities in general to motion in particular? One way is to consider a body, the different parts of which move with different velocities; a rod held by a pin through one end and rotated about that pin would be a good example. In such a case, we can draw the rod horizontally and erect a perpendicular at any point, indicating the velocity of that point. The result will be a distribution of velocities in a subject, as in figure 12.5.

But there is another case, more difficult because it requires more abstract treatment. Suppose we have a body that moves as a unit, all of its parts having the same velocity, but a velocity that varies over time. The way to understand this, Oresme explained, is to see that here the subject line is not the extension of a corporeal object, as in the examples above, but the duration of a local motion. Time becomes the subject. This gives us a primitive coordinate system in which velocity can be plotted as a function of time (see fig. 12.6). Oresme proceeded to discuss various configurations of velocity with respect to time. Uniform velocity will be represented by a figure in which all the vertical lines are of equal length—that is, a rectangle. Nonuniform velocity requires verticals of variable length. Within this category of nonuniform velocity, we have uniformly nonuniform ve-

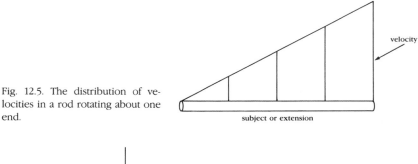

velocity

Fig. 12.5. The distribution of velocities in a rod rotating about one end.

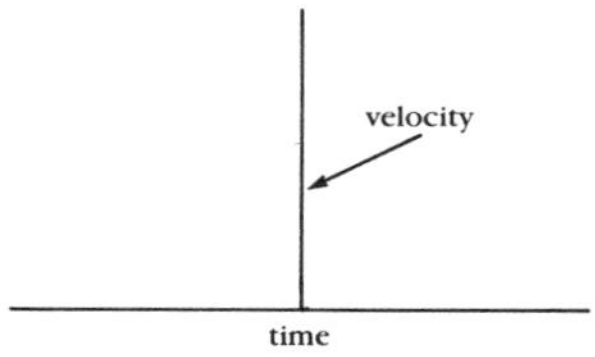

subject or extension

velocity

time

Fig. 12.6. Velocity as a function of time.

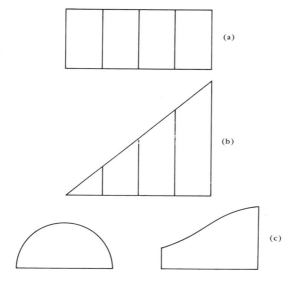

Fig. 12.7. The representation of various motions.
(a) Uniform velocity.
(b) Uniformly nonuniform velocity (uniformly accelerated motion).
(c) Nonuniformly nonuniform velocity.

locity (uniformly accelerated motion), represented by a triangle, and non-uniformly nonuniform velocity, represented by a variety of other figures (see fig. 12.7). Finally, how did Oresme deal with that other feature of qualities noted above—their total quantity? He identified the total quantity of motion with the distance traversed; he argued, moreover, that in the velocity-time diagram this must be represented by the area of the figure.

Oresme has thus developed a very clever geometrical system for the representation of motion. Did he and those who followed him merely sit and admire it, or could they do something with it? The fact is they succeeded in developing kinematic theorems that revealed some notable mathematical characteristics of uniform or uniformly accelerated motion. The most important case was the latter, represented in figure 12.7(b). This case was of special interest in the fourteenth century, not because it was identified with any particular motion in the real world but because it offered a substantial mathematical challenge. Let us examine two important theorems applicable to uniformly accelerated motion.

The first had already been stated, without geometrical proof or illustration, by the Merton scholars; it is now known as the "Merton rule" or the "mean-speed theorem." This theorem seeks to find a measure for uniformly accelerated motion by comparing it with uniform motion. The theorem claims that a body moving with a uniformly accelerated motion covers the same distance in a given time as if it were to move for the same duration with a uniform speed equal to its mean (or average) speed. Expressed in numerical terms, the claim is that a body accelerating uniformly from a velocity of 10 to a velocity of 30 traverses the same distance as a body moving uniformly for that same period of time with a velocity of 20. Now Oresme provided a simple but elegant geometrical proof of this theorem. The uniformly accelerated motion can be represented by triangle ACG (fig. 12.8) and its mean speed by line BE. The uniform motion that is to be compared with this must therefore be represented by rectangle ACDF (the altitude of which is BE, the mean speed of the uniformly accelerated motion). The Merton rule claims simply that the distance traversed by the one motion is equal to the distance traversed by the other. Since, in Oresme's diagrams, distance traversed is measured by the area of the figure, we can prove the theorem by showing that the area of triangle ACG equals the area of rectangle ACDF. A glance at the two figures will reveal that this is so.[26]

The second theorem, like the first, aimed to elucidate the mathematical properties of uniformly accelerated motion by means of a comparison involving distances traversed. In this case, the distance covered in the first half of a uniformly accelerated motion was compared to the distance covered in the second half of the same motion; the claim was that the latter is three times the former. To prove this theorem geometrically, we need merely show that the area of quadrangle BCGE (fig. 12.8), which represents the distance covered in the second half of the time, BC, is three times the area of triangle ABE, representing the distance traversed in the first half of the time, AB. Once again, inspection will establish that this is the case.[27]

Finally, two general points need to be made. First, we must remind ourselves that medieval kinematics was a totally abstract endeavor—much like modern mathematics. It was claimed, for example, that *if* a uniformly ac-

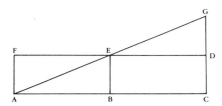

Fig. 12.8. Nicole Oresme's geometrical proof of the Merton rule.

celerated motion were to exist, *then* the Merton rule would apply to it. Never did a medieval scholar identify an instance of such motion in the real world. Is there a satisfactory explanation of such seemingly odd behavior? Yes, there is. Given the technology available in the Middle Ages (particularly for the measurement of time), demonstrating that a particular motion was uniformly accelerated would have been a considerable feat; even in the twentieth century, imagine the difficulty of producing or identifying uniformly accelerated motion without the kind of equipment available in a physics laboratory. But perhaps more importantly, the medieval scholars who developed this kinematic analysis were mathematicians and logicians; and no more than modern mathematicians and logicians would they have thought of moving their place of labor from the study to the workshop.

Second, out of this purely intellectual labor came a new conceptual framework for kinematics and a variety of theorems (the Merton rule, for example) that figured prominently in the kinematics developed in the seventeenth century by Galileo—through whom they entered the mainstream of modern mechanics.[28]

THE DYNAMICS OF LOCAL MOTION

Having dealt at length with medieval kinematics—the effort to describe motion mathematically—I must conclude this discussion of medieval mechanics with a brief account of contributions to the *causal* analysis of motion. The starting point of all dynamical thought in the Middle Ages was the Aristotelian principle that moved things are always moved by a mover. We must first get clear on what this principle was taken to mean in the Middle Ages. We will then look at attempts to identify the mover in several particularly difficult cases of motion. And finally we will examine attempts to quantify the relationship between the force or power of a mover and the resulting velocity of the moved body.

Aristotle, the reader will recall, divided motion into two categories: natural and forced. A natural motion, by which an object moves toward its natural place, apparently arises from an internal principle, the nature of the body. A motion in any other direction must be a forced motion, produced by the application of an external force in continuous contact with the moved body. This seems clear enough in broad outline, but problems arose when medieval scholars attempted to bring precision to the identification of the mover in natural motion and in one particularly troublesome case of forced motion.

In his *Physics,* where he gave an account of the mover for natural motion, Aristotle vacillated, suggesting first that natural motion may result
from an internal cause, the nature of the body, but arguing later that the
nature of the body cannot be the whole story and that the participation of
an external mover is also required. Aristotle's ambivalence posed an obvious problem for his medieval followers, who felt compelled to inquire
whether or not it is sufficient to affirm that the body is moved by its own
nature. Avicenna and Averroes considered this explanation unacceptable
on the grounds that it did not distinguish sufficiently between that which is
moved (the body) and that which moves it (the nature of the body). They
discovered what seemed to them an adequate alternative in the form-
matter distinction, proposing that the form of the body is the mover, while
its matter is the thing moved. In the West, Thomas Aquinas repudiated this
solution, reminding his readers that matter and form are inseparable and
cannot be treated as distinct things. Aquinas argued instead (reviving one
of Aristotle's proposals) that the mover in the case of natural motion is
whatever generated the body outside its natural place to begin with; thereafter the body requires no mover but simply does what comes naturally.
The debate over this issue continued through the later Middle Ages, with
no clear victor.[29]

The particular case of forced motion that proved troublesome was that
of projectiles; the problem was to explain their continued motion after
they lose contact with the original projector. Aristotle had assigned causation to the medium, arguing that the projector simultaneously projects the
projectile and endows the surrounding medium with the power to
produce motion; this power is transmitted from part to part in such a way
that the projectile is always surrounded by a portion of the medium capable of moving it. It was clear from this account that an external force,
continuously in contact with the projectile, is required.

The first major opposition to Aristotle's explanation came in the commentary on Aristotle's *Physics* by the sixth-century Alexandrian philosopher John Philoponus (d. after 575), to whom it seemed that the medium
serves as resistance rather than mover and who doubted that it could serve
both functions simultaneously. As a Neoplatonist and a dedicated anti-
Aristotelian, Philoponus launched a broad attack on Aristotelian natural
philosophy, including the notion that forced motions require external
movers. He proposed, rather, that all motions, natural and forced alike, are
the result of internal movers. Therefore, when a projectile is hurled, the
projector impresses on the projectile an "incorporeal motive force," and
this internal force is responsible for its motion.[30]

Although Philoponus's impressed motive force had radically anti-Aristotelian origins, it was eventually absorbed into the medieval Aristotelian tradition. Philoponus's commentary on Aristotle's *Physics* had an influential career in Arabic translation and seems to have had an indirect impact on medieval Latin thought, although the details of transmission remain to be fully traced.[31] In the thirteenth century, theories bearing a close resemblance to that of Philoponus were discussed and rejected by Roger Bacon and Thomas Aquinas. In the fourteenth century, the theory of impressed force was defended, first by the Franciscan theologian Franciscus de Marchia (fl. 1320), subsequently by John Buridan (ca. 1295–ca. 1358) and others. Let us examine Buridan's version of the theory, often considered its most advanced form.

Buridan employed a new term, "impetus," to denote this impressed force—terminology that remained standard down to the time of Galileo. Buridan described impetus as a quality whose nature it is to move the body in which it is impressed, and took pains to distinguish this quality from the motion it produces: "Impetus is a thing of permanent nature distinct from the local motion in which the projectile is moved. . . . And it is probable that impetus is a quality naturally present and predisposed for moving a body in which it is impressed." In defense of the impetus theory, Buridan pointed to the analogous case of a magnet, which is able to impress in iron a quality capable of moving that iron toward the magnet. Like any quality, impetus is corrupted by the presence of opposition or resistance, but otherwise retains its original strength. Buridan took a first step toward quantifying impetus by declaring its strength to be measured by the velocity and the quantity of matter of the body in which it inheres. Finally, Buridan extended the explanatory range of the impetus theory beyond simple projectile motion, arguing that motion in the heavens might plausibly be explained by God's imposition of an impetus on the celestial spheres at the moment of creation; because the heavens offer no resistance, this impetus would not be corrupted, and the celestial spheres would be moved (as observation reveals they are) with an eternally unchanging motion. And he explained the acceleration of a falling body by the assumption that as the body falls its gravity continually generates additional impetus in the body; as the impetus increases, it gives rise to increasing velocity.[32]

The theory of impetus became the dominant explanation of projectile motion until the seventeenth century, when a new theory of motion, which denied that force (either internal or external) is required for the continuation of unresisted motion, gradually won acceptance. There have been many attempts to view the theory of impetus as an important step in the direction of modern dynamics; for example, attention has often been called to the quantitative resemblance between Buridan's impetus (velocity × quantity of matter) and the modern concept of momentum (velocity × mass). No doubt there are connections, but we must note that Buridan's impetus was the *cause* of the continuation of projectile motion, whereas our momentum is the measure of a motion that requires no cause for its continuation so long as no resistance is encountered. In short, Buridan was still working within a conceptual framework that was fundamentally Aristotelian; and this meant that he was a world (or worldview) away from those natural philosophers in the seventeenth century who formulated a new mechanics based on a new conception of motion and inertia.

THE QUANTIFICATION OF DYNAMICS

One question remains: is it possible to quantify the dynamic relations between force, resistance, and velocity? Many medieval scholars believed that it was. The problem went back as far as Aristotle, who had made a brief and preliminary stab at quantitative analysis, defending a variety of propositions such as the following: the greater the weight (of a falling body), the swifter its motion; the greater the resistance (encountered by a falling body), the slower its motion; and the smaller a moved object, the more rapidly a given force will move it. Historians have managed, by concerted effort, to extract a mathematical relationship from these claims, attributing to Aristotle the view that velocity is proportional to the force and inversely proportional to the resistance. Expressed in modern terms, this becomes:

$$v \propto F/R$$

This relationship is unquestionably useful as an economical means of conveying substantial pieces of Aristotelian dynamics, which explains why it continues to be repeated. But it is also potentially misleading and must be employed with great caution. Aristotle would certainly not have agreed that velocity is proportional to force and inversely proportional to resistance for all values of F and R, as the mathematical form of the relationship might suggest. Moreover, he had no clear conception of velocity as a technical, quantifiable philosophical or scientific term.

Aristotle's dynamic ideas had clear implications for the possibility of motion in a void. If it is true that the swiftness of a falling body is a function of the resistance it encounters, then in a vacuum, where there is no resistance at all, there would be nothing to retard the motion of the body; in that case it would move with infinite swiftness. Since infinitely swift motion is absurd, Aristotle argued,[33] it is plain that a vacuum is impossible. Now it was this use of his theory of motion to prove the impossibility of a void that provoked a broad attack from the Alexandrian Neoplatonist John Philoponus. Philoponus appealed to everyday observation to refute the basic Aristotelian claim that the time of descent for a body falling through a medium is inversely proportional to the weight:

> But this [view of Aristotle] is completely erroneous, and our view may be corroborated by actual observation more effectively than by any sort of verbal argument. For if you let fall from the same height two weights, one many times as heavy as the other, you will see that the ratio of the times required for the motion does not depend [solely] on the ratio of the weights, but that the difference in time is very small. And so, if the difference in the weights is not considerable, that is, if one is, let us say, double the other, there will be no difference, or else an imperceptible difference, in time. . . .[34]

If Aristotle's theory is false, what is the truth? Philoponus urged his reader to think about falling bodies in the following way. The efficient cause of the descent of a falling body is weight. In a void, where there is no resistance, the sole determinant of motion will be the weight of the body; consequently, heavier bodies will traverse a given distance more rapidly (that is, in less time) than will lighter bodies; and, of course, none of them will move with infinite speed, as Aristotle had supposed. (Philoponus did not state that the rapidity of motion in the void would be directly proportional to the weight, but perhaps he expected this to be assumed.) Now in a medium, the resistance of the medium slows the motion by a certain amount, and the net effect of this slowing is to close the gap in swiftness between heavier and lighter bodies, leading to the observed results described in the quotation above.

Philoponus's point of view was developed and defended in Islam by Avempace (Ibn Bājja, d. 1138). Avempace, in turn, was attacked by Averroes; and through Averroes the controversy was transmitted to the West, where it was resumed by the fourteenth-century Mertonian Thomas Bradwardine. But with Bradwardine there was a difference. Whereas all of his predecessors had been interested primarily in the nature and causes of motion, Bradwardine was determined to view the problem in mathematical terms. This meant that he had to begin by giving a mathematical formulation of each of the alternatives—three of which he was able to identify. Bradwardine expressed these alternatives in words rather than mathematical symbols, but the following formulas adequately capture his intent.

First theory (no doubt meant to represent the opinion of Philoponus and Avempace):

$$V \propto F - R$$

Second theory (suggested by a passage in Averroes);

$$V \propto \frac{F - R}{R}$$

Third theory (representing the traditional interpretation of Aristotle):

$$V \propto \frac{F}{R}$$

Bradwardine was able to refute each of these theories by calling attention to its absurd or unacceptable consequences. The first theory, for example, fails on the grounds that it contradicts Aristotle's claim that doubling both the force and the resistance will leave the velocity unchanged. And the third theory fails on the grounds that it does not predict zero velocity when the resistance is equal to or larger than the force.

In place of these discredited theories Bradwardine proposed an alternative "law of dynamics." There is no easy way of stating Bradwardine's "law." To remain close to Bradwardine's own account would lead us more deeply into the medieval theory of the compounding of ratios than we can afford to go. Perhaps the simplest *modern* way of expressing the mathematical relationship that Bradwardine had in mind is to state that according to his "law" velocity increases arithmetically as the ratio F/R increases geometrically. That is, in order to double the velocity, we must square the ratio F/R; to triple the velocity, we must cube the ratio F/R; and so on. Or consider the following numerical example:

Apply first a force (F_1) of 4, then a force (F_2) of 16, to a body that offers a resistance (R) of 2. First calculate the F/R ratios:

$$\frac{F_1}{R} = \frac{4}{2} = 2$$

$$\frac{F_2}{R} = \frac{16}{2} = 8$$

What will be the ratio of the velocities produced? Since 8 is
the *cube* of 2, the velocity produced by the force of 16 will be
three times the velocity produced by the force of 4.[35]

Three points need to be made about Bradwardine's achievement. First,
we make Bradwardine's "law" more complicated than it really was by
expressing it, as we have done above, in modern terms. We need to understand that in the medieval mathematical tradition within which Bradwardine was working, the way to talk about the compounding or increase of
ratios was through the language of addition. Therefore, the operation that
we refer to as the multiplication of two ratios would have been, in Bradwardine's terminology, the addition of one ratio to the other; and what we
refer to as the squaring of the ratio F/R would have been, in his terms, the
doubling of F/R. Consequently, instead of relating geometrical increases in
the ratio F/R to arithmetical increases in the velocity (as we did above),
Bradwardine would merely have stated that to "double" the velocity you
must "double" the ratio of F to R. In short, Bradwardine proposed not
some esoteric mathematical relationship, but (as one historian has recently put it) "the least complicated expression available to him."[36]

Second, Bradwardine's formulation of a "law of dynamics" proved influential. Its implications were brilliantly worked out by Richard Swineshead
and Nicole Oresme in the fourteenth century, and the law continued to be
discussed as late as the sixteenth century.[37] Third, whatever our precise
evaluation of Bradwardine's achievement, we must acknowledge that his
enterprise was unmistakably a mathematical one. It is true that his refutation of the alternatives included appeal to everyday experience, but it is
clear that his primary aim was to satisfy the criteria of mathematical coherence. In short, Bradwardine neither discovered nor defended his "law" by
experimental means; nor is it clear what benefits an experimental approach, if he had been inclined to adopt it, could have rendered. The task
undertaken by medieval scholars was the formulation of a conceptual and
a mathematical framework suitable for analyzing problems of motion.
Surely this was the first order of business, and medieval scholars carried it
out brilliantly. The further task of interrogating nature, in order to find out
whether it would accept the conceptual framework thus formulated, was
left to future generations.

THE SCIENCE OF OPTICS

I conclude this analysis of sublunary physics with a brief account of the
science of optics (or *perspectiva,* as it came to be called in Latin

Christendom). The decision to treat optics in this chapter is somewhat arbitrary, for optics was a discipline of exceedingly broad reach, affiliated in one way or another with many subjects, including mathematics, physics, cosmology, theology, psychology, epistemology, biology, and medicine.[38] But it will fit here well enough.

The works of Aristotle, Euclid, and Ptolemy, which had dominated Greek thought about light and vision, were all translated into Arabic and gave rise to a substantial Islamic tradition of optical studies. The various Greek approaches to optical phenomena were taken seriously, defended, and extended. But the major achievement of Islamic optics was the successful integration of these separate and incompatible Greek optical traditions into a single comprehensive theory.

Most Greek optical thought was narrowly focused, guided by one or another relatively narrow set of criteria. Aristotle, for example, was concerned almost exclusively with the physical nature of light and the physical mechanism of contact in visual perception between the observed object and the observing eye; neither mathematical analysis nor anatomical or physiological issues occupied a significant place in his theory. Specifically, he argued that the visible object produces an alteration of the transparent medium; the medium instantaneously transmits this alteration to the observer's eye, with which it is in contact, to produce sensation. This is an "intromission" theory—so called because the agent responsible for vision passes from the observed object to the eye. The Greek atomists, who also demanded a physical account of vision, identified a different causal agent—a thin "skin" or "simulacrum" of atoms "peeled" from the outer surface of the object, rather than an alteration of the transparent medium—but joined Aristotle in the belief that a causal theory must be an intromission theory.

Euclid's concerns, by contrast, were almost exclusively mathematical; the aim of his *Optics* was to develop a geometrical theory of the perception of space, based on the visual cone, with only minimal concern for the nonmathematical aspects of light and vision. According to his theory of vision, radiation emanates from the eye in the form of a cone; perception occurs when the rays within the cone are intercepted by an opaque object. The perceived size, shape, and location of the object are determined by the pattern and location of intercepted rays. Because it holds that radiation issues from the eye, we refer to this as an "extramission" theory.

Finally, physicians such as Herophilus and Galen were preoccupied with the anatomy of the eye and the physiology of sight. Galen revealed a firm grasp of the mathematical and causal issues, but made his principal contribution to visual theory through an analysis of the anatomy of the eye and the participation of the various organs forming the visual pathway in the process of vision.

The Islamic contribution, as I have indicated, was to produce a merger of these disparate Greek theories. The primary architect of the merger was the brilliant mathematician and natural philosopher Alhazen (Ibn al-Haytham, ca. 965–ca. 1040)—although Ptolemy, the last great optical writer of antiquity, had pointed the way. Our analysis of Alhazen's achievement will be simplified if, for the moment, we lay aside the anatomical and physiological concerns of the medical tradition and restrict our attention to the mathematical and physical aspects of vision.

To begin, it is important to notice that ancient theories of vision with mathematical aims (those of Euclid and Ptolemy) invariably assumed the extramission of light from the eye, while theories with physical plausibility as their primary concern (if we may judge from the works of Aristotle and the atomists) tended to assume the intromission of light into the eye.[39] If there was any doubt about this correlation, it would have been removed for the attentive reader of Aristotle's works by the discovery that on the one occasion when he attempted a mathematical analysis of optical phenomena (in his theory of the rainbow), Aristotle employed an extramission theory of vision.[40]

What Alhazen achieved, then, was twofold. First of all, he demolished the extramission theory with a compelling set of arguments. For example, he called attention to the ability of bright objects to injure the eye (noting that it is the nature of injury to be inflicted from without) and inquired how it would be possible, when we observe the heavens, for the eye to be the source of a material emanation that fills all of the space up to the fixed stars. Having refuted the extramission theory, he proceeded to formulate and defend a new version of the intromission theory, which appropriated the visual cone of the extramissionists. Along with the visual cone came the mathematical power of the extramission theory, which was thereby coupled for the first time with the satisfying physical explanations provided by the intromission theory. This might seem a simple step, but consider some of the obstacles.[41]

In the first place, ancient writers offered no theory of radiation adequate for Alhazen's purposes. In ancient sources, radiation was generally presented as a holistic process in which the visible object radiates as a coherent unity. Radiation was not thought to proceed independently from individual points (as in modern optical theory); rather, the object as a whole was held to send a coherent image or power through the medium

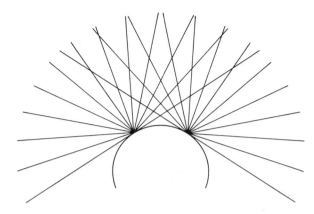

Fig. 12.9. Incoherent radiation from two points of a luminous body.

to the eye (as in the atomists' theory of films or simulacra).[42] There was no way of imposing a visual cone on such a conception of the process of radiation. However, a new conception of radiation was formulated by the philosopher al-Kindī (d. ca. 866) and adopted (or independently invented) by Alhazen. Al-Kindī and Alhazen conceived radiation as an incoherent process, in which individual points or small parts of the luminous body radiate not as a coherent group but each one independently of the others and in all directions (see fig. 12.9).

This was an important innovation, but it raised new problems for anybody hoping to defend an intromission theory of vision. Can an incoherent process of radiation from visible objects account for the coherent visual perception that all normal-sighted people experience? If every point of the visible object radiates in all directions, then surely every point in the eye will receive radiation from every point in the visual field (see fig. 12.10). This should lead to total confusion rather than clear perception. What we require to explain our perceptual experience is a one-to-one correspondence, each point of the principal sensitive humor or organ of the eye (identified by Galen and his followers as the crystalline humor or lens) responding to radiation from one point in the visual field; and if possible, the pattern of recipient points in the eye should exactly replicate the pattern of radiating points in the visual field, thus explaining the correspondence between the world out there and the world as we see it.

Alhazen's solution to this problem was to argue that although every point in the visual field does indeed send radiation to every point in the

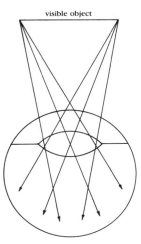

Fig. 12.10. Rays issuing from the endpoints of the visible object mixing within the eye. For the sake of simplicity, the bending of rays by refraction at the various interfaces has not been shown.

eye, not all of this radiation is capable of making itself felt. Only one ray from each point in the visual field, he pointed out, falls on the eye perpendicularly (see fig. 12.11); all others fall obliquely and are refracted. As a result of refraction these other rays are weakened to the point where they play only an incidental role in the process of sight. The primary sensitive organ of the eye, the crystalline humor or lens, pays attention to the perpendicular rays, and these happen to form a visual cone, with the visual field as base and the center of the eye as vertex. Alhazen has thus achieved his aim: by successfully importing the visual cone of the extramissionists into the intromission theory, he has combined the advantages of extramissionism and intromissionism; he has united the mathematical and physical approaches to vision in a single theory. It is important to add, though we do not have space to delve into it, that he also incorporated the anatomical and physiological ideas of the Galenic tradition (fig. 12.11 conveys his basic conception of the anatomy of the eye), thus producing a unified visual theory, answering to all three kinds of criteria.

Visual theory may have been the centerpiece of Alhazen's optics, but his interests extended to the whole range of optical phenomena. He analyzed the nature of the radiation associated with light and color, distinguishing between naturally luminous objects and those that shine with derived or secondary light. He considered the physics of reflection and refraction. He continued and extended the mathematical analysis of radiating light and color, dealing in sophisticated fashion with problems of image-formation

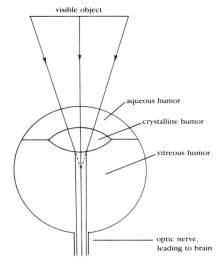

visible object

aqueous humor

crystalline humor

vitreous humor

optic nerve,
leading to brain

Fig. 12.11. The visual cone and the eye in Alhazen's intromission theory of vision. Rays from the object that fall obliquely on the eye (and undergo refraction) are not shown, since they enter only incidentally into the process of vision.

by reflection and refraction. And he offered a serious and influential discussion of the psychology of visual perception.

Alhazen's *Optics* exercised a powerful influence on Western optics after its translation into Latin late in the twelfth century or early in the thirteenth. But it was not the only influential source. Plato's *Timaeus* had long been available; not only did it address vision, but it had also given rise to a substantial tradition of Neoplatonic optical thought. The optical works of Euclid, Ptolemy, and al-Kindī, translated in the second half of the twelfth century, revealed the promise of a mathematical approach to optics at a time when Alhazen's *Optics* was not yet available. Writings by Aristotle, Avicenna, and Averroes left the definite impression that the real problems were physical and psychological, rather than mathematical. And a variety of sources, including a small work by Ḥunayn ibn Isḥāq, conveyed the anatomical and physiological content of the Galenic tradition. As in so many other areas, then, Western scholars found themselves suddenly enriched with a splendid new body of knowledge—but one that was complex rather than simple, containing conflicting ideas and tendencies. The problem confronting Western scholars was to reconcile and harmonize, reworking this perplexing intellectual heritage into a coherent and unified philosophy of nature.[43]

Among the first to undertake such efforts were two distinguished Oxford scholars: Robert Grosseteste in the 1220s and 1230s and Roger Bacon in the 1260s. Working early in the century, Grosseteste (ca. 1168–1253) was handicapped by an imperfect knowledge of the optical sources listed above, and his optical writings were valuable primarily as inspiration. It was Roger Bacon (ca. 1220–ca. 1292), inspired by Grosseteste but with the advantage of a full mastery of the optical literature of Greek antiquity and medieval Islam, who determined the future course of the discipline.

Following the broad outlines of optical theory as it was developed by Alhazen, Bacon adopted Alhazen's intromission theory of vision in almost all of its details. He was extraordinarily impressed by Alhazen's successful mathematical analysis of light and vision, and in his own works he effectively communicated the promise of the mathematical approach to future generations. But Bacon (like many of his generation) was convinced that all of the ancient and Islamic authorities were in fundamental agreement, and he was therefore committed to showing that all (or almost all) who had written about light and vision were of one mind. This meant that he would have to reconcile the optical teachings of such a diverse group as Aristotle, Euclid, Alhazen, and the Neoplatonists. Two examples will serve to illustrate how he managed this feat.[44]

Regarding the direction of radiation (from or toward the eye—the point of contest between the extramissionists and the intromissionists), Bacon agreed with Alhazen and Aristotle that vision occurs only through intromitted rays. What then of the extramitted rays advocated by Plato, Euclid, and Ptolemy? Clearly they could not be responsible for vision, but they could still exist and play an auxiliary role in the visual process—that of preparing the medium to receive the rays emanating from the visible object and of ennobling the incoming rays to the point where they could act on the eye. Regarding the nature of the radiation, Bacon accepted the Neoplatonic conception of the universe as a vast network of forces, in which every object acts on objects in its vicinity through the radiation of a force or likeness of itself. Moreover, he conceived this universal force to be the instrument of all causation and, on this basis, developed (what proved to be) an influential philosophy of nature. As for light and color, Bacon argued that they (and any other visible agents discussed by the optical authors) were merely particular manifestations of this universal force.[45]

Bacon's was not the only voice addressing optical problems in the second half of the thirteenth century, but it was in large part through his influence and that of two younger contemporaries—another English Franciscan, John Pecham (d. 1292), and a Polish scholar named Witelo

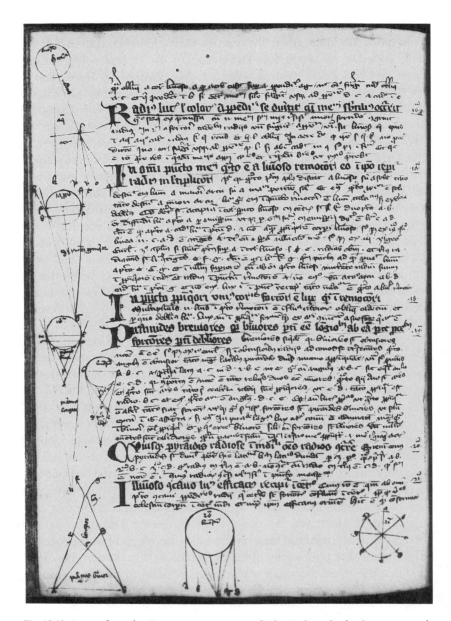

Fig. 12.12. A page from the *Perspectiva communis* of John Pecham, by far the most popular optical text in the medieval universities. Kues, Bibliothek des St. Nikolaus-Hospitals, MS 212, fol. 240v (early 15th c.)—a manuscript that once belonged to Nicholas of Cusa. Refraction of light is represented in the upper left-hand corner, various patterns of radiation in the remaining diagrams.

(d. after 1281), connected with the papal court—that Alhazen's optical theories, including his combined physical-mathematical-physiological approach, became dominant in Western thought. When theories of light and vision appeared in fourteenth-century natural philosophy (as they frequently did, especially in epistemological discussions), they were almost always derived from the tradition of Alhazen and Bacon. When Johannes Kepler began to think about visual theory in the year 1600 (efforts that led eventually to his invention of the theory of the retinal image), he took up the problem where Bacon and Pecham and Witelo had left it.[46]

THIRTEEN

Medieval Medicine and Natural History

THE MEDICAL TRADITION OF THE EARLY MIDDLE AGES

Medieval medicine was an outgrowth and continuation of the ancient medical tradition (examined above, chap. 6). Medieval medical practitioners were heirs to Greek and Roman theories of health and disease, diagnostic techniques, and therapeutic procedures. But access to this ancient legacy was partial and sometimes precarious, and the portions of it that were available in medieval Islam and Christendom had to be adapted to new cultural circumstances that profoundly shaped their development and use.[1]

It is difficult to get a clear picture of early medieval medicine in the West.[2] The social and economic chaos that accompanied the disintegration of the Roman Empire probably did not seriously affect the *craft* side of healing—the treatment of wounds and common ailments, midwifery, bone-setting, the preparation and distribution of familiar remedies, and the like. Especially in rural areas and on the domestic front, people skilled in the healing arts continued to practice their craft more or less as local healers had always done. What suffered from Roman collapse was the learned, and especially the *theoretical or philosophical,* component of medicine. The decline of schools and the gradual disappearance of facility in Greek increasingly deprived the West of the learned aspects of the Greek medical tradition, so that the number of medical practitioners with a command of the learned traditions of ancient medicine declined precipitously.

This is not to suggest that the West was totally cut off from Greek medical knowledge. Medicine received a certain amount of coverage in early Latin encyclopedias—those of Celsus, Pliny, and Isidore of Seville, for example.[3] Moreover, by the middle of the sixth century, a small collection of Greek medical writings was available in Latin translation. But Greek medical literature covered a broad spectrum of medical interests, from theoretical to practical, and the translated works tended toward the practical. Included were several works by Galen and Hippocrates, a collection of excerpts from Greek medical sources assembled by Oribasius (fl. 4th c. A.D.), a handbook for midwives by Soranus (1st c. A.D.), and the great pharmacopeia (*De materia medica*) of Dioscorides (fl. 50–70 A.D.).

The practical, therapeutic orientation of early medieval medicine is nicely illustrated by Dioscorides' *Materia medica* and the pharmaceutical tradition it spawned. Containing descriptions of some nine hundred plant, animal, and mineral products alleged to have therapeutic value, Dioscorides' work was one of the monumental achievements of Hellenistic medicine. Translated into Latin in the sixth century, it enjoyed only a limited circulation during the early Middle Ages, perhaps because it was too comprehensive to be useful—containing, as it did, descriptions of many substances unavailable to early medieval Europeans. Far more popular was a shorter, illustrated herbal entitled *Ex herbis femininis,* based on Dioscorides but containing descriptions of only seventy-one medicinal plant substances, all available in Europe. Many additional collections of medical recipes were produced in the course of the early Middle Ages.[4]

Who were the medical practitioners able to make use of these texts? In Italy, the Roman pattern of secular, nonreligious medicine persisted, although it no doubt experienced quantitative decline. Publicly salaried physicians could still be found in early sixth-century Italy under Ostrogothic rule. Alexander of Tralles (a Greek physician) is known to have practiced in Rome in the second half of the sixth century. And a variety of evidence points to the continued existence of lay medical practice at royal courts (for example, that of the Frankish king Clovis at the end of the fifth century) and in major cities (Marseilles and Bordeaux) outside Italy.[5]

But increasingly the most hospitable settings for medical practice seem to have been religious ones, particularly monasteries, where the care of sick members of the community was an important obligation. Our earliest evidence comes from Cassiodorus (ca. 480–ca. 575), founder of a monastery at Vivarium, who instructed his monks to read Greek medical works in Latin translation, including the works of Hippocrates, Galen, and Dioscorides (possibly a reference to *Ex herbis femininis*). Other evidence reveals a high level of medical practice, including the use of secular medical literature, in such monastic centers as Monte Cassino, Reichenau, and St. Gall.[6] It is probable that substantial medical expertise could be found in most monasteries, except the very smallest, throughout the Middle Ages. And although the medicine practiced within the monastery was intended pri-

Fig. 13.1. A page from a Greek manuscript of Dioscorides' *Materia medica*. Paris, Bibliothèque Nationale, MS Gr. 2179, fol. 5r (9th c.).

marily for the members of the monastic community, there is no doubt that on occasion it was made available to others—pilgrims, visitors, and the surrounding population.

The presence, in a monastic environment, of secular medical literature and the medical practices linked with it raises an obvious question, which we must now consider: how did the traditions of Greek and Roman secular medicine interact with Christian ideas about healing? There is no simple answer, but we can begin to make sense of the complex reality if we keep in mind (1) that a philosophical tension did emerge between the naturalism of the medical tradition (the assumption that only natural causes are at work) and supernaturalist traditions (miraculous healing) within Christianity; (2) that most people (including literate people) were not philosophically inclined, and therefore few ever noticed the tension; and (3) that for those who did, there were various ways of easing or resolving the tension, short of repudiating one kind of healing or the other.

The sources of tension are obvious enough. As medieval Christianity matured, it became common for sermons and religious literature to teach that sickness is a divine visitation, intended as punishment for sin or a stimulus to spiritual growth. The cure, in either case, would seem to be spiritual rather than physical. Moreover, within medieval Christianity there developed a widespread tradition of miraculous cures, associated especially with the cult of saints and relics. And to complete the picture, we have the concrete evidence of religious leaders denouncing secular medicine for its inability to produce results.[7]

It is fairly easy to inflate such beliefs and attitudes into a general portrayal of the Christian church as an implacable opponent of Greek and Roman medicine, resolutely committed to belief in supernatural causation and to the exclusive use of supernatural remedies. Unfortunately, such attempts seriously misrepresent the real historical situation. Although it is true that sickness was widely understood to be of divine origin, this did not rule out natural causes, for most medieval Christians shared the view, common since the Hippocratic writers, that an event or a disease could be simultaneously natural and divine (see above, chap. 6). Within a Christian context, it made perfectly good sense to believe that God customarily employs *natural* powers to accomplish *divine* purposes. For example, plague could be explained both as divine retribution for sin and as the result of an unfavorable conjunction of planets or corruption of the air.[8] As for the practice of medicine and the use of natural remedies, all Christian writers would have agreed that cure of the soul is more important than cure of the body, and a few spoke out against any use of secular medicine. Bernard of

Clairvaux (1090–1153), writing to a group of monks in the twelfth century, expressed views that had existed for centuries:

> I fully realize that you live in an unhealthy region and that many of you are sick. . . . It is not at all in keeping with your profession to seek for bodily medicines, and they are not really conducive to health. The use of common herbs, such as are used by the poor, can sometimes be tolerated, and such is our custom. But to buy special kinds of medicines, to seek out doctors and swallow their nostrums, this does not become religious [i.e., monks].[9]

But the vast majority of Christian leaders looked favorably on the Greco-Roman medical tradition, viewing it as a divine gift, an aspect of divine providence, the use of which was legitimate and perhaps even obligatory. Basil of Caesarea (ca. 330–79) spoke for many of the church fathers when he wrote that "we must take great care to employ this medical art, if it should be necessary, not as making it wholly accountable for our state of health or illness, but as redounding to the glory of God." Even a writer as hostile to Greco-Roman learning as Tertullian (ca. 155–ca. 230) revealed his appreciation of the value of Greco-Roman medicine. The denigrating accounts of conventional medicine that appear in saints' lives served an obvious polemical function—namely, to authenticate and magnify the power of the saint in question by demonstrating how he or she had healing abilities that transcended those of the secular healer. That we cannot take such denunciations as representative of the views of the author (let alone the remainder of medieval society) toward secular medicine is evident from the fact that many of these same authors, in other contexts or even in the same context, reveal a large measure of respect for conventional healing practices. What the church fathers were eager to denounce was not the use of secular medicine, but the tendency to overvalue it and the failure to recognize and acknowledge its divine origin.[10]

While defending the church against the charge of having repudiated the medical tradition, we must be careful to avoid the opposite error. There is no question that early medieval Christians believed in healing miracles and that they availed themselves of both religious healing and secular medicine, sometimes simultaneously, sometimes sequentially. In the fourth and fifth centuries the cult of saints became a dominant feature of European culture. Shrines were established around the tomb or some relic (perhaps a bone) of a saint; and these became pilgrimage sites of enormous drawing power. One of the features of these sites that contributed

Fig. 13.2. The miraculous healing of a leg. Paris, Bibliothèque Nationale, MS Fr. 2829, fol. 87r (late 15th c.). For discussion of this illustration, see Marie-José Imbault-Huart, *La médecine au moyen âge à travers les manuscrits de la Bibliothèque Nationale,* p. 182.

most powerfully to their attraction was the report of miraculous cures produced there. A single example will serve to illustrate: Bede (d. 735), in his *Ecclesiastical History of the English People,* recounted many stories of miraculous healings, including that of a monk on the island of Lindisfarne (off the northeastern coast of England), suffering from palsy, who was brought to the tomb of Cuthbert:

> falling prostrate at the corpse of the man of God, he prayed
> with godly earnestness that through his help the Lord would
> become merciful unto him: and as he was at his prayers, . . .
> he felt (as he himself was afterwards wont to tell) like as a
> great broad hand had touched his head in that place where
> the grief was, and with that same touching passed along all
> that part of his body, which had been sore vexed with sick-

ness, down to his feet, and by little and little the pain passed away and health followed thereon.[11]

Similar tales from the medieval period could be multiplied without end.

If the church was neither the enemy of the Greco-Roman medical tradition nor its single-minded supporter, how are we to characterize its attitude and influence? A familiar approach would be to weigh the factors on each side of the equation—both the opposition and the support offered by the church—and to argue that *on balance* the church was a force for good or ill, as the case might be. But such a conclusion would be simplistic. We will come closer to the truth if we avoid the categories of opposition and support altogether and see the church as a powerful cultural force that *interacted* with the secular medical tradition, appropriating and transforming it. Churchmen neither simply repudiated nor simply adopted secular medicine, but put it to use; and to use it was to adapt it to new circumstances, thereby subtly (or, in some respects, radically) altering its character. It is not too strong to claim that within Christendom there was a fusion of secular and religious healing traditions. In its new context, Greco-Roman medicine would have to be accommodated to Christian ideas of divine omnipotence, providence, and miracles. In the radically new institutional setting provided by the monasteries, it was not only nurtured and preserved through a dangerous period in European history, but it was also pressed into service on behalf of Christian ideals of charity (one important outcome of which was the development and spread of hospitals). And eventually, its institutionalization in the universities restored its contact with various branches of philosophy and elevated its status as a science.

There is one further development, of critical importance, that requires our attention before we leave the early medieval period. The translation of Greek medical works into Arabic began in the eighth century and continued through the tenth. When it was finished, most of the major Greek medical sources were available in Arabic, including Dioscorides' *De materia medica,* many Hippocratic works, and nearly all the works of Galen. The magnitude of the gap between Islamic and Western access to this Greek medical literature can be illustrated by reference to the Galenic corpus: only two or three of Galen's works were available in Latin before the eleventh century, whereas Ḥunayn ibn Isḥāq (808–73) listed 129 Galenic works known to him in Baghdad, forty of which he claimed personally to have translated into Arabic.

This Greek medical literature served as a foundation on which a sophis-

Fig. 13.3. Arabic surgical instruments from the treatise by Abu-l-Qāsim az-Zahrāwī (Abulcasis), *On Surgery and Instruments.* Oxford, Bodleian Library, MS Huntington 156, fol. 85v.

ticated Islamic medical tradition would be built. Several features of this medical tradition require brief mention. First, the Islamic medical tradition was built on a full mastery of Greek medical literature and an assimilation of many of the aims and much of the content of Greek medicine. Second, central to the medical thought that emerged were Galenic anatomy and physiology and Galenic theories of health, disease (including epidemic disease), diagnosis, and therapy. An important aspect of Galenic influence was the linkage it revealed between medicine and philosophy—a linkage that became characteristic of much Islamic medical thought.

Third, Galenic medical theory did not rigidly constrain medical thought and practice in Islam, but functioned as a framework to be extended, modified, and integrated with other medical and philosophical systems; medicine in Islam was a dynamic, rather than a static, enterprise. Fourth, not only did Greek medical works circulate in translation, but along with them a large native medical literature was produced by Islamic physicians. This original Arabic literature contained a great deal of variety, of course, but particularly prominent was a series of comprehensive, encyclopedic works that surveyed large segments, or even the whole, of medical theory and practice. Three such encyclopedic works that were to have a profound influence on later Western medicine were the *Almansor* of Rhazes (al-Rāzī, d. ca. 930), the *Pantegni* (or *Universal Art*) of Haly Abbas ('Alī ibn 'Abbās al-Majūsī, d. 994), and the *Canon of Medicine* of Avicenna (Ibn Sīnā, 980–1037). These, along with many other translated works, helped to shape and redirect Western medicine in the later Middle Ages.[12]

THE TRANSFORMATION OF WESTERN MEDICINE

In the eleventh and twelfth centuries, a number of influences began to impinge on the European medical tradition and alter its character. The political and economic renewal of the period, accompanied by dramatic population increase, led to far-reaching social change, including the urbanization and expansion of educational opportunity. In the new urban schools the curriculum was broadened, as emphasis came to be placed on subjects that had been of minor significance, or even totally absent, in the monastic setting. Meanwhile, reform movements within monasticism were attempting to diminish monastic involvement in secular culture (see above, chap. 9). The convergence of these movements brought about a shift in the location of medical education from the monasteries to the urban schools, with a corresponding shift toward professionalization and secularization. At the same time there was a growing demand among urban elites for the services of skilled medical practitioners, which contributed to the emergence of medical practice as a lucrative (and sometimes prestigious) career.

The earliest example of renewed urban medical activity is at Salerno, in southern Italy, in the tenth century. By the end of the century, Salerno had acquired a reputation for its numerous and skilled medical practitioners, including clergy and women. There seems to have been no school in any formal sense, but simply a center (increasingly a famous center) of medical activity, with ample opportunities for men and women to master the healing arts through apprenticeship. What flourished at Salerno in the tenth

Fig. 13.4. Constantine the African practicing uroscopy. Oxford, Bodleian Library, MS Rawlinson C.328, fol. 3r (15th c.). For commentary, see Loren C. MacKinney, *Medical Illustrations in Medieval Manuscripts,* pp. 12–13.

century and into the eleventh was not medical learning but skill in the healing arts. In the course of the eleventh century, however, some of the practitioners at Salerno began to produce medical writings of a practical sort. Early in the twelfth century, the literature emanating from Salerno began to broaden and become more theoretical, reflecting the philosophical orientation of the Arabic medical texts beginning to circulate in Latin translation. Many of the new texts were teaching texts, connected (apparently) with the emergence of organized medical instruction at Salerno.[13]

The translations from the Arabic that influenced medical activity at Salerno in the twelfth century soon transformed medical instruction and medical practice throughout Europe. The earliest translations appear to have been those of Constantine the African (fl. 1065–85), a Benedictine monk at the monastery of Monte Cassino in southern Italy, who had close ties with Salerno. Constantine, whose knowledge of Arabic was undoubtedly connected with his north African origins, translated works of Hippocrates and Galen, the *Pantegni* of Haly Abbas, medical works by Ḥunayn ibn Isḥāq, and other sources. He was followed by other translators over the next hundred and fifty years, in southern Italy, Spain, and elsewhere, who little-by-little rendered from Arabic to Latin much of the corpus of Greco-Arabic medicine. At Toledo, Gerard of Cremona (ca. 1114–87) translated nine Galenic treatises, Rhazes' *Almansor* (from the name of Rhazes' patron, Manṣūr ibn Isḥāq, to whom it was dedicated), and Avicenna's great *Canon of Medicine*. These new texts vastly broadened and deepened Western medical knowledge, giving it a much more philosophical orientation than it had possessed during the early Middle Ages and ultimately shaping the form and content of medical instruction in the newly founded universities.[14]

MEDICAL PRACTITIONERS

Today we generally think of medicine as a learned profession, which can be practiced only by those who have undergone a long period of schooling and acquired the appropriate professional credentials. But if we project such a model onto the Middle Ages, we will be sorely misled. A far more useful modern analogue would be that of carpentry. Carpentry covers a continuum from elementary home-maintenance through the professional carpentry of the building trades to civil engineering and architecture. Carpentry of the simplest kind falls into the realm of general knowledge (almost everybody knows, or is willing to learn, something about elementary home-repair); the weekend amateur, who (for example) restores antiques for a hobby, may command considerable knowledge and skill; the building trades are staffed by expert professionals who have, for the most part, learned their trade through apprenticeship; and finally, the civil engineer and the architect bring theoretical knowledge to bear on the subject.

So it was with the practice of medicine in the Middle Ages. Simple domestic medicine, practiced in the home, was the property of almost everybody. If more expertise was required, in every community there were people known to have the knack of treating certain kinds of ailments, and we begin here to move up the ladder of medical expertise and specialization. Most villages would have midwives, bone-setters, and people knowledgeable in herbs and herbal remedies. In the cities, one would find a variety of "empirics" with such specialties as the treatment of

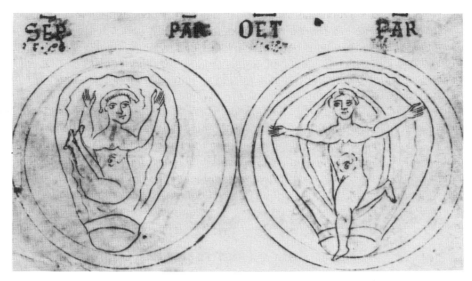

Fig. 13.5. Fetuses in the womb. Copenhagen, Kongelige Bibliotek, MS Gl. kgl. Saml. 1653 4°, fol. 18r (12th c.).

Fig. 13.6. Trotula, a twelfth-century Salernitan medical practitioner. London, Wellcome Institute Library, MS 544, p. 65 (12th c.).

wounds, dental problems, and certain kinds of surgery (for example, lancing boils, repair of hernia, or removal of kidney stones). At a higher level of professionalization, there were apothecaries, trained surgeons, skilled professional medical practitioners educated through apprenticeship, and finally university-educated physicians. This was by no means a static or strictly linear hierarchy, nor was it invariable from place to place; it was also complicated by the existence of both secular and religious practitioners (clerics, for example, who frequently combined conventional medical practice with religious duties) at many of the levels; moreover, the lines of demarcation were rarely clear, because the regulation or licensing of medical practitioners, which would have demanded relatively clear categories, was only slowly instituted in the course of the later medieval period and never became universally effective. But some semblance of this classification scheme was generally characteristic of the medieval medical scene.[15]

We have only the sketchiest data on the numbers of medical practitioners in medieval Europe. We can learn something, however, from the fragments of data in our possession. In 1338, Florence (which was undoubtedly blessed with far more physicians per capita than the average European city) had approximately 60 licensed medical practitioners of all kinds (including surgeons, and unlettered "empirics") for a population of 120,000. Twenty years later, after the population had been decimated by the black death, Florence had 56 licensed medical practitioners for a population of about 42,000; and this ratio of twelve or thirteen physicians for every 10,000 residents held for the remainder of the century.[16] Access in rural areas to a trained physician must have been far less common.

Included among medieval medical practitioners were substantial numbers of women, active in obstetrics and gynecology but also in other medical specialties. The most famous of these is Trota or Trotula, from twelfth-century Salerno, who may not have written the gynecological work usually attributed to her but seems to have produced a more general work of practical medical remedies and advice. In certain parts of Europe Jewish medical practitioners were also numerous.[17]

MEDICINE IN THE UNIVERSITIES

The medical practitioners about whom we know the most are those who studied or taught in the formally organized medical schools of medieval Europe. Because these physicians were literate and left written records that still survive, we can learn something about their identities, their studies, and the kind of medical practice in which they engaged.[18]

Formal medical studies seem to have appeared first in the cathedral schools of the tenth and eleventh centuries—not for the purpose of educating professional physicians, but as an aspect of general education. At Chartres, for example, medical studies appeared by about 990, and in the next century medical instruction could be found in similar schools elsewhere.[19] However, it was at Salerno in the twelfth century that the newly translated medical works of the Greco-Arabic tradition were first assimilated, and it was there that medicine began to emerge as a learned profession. The driving force behind these developments was not mere intellectual curiosity or medical altruism (though a measure of both no doubt existed), but the desire for status and professional advancement. Physicians already at the top of the medical hierarchy outlined above, and therefore already literate, perceived the possibility of elevating their status by imitating other learned professions, such as law, in demanding that practitioners acquire formal intellectual credentials. The aim was to elevate the status of medicine from art or craft to science. Developments at Salerno were influential, and in the thirteenth century medical faculties became prominent at the universities of Montpellier, Paris, and Bologna. Medical faculties of lesser significance were created at Padua, Ferrara, Oxford, and elsewhere.

The institutionalization of medicine in the medieval universities was of enormous importance for the course of medical theory and practice. In the first place, it assured the continuation and the continuity of medical studies and the existence, from the Middle Ages to the present, of an influential community of university-educated physicians. Second, the establishment of medical studies in the university (as opposed to some other possible institutional home) created a linkage between medicine and other branches of knowledge that profoundly shaped the development of medicine. Specifically, a degree in the faculty of arts came to be a typical (if not quite universal) prerequisite for medical studies; and this meant that medical students came equipped with the logical and philosophical tools that would transform medicine (for better or for worse) into a rigorous, scholastic enterprise. It also gave medicine access to Aristotelian natural philosophy, which would provide medicine with some of its important principles, and to astrological theory (and its companion, astronomy), which would become a universal part of the physician's diagnostic and therapeutic armory. Let us give the medical curriculum a brief examination.

Fig. 13.7. Medical instruction. From a copy of Avicenna's *Canon of Medicine,* Paris, Bibliothèque Nationale, MS Lat. 14023, fol. 769v (14th c.).

Teaching, first at Salerno and later in the other medical schools, co-alesced, for a time, around a collection of brief treatises known collectively as the *Articella.* This collection included an introduction to medicine by Ḥunayn ibn Isḥāq (known in the West as Johannitius), several short works from the Hippocratic corpus, and books on urinalysis and diagnosis by pulse. In the fourteenth and fifteenth centuries, these were supplemented by the works of Galen, Rhazes, Haly Abbas, Avicenna, and others. This curriculum had a marked philosophical orientation—medical theory being required to conform to broader principles of natural philosophy. And the teaching methods employed were the typical scholastic ones of commentary on authoritative texts and debate over disputed questions. But that did not mean (as has sometimes been alleged) that university medicine was a purely theoretical, textbook activity. In fact, many university professors of medicine engaged in private practice on the side, and medical students were frequently required to obtain practical experience.[20]

Finally, do we have any idea of the numbers of students involved? In fact, there are scraps of data. During a period of fifteen years early in the fifteenth century, the University of Bologna (one of the foremost medical schools in Europe) granted sixty-five degrees in medicine and one in surgery. During a thirty-six year period a little later in the same century, the University of Turin (also in northern Italy) awarded a total of thirteen medical doctorates. And during its first sixty years of existence (beginning in 1477) the University of Tübingen awarded medical degrees at the rate of about one every other year. The number of medical students, of course, was far higher than the number of degree recipients, since most students did not complete the course of studies: the ratio 10:1 has been suggested as a possible multiplier. About all we learn from these numbers is that university-trained physicians, and especially physicians with doctorates in medicine, were rare creatures, members of an urban elite and accessible, for the most part, only to the rich and powerful.[21]

DISEASE, DIAGNOSIS, PROGNOSIS, AND THERAPY

The medical theories held and the diagnostic measures and therapies employed by a medieval medical practitioner varied with the practitioner's level of education, specialty, and professional circumstances. We know most, of course, about the views and procedures of learned physicians; but there is reason to believe that their beliefs and practices filtered downward and therefore influenced other kinds of healers. For example, there is ample evidence that Latin medical treatises were translated into vernacular languages, or translated and excerpted, for the benefit of medical practitioners who were literate but could not read Latin.[22] At the same time, it is clear that folk medicine and folk remedies had a tendency to filter upward and influence professional and even (to some extent) learned medicine. We will not be far off, therefore, if we judge the following elements of medical belief and practice to have been present, to varying degrees, in much medieval healing activity.

Fundamental to medieval theories of disease was the idea that every person has a characteristic *complexion* or temperament, determined by the balance of the four elements and their corresponding qualities (hot, cold, wet, dry) in the person's body. It was understood that complexion was peculiar to the individual; the balance that was normal for one would be abnormal for another. Closely associated with the theory of complexion was the idea, stemming from Galen and the Hippocratics, that the body contains four principal, physiologically significant fluids or humors—blood, phlegm, black bile, and red or yellow bile—and that these humors are the vehicle by which the proper balance of qualities is maintained. And it was understood that health is associated with proper balance, illness with imbalance. For example, fever was conceived to be the result of abnormal heat emanating from the heart. Finally, health and disease were thought to be influenced by a set of conditions called the "non-naturals": the air breathed, food and drink, sleep and wakefulness, activity and rest, retention and elimination (of nutrients), and state of mind.[23]

If sickness is the result of deviation from a person's normal complexion, then therapy must be directed toward the restoration of balance. Various techniques were available for the achievement of this end. The first was dietary; since the humors are the end products of the food consumed, a suitable diet was absolutely essential to the maintenance of health. Drugs, classified according to their predominant qualities, could also be prescribed to help restore balance. And if more heroic treatment seemed called for, it was possible to eliminate excess bodily fluids through purging, "puking," and blood-letting. In order to determine which of these measures to employ, the physician would need to inquire into the patient's life-style or regimen (such matters as diet, exercise, sleep, sexual activity, and bathing) in order to ascertain his or her specific complexion and the regimen required to maintain it. Indeed, for maximum effect the physician should closely monitor the patient's activities over an extended period of time—a realistic aim only for a physician (presumably learned) in the employ of a wealthy patron. Having observed his patron-patient over a period of time, the learned physician would (in theory) be in a position to offer the advice needed for the maintenance or recovery of health. The ideal that governed learned medicine (and, to some extent, less learned varieties of medical practice) thus portrayed the physician as medical advisor, with a primary responsibility for what we would call preventive medicine, but capable of following up with suitable remedies when preventive measures failed.[24]

The most common form of medical intervention was drug therapy, and the ability to identify and prepare drugs, along with knowledge of their therapeutic properties, was therefore an essential part of the repertory of most medieval healers. Drugs could be simple or compound; the most common ingredients were herbal, but animal and mineral substances were also employed. Many drugs were folk remedies, sanctioned by apparently successful use over many generations; for example, long experience had taught local healers that certain plant substances were effective as laxatives or pain-killers. There is no question that some medieval drugs were effective; the majority, however, were simply harmless, while a few may have been dangerous. And some were downright disgusting—for ex-

Fig. 13.8. An apothecary shop. London, British Library, MS Sloane 1977, fol. 49v (14th c.). By permission of the British Library.

ample, the belief that pig manure was an effective cure for nose-bleed. In this case, the cure might well seem worse than the ailment.[25]

But if there was a substantial empirical (frequently folk) component in medieval drug therapy, there was also a strong theoretical component emanating from the Greek and Arabic medical traditions. Dioscorides' *De materia medica* (in a revised and augmented version) had a very modest circulation in the West; in the twelfth century new and more influential collections of medical recipes appeared; and finally, fresh translations of works by Galen, Avicenna, and others supplied the theoretical underpinnings needed to organize and systematize pharmaceutical knowledge. The basic theoretical assumption was that natural substances have therapeutic properties, associated with their primary qualities: hot, cold, wet, dry. To this theory, Avicenna added the idea that medicinal substances may also have a "specific form," independent of their primary qualities, which explains therapeutic effects not readily accounted for by the four primary qualities. It was thus through its specific form that theriac (a drug known since antiquity, made from viper's flesh and other ingredients) acquired the remarkable curative properties assigned to it in the twelfth-century *Antidotarium Nicolai:*

> Theriac . . . is good for the most serious afflictions of the entire human body: against epilepsy, catalepsy, apoplexy, headache, stomach ache, and migraine; for hoarseness of voice and constriction of the chest; against bronchitis, asthma, spitting of blood, jaundice, dropsy, pneumonia, colic, intestinal wounds, nephritis, the stone, and choler; it induces menstruation and expels the dead fetus; it cures leprosy, smallpox, intermittent chills, and other chronic ills; it is especially good against all poisons, and the bites of snakes and reptiles . . . ; it clears up every failing of the senses[?], it strengthens the heart, brain, and liver, and makes and keeps the entire body incorrupt.[26]

Another area of theoretical concern was the problem of determining how the properties of compound medicines depended on the qualities of their simple ingredients. Elaborate theoretical discussions (including mathematical analysis) of this problem were undertaken by both Islamic and European authors. Indeed, the doctrines of the intensification and remission of forms and qualities discussed above (chap. 12) were developed in part because of their applicability to pharmaceutical theory.[27]

We cannot conclude this discussion of medieval disease and treatment without discussing two prominent diagnostic techniques—urinalysis and the examination of pulse. Both had been recommended in antiquity, by various writers including Galen; the further influence of two short treatises, one on pulse and one on urine, contained in the *Articella* collection, as well as longer discussions in Avicenna's *Canon of Medicine,* assured their centrality in later medieval diagnosis. It was held that urinalysis could reveal the state of the liver, while pulse reflected the state of the heart. The critical features of the urine were color, consistency, odor, and clarity. For example, an early thirteenth-century medical writer, Giles of Corbeil, maintained that "thick urine, whitish, milky, or bluish-white, indicates dropsy, colic, the stone, headache, excess of phlegm, rheum in the members, or a flux."[28] Charts revealing the connection between different colors of urine and various ailments were a common feature of medieval medical writing (see fig. 13.9).

In taking a patient's pulse, the physician attempted to determine its

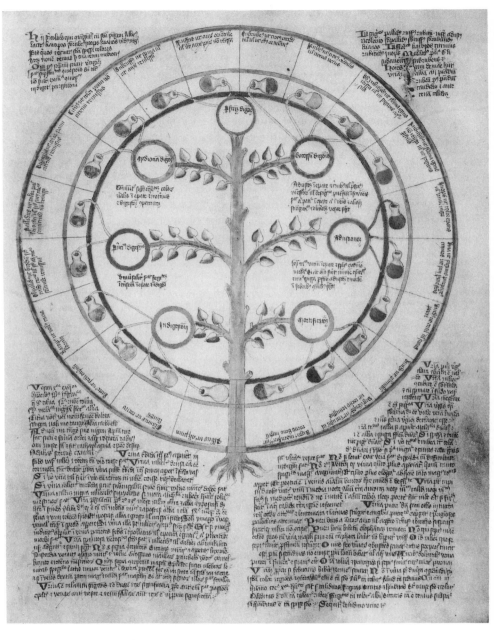

Fig. 13.9. A urine color chart, which connects variations in the color of urine with various stages of digestion. London, Wellcome Institute Library, MS 49, fol. 42r (15th c.). For further commentary, see Nancy G. Siraisi, *Medieval and Early Renaissance Medicine*, p. 126.

strength, duration, regularity, breadth, and so forth. Many varieties of pulse were differentiated and various classification schemes developed. An anonymous treatise of the thirteenth century offered the following scheme:

> The varieties of pulses are differentiated by the physician in a number of ways, in particular according to five considerations: (1) motion of the arteries; (2) condition of the artery; (3) duration of diastole and systole; (4) strengthening or weakening of pulsation; (5) regularity or irregularity of the beat. Ten varieties of pulse derive from these considerations.[29]

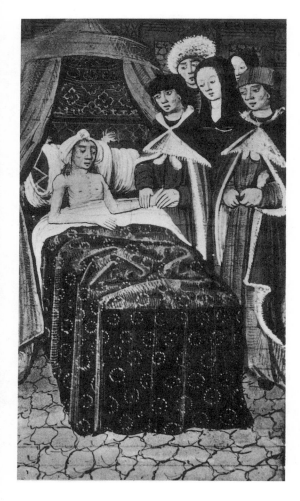

Fig. 13.10. Diagnosis by pulse. Glasgow University Library, MS Hunter 9, fol. 76r (15th c.). For a discussion of this illustration, see MacKinney, *Medical Illustrations from Medieval Manuscripts,* pp. 16–17. By permission of the Librarian, Glasgow University Library.

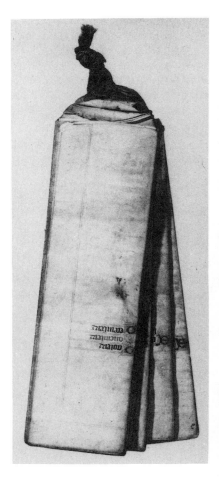
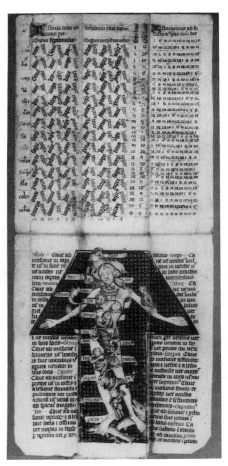

Fig. 13.11. A physician's girdle book. London, Wellcome Institute Library. A handy guide for the physician, meant to be suspended from the belt. The left-hand illustration shows the book in its folded form. The right-hand illustration shows one of its leaves, this one containing astrological information. For a full discussion, see John E. Murdoch, *The Album of Science: Antiquity and the Middle Ages,* pp. 318–19.

Failing pulse could be used to foretell the time of death and was therefore useful for prognosis as well as diagnosis.

Thus far we have sidestepped one pervasive element in medical theory and practice, which hovered over and shaped what the medieval healer believed and the therapeutic measures he or she prescribed. This was medical astrology—the belief that planetary influence is implicated both

in the cause and in the cure of disease. There were good reasons for believing in such planetary influence. For one, there was medical authority:
several of the Hippocratic works contained passages that could be interpreted as affirmations of celestial influence, and during the later Middle
Ages a treatise on astrological medicine circulated under Hippocrates'
name. But far more importantly, anybody who had grasped the fundamentals of natural philosophy knew that the heavens exercised an influence on
the human body and its environment; and there was no reason at all to
doubt that this would have an effect on health and the course of disease.[30]

Celestial influence began at conception, contributing to the temperament or complexion of the newly conceived embryo. After birth, every
human was the recipient of a continuous flow of celestial forces, either
directly or through the surrounding air, and these influences affected temperament, health, and disease. Indeed, astrological influence was frequently
invoked to explain major epidemics, such as the black death of 1347–51.
Pressed for an explanation of this particular plague, the medical faculty at
the University of Paris concluded that it resulted from corruption of the air
caused by a conjunction of Jupiter, Saturn, and Mars in 1345.[31]

If illness struck, the physician needed to take account of the planetary
configuration in order to prescribe effective treatment. The preparation
and administration of drugs had to be properly timed to coincide with
favorable planetary configurations, and proper dosage depended on astrological factors. It was also necessary to determine propitious times for surgical procedures, such as blood-letting. Surgical treatises often contained
"blood-letting figures" that instructed the user on the appropriate times
for bleeding from specific bleeding points. Finally, the Hippocratic theory
of "critical days," which held that the course of acute diseases is marked
by crises or turning points, became linked to astrology; among the factors
that were believed to determine the outcome of a crisis was its timing—
whether or not it occurred on an astrologically favorable day.

ANATOMY AND SURGERY

Medieval healers were no doubt inclined toward moderate forms of medical intervention, such as control of diet and prescription of drugs. But
there were ailments and medical emergencies that demanded more intru-
sive measures, and Europe always had medical practitioners willing to
invade the body surgically. There were many kinds of surgeons, with differing specialties and levels of education, from itinerant empirics specializing in a particular surgical procedure to university-educated surgeons in

Fig. 13.12. Operation for cataract (above) and nasal polyps (below). Oxford, Bodleian Library, MS Ashmole 1462, fol. 10r (12th c.). For commentary on this figure, see MacKinney, *Medical Illustrations from Medieval Manuscripts,* pp. 70–71.

Fig. 13.13. Operation for scrotal hernia. Note that the patient is both tied and held down. Montpellier, Bibliothèque Interuniversitaire, Section Médecine, MS H.89, fol. 23r (14th c.). This illustration is discussed by MacKinney, *Medical Illustrations from Medieval Manuscripts,* pp. 78–80.

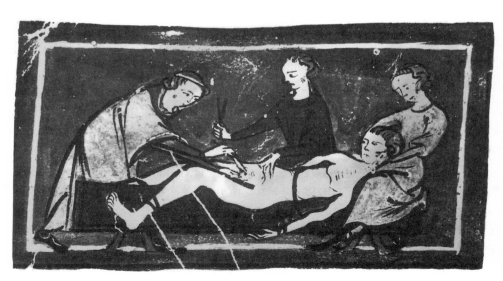

the employ of king or pope. There was always a tendency to view surgery as a craft, beneath the dignity of the university-educated physician; however, in southern Europe surgeons managed to institutionalize their enterprise in the universities (Montpellier and Bologna, for example), thereby acquiring intellectual status. A substantial Arabic surgical literature was made available in the West through the translations of the twelfth and thirteenth centuries, and this stimulated a European tradition of surgical writing. Among the most influential of the European treatises were the *Surgery* of Roger Frugard (twelfth century), which frequently circulated in short sections, and the *Chirurgia magna* (or *Great Surgery*) of Guy de Chauliac (ca. 1290–ca. 1370), physician and surgeon to three popes. Guy's work not only circulated widely in Latin, but was also translated into English, French, Provençal, Italian, Dutch, and Hebrew.[32]

There is no doubt that most surgery was not particularly heroic—the setting of a broken bone, reduction of a dislocated joint, dressing of an ulcer or sore, cleaning and suturing of a wound, or lancing of a boil. Blood-letting and cautery (application of hot irons to various parts of the body in order to create ulcers through which unwanted fluids could drain) were also common procedures.[33] Removal of external hemorrhoids may also have been fairly routine. But some medieval surgeons undertook much more ambitious procedures. Operation for removal of cataract, by inserting a sharp instrument through the cornea and forcing the lens of the eye out of its capsule and down to the bottom of the eye, is one example. Removal of bladder stone and surgical correction of hernia are others. A text describing the removal of bladder stone will serve to illustrate:

> If there is a stone in the bladder make sure of it as follows: have a strong person sit on a bench, his feet on a stool; the patient sits on his lap, legs bound to his neck with a bandage, or steadied on the shoulders of the assistants. The physician stands before the patient and inserts two fingers of his right hand into the anus, pressing with his left fist over the patient's pubes. With his fingers engaging the bladder from above, let him work over all of it. If he finds a hard, firm pellet it is a stone in the bladder. . . . If you want to extract the stone, precede it with light diet and fasting for two days beforehand. On the third day, . . . locate the stone, bring it to the neck of the bladder; there, at the entrance, with two fingers above the anus incise lengthwise with an instrument and extract the stone.

As a final example of dangerous surgery, fracture of the skull sometimes required trephining (the making of small holes in the skull with a saw) in order to reduce pressure and drain blood and pus. And all of this surgery was performed with only the most modest use of sedatives or anesthetics; if there was anything obviously heroic about medieval surgery, it was the patient.[34]

How much human anatomy did the medieval surgeon or physician know, and what place did anatomical instruction and firsthand anatomical investigation have in the education of medical practitioners? Despite Galen's stress on the importance of anatomical knowledge for the successful treatment of disease, the connection between anatomical knowledge and the clinical side of medical practice remained as tenuous during the Middle Ages as it had been in antiquity. Most medieval practitioners no doubt found that they could get along quite nicely with a minimum of anatomical knowledge, for the advice they dispensed and the dietary and herbal remedies they prescribed rarely, if ever, depended on detailed structural knowledge of the human body. The surgeon's requirements were undoubtedly greater but still modest; much of the required knowledge was common property through such daily experience as animal butchery, and the rest could be obtained by experience in the course of apprenticeship or surgical practice.

Nonetheless, the translations of the twelfth century provoked new interest in anatomical questions. The translation of Galen's anatomical writings and Arabic works based on them (works of Avicenna, Haly Abbas, Rhazes, and later Averroes) brought to the West a body of anatomical literature that demanded attention—not because it promised a large, immediate impact on healing practices, but because it belonged to the body of medical theory that learned physicians were attempting to assimilate in their quest for intellectual status. The new interest in anatomical knowledge first found expression in the form of actual anatomical dissections in twelfth-century Salerno; the object of dissection was the pig, considered anatomically analogous to humans.

Human dissection appears to have begun in certain Italian universities, especially Bologna, late in the thirteenth century. The picture is murky, but the purpose seems originally to have been legal—autopsies within the law faculty for the purpose of determining the cause of death—the practice spreading subsequently, by steps we know nothing about, to include dissections for medical instruction. By 1316, Mondino dei Luzzi (d. ca. 1326), who taught at Bologna, had become sufficiently skilled in human dissec-

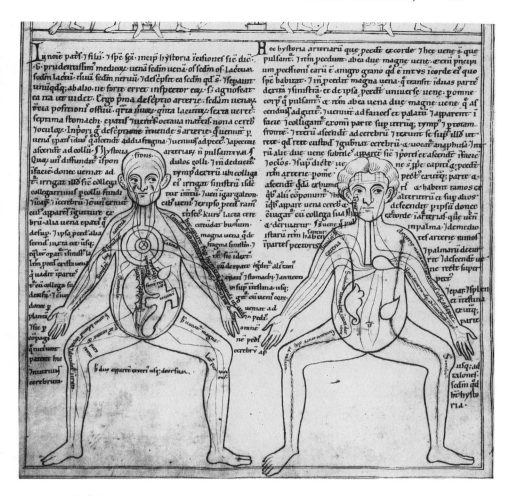

Fig. 13.14. Human anatomy, illustrating the Galenic conception of veins (left) and arteries (right). Munich, Bayerische Staatsbibliothek, CLM 13002, fol. 2v (12th c.). For commentary and additional anatomical drawings, see Siraisi, *Medieval and Early Renaissance Medicine,* pp. 92–95.

tion to write a dissection manual entitled *Anatomia,* which became the standard guide to human dissection for the next two centuries.[35]

In the course of the fourteenth century, dissection became a regular part of medical instruction at Padua, Bologna, and a few other universities. In his *Chirurgia magna,* Guy de Chauliac described the procedures of his master at Bologna, Nicolaus Bertrucius:

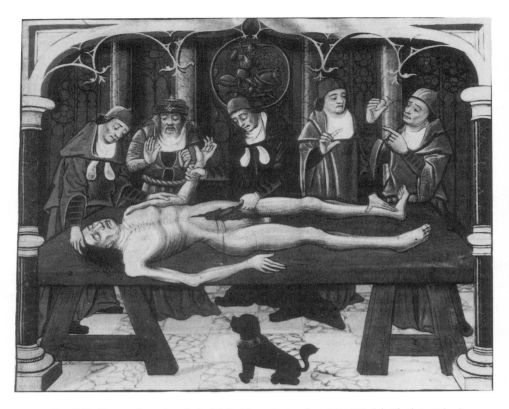

Fig. 13.15. Human dissection. Paris, Bibliothèque Nationale, MS Fr. 218, fol. 56r (late 15th c.).

Having laid the dead body on the table, he made four lessons
on it. In the first the nutritive members [stomach and intes-
tines] were treated, since they decay the soonest. In the
second, the spiritual members [heart, lungs, and trachea], in
the third the animal members [skull, brain, eyes, and ears],
and in the fourth the extremities were treated. And following
the commentary on the book of Sects [of Galen], in each there
are nine things to see: that is, to know the situation, the substance, the constitution, the number, the figure, the relations
of connections, the actions and uses, and the diseases which
affect them. . . . We make anatomies also on bodies dried in
the sun, or consumed in the earth, or submerged in running
or boiling water. This shows us the anatomy at least of the
bones, cartilage, joints, large nerves, tendons, and ligaments.[36]

Such dissections were generally performed on the corpses of criminals, whose execution might be timed to meet the needs of the medical school. They were infrequent, an annual dissection being perhaps the most common pattern. And it is important to understand that the medical student was an observer rather than an experimenter; the function of dissection was to illustrate the Galenic text; this was not research, but pedagogy.

Medieval physicians have been severely criticized by modern historians for adopting a methodology that made texts, rather than cadavers, the primary anatomical authority. The unfortunate result of this methodology, it has frequently been argued, was the continued propagation of a variety of errors in Galen's account of human anatomy. What are we to think of such criticisms? There is no question that medieval physicians found Galenic anatomy an awesome achievement and were therefore inclined to attach great (though not absolute) authority to Galenic texts, but it does not follow that they were fools. Consider a modern parallel: the modern anatomical textbook is also a remarkable achievement, and when a medical student taking the obligatory anatomy course finds a discrepancy between text and cadaver, he or she interprets this discrepancy as a variation in the cadaver rather than a mistake in the textbook. We should not be surprised to see medieval physicians and surgeons behaving similarly. They had every reason to believe that Galen had gotten it right (as, for the most part, he had) and to view the study of Galenic texts as the surest and most efficient, not to mention cleanest, way of acquiring anatomical knowledge.

Despite the secondary importance of anatomical dissection within medical education, we have seen that a tradition of anatomical dissection did develop late in the thirteenth and early in the fourteenth century. It grew in strength and sophistication over the next two hundred years, while maintaining a continuous dialogue with the textual tradition of anatomical knowledge. In the fifteenth century it became allied with the technology of printing, which made possible the cheap production of texts and the faithful reproduction of anatomical drawings. The quality of anatomical drawings was further enhanced by contributions from the growing company of talented artists. And in the sixteenth century, these factors combined with renewed access to the Greek text of Galen to produce the stunning anatomical achievements of Andreas Vesalius (1514–64) and others.

DEVELOPMENT OF THE HOSPITAL

I conclude the discussion of medieval medicine on an institutional note, with a brief account of one of the most celebrated medieval medical achievements—the invention of the hospital. One of the difficulties in tracing the origin of the hospital is deciding what the term means. If, by "hospital," we mean anything called "hospice" or "hospital," then we include many institutions that offered food and shelter to paupers and pilgrims, including the sick, but which provided little or no specialized medical care. If, however, we wish to reserve the term for institutions dedicated to the treatment of the sick, including the provision of skilled medical care, then we are applying a much narrower criterion. The former sort of hospital, which was common throughout medieval Europe (often maintained by monasteries or communities of lay brethren), will not interest us. It is the latter kind of institution that will be the object of our attention.[37]

Where, then, did the hospital as a medical institution come from? Its origins seem to lie in the Byzantine Empire, where, by the sixth century and perhaps well before, ideals of Christian charity led to the establishment of hospitals that provided specialized medical care. One of the earliest for which we have hard evidence was the Sampson hospital (named after a saint of the fourth century) in Constantinople; here, early in the seventh century, for example, a church official suffering from a groin infection was hospitalized for surgery and convalescence. Other Byzantine hospitals were organized along the same lines: in the twelfth century, the Pantokrator hospital, also in Constantinople, had space for fifty patients (thirty-eight male and twelve female); to meet their medical and other needs, the hospital employed a staff of forty-seven, including physicians and surgeons.[38]

This Byzantine model became known in both Islam and the West, where it interacted with, and helped to shape, indigenous traditions of health care. In Islam, we find comparable institutions early in the ninth century, perhaps owing to the influence of the Barmak family, which occupied a position of power under the Khalif Hārūn ar-Rashīd (786–809). There were no doubt many strands in the transmission of the Byzantine model to the West; one of them seems to have come as a byproduct of the conquest of Jerusalem in 1099, during the First Crusade. Shortly after the fall of Jerusalem, the lay brothers (subsequently known as "Hospitallers") who operated the hospital of Saint John in Jerusalem reorganized it on the Byzantine model. Because of its prominent location and large size, it became renowned throughout Europe: visitors a century later reported that it housed a thousand patients or more. The Hospitallers eventually established a string of hospitals in Italy and southern France. Through the promulgation of various statutes regulating these hospitals (requiring, in one version, the hiring of four physicians to treat patients), the Jerusalem pattern became familiar in the West, where it influenced the conception of charitable care

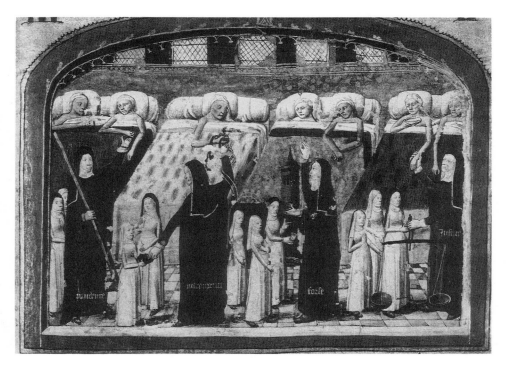

Fig. 13.16. A medieval hospital. From Jean Henry, *Le livre de vie active des religieuses de l'Hôtel-Dieu* (late 15th c.). Paris, Centre de l'Image de l'Assistance Publique. This illustration is discussed by Imbault-Huart, *La médecine au moyen âge,* p. 168.

for the ill and the indigent, encouraging the development of the hospital as a specialized medical institution.[39]

This is, to be sure, a very sketchy picture, with many uncertainties remaining. Whatever the precise details of transmission and assimilation, the model of the hospital as a medical institution spread rapidly in the West in the twelfth and thirteenth centuries, so that hospitals could be found in cities and towns throughout Europe. They might be large or small, containing anywhere from hundreds of beds to half-a-dozen. Their sponsorship could be either religious or secular. Their clientele was principally from the lower classes, though there were exceptions. They were typically staffed by professional physicians, paid an annual salary for their labors. Considerable thought was given to the needs of the patients—cleanliness and diet, for example. Beds consisted of straw mattresses suspended on ropes from bedposts, designed to hold two, or even three, patients. An account of medical facilities in the city of Milan, written about 1288, is instructive:

> In the city, including the suburbs . . . thére are ten hospitals for the sick. . . . The principal one is the Hospital of the Brolo, very rich in possessions, and founded by Geoffrey de Bussero in 1145. In it . . . there are found, particularly in bad times, more than five hundred bed patients and as many more not lying down. All these receive food at the expense of the hospital itself. Besides them, there are 350 babies or more, placed with individual nurses after their birth, under the hospital's care. Every kind of poor person, except lepers, for whom another hospital is reserved, is received there, and kindly and bountifully restored to health, bed as well as food being provided. Also, all the poor needing surgical care are diligently cared for by three surgeons who are assigned to this particular task. . . .[40]

Though surely putting the best face on things, this account reveals the impressive level of care to which a medieval hospital might aspire.

NATURAL HISTORY

Medicine was no doubt the principal repository of biological knowledge during the Middle Ages, but it was not the only one. Aristotelian natural philosophy included a large component of zoological and botanical information. Encyclopedias almost always contained sections on plants and animals. Herbals and bestiaries specialized, of course, in the plant and animal kingdoms, respectively. And finally, medieval people had intimate firsthand knowledge of the local flora and fauna. We will conclude this chapter with a brief examination of medieval botanical and zoological knowledge.

Medieval botanical knowledge was closely linked with medicine, since the principal use of plants (if we ignore those that formed part of the European diet) was for herbal remedies. If the medicinal use of herbs was to be effective, manuals were needed, which would describe the various herbs and their therapeutic uses. A significant herbal literature thus developed, most of it designed for practical purposes. The model was Dioscorides' *De materia medica,* in its revised Latin translation, which arranged medicinal substances in alphabetical order so as to facilitate use. A typical entry in an herbal would include the name or names of the plant, an account of its identifying features, including habitat, a description of the medicinally sig-

Fig. 13.17. A page from the *Herbal* of Pseudo-Apuleius, describing and illustrating couchgrass, sword lily, and rosemary. Oxford, Bodleian Library, MS Ashmole 1431, fol. 21r (12th c.). Described in Joan Evans, ed., *The Flowering of the Middle Ages,* pp. 190, 352.

nificant parts and their therapeutic properties, and instructions regarding preparation and use. The alphabetical arrangement of the herbal reveals that practical aims (the ability to look up a medicinal substance by name) prevailed over classification according to biological type or any other theoretical consideration.[41]

But alongside these practically oriented herbals, there was also a more theoretical or philosophical literature, which placed plant life within the context of natural philosophy. Most of this literature descended in one way or another from the book *On Plants,* attributed to Aristotle and believed by medieval scholars to be his, but probably written by Nicholas of Damascus (1st century B.C.). A few commentaries were written on this treatise (perhaps a dozen are known), by far the most impressive of which was the *On Vegetables* of Albert the Great (ca. 1200–1280). Albert's *On Vegetables* contains a paraphrase of *On Plants,* accompanied by Albert's own attempt to bring intellectual order to the natural philosophy of plants, and finally a traditional alphabetical list of herbs and their uses. A reading of this work reveals the extraordinary skill, unmatched by any contemporary, with which Albert observed and described botanical phenomena.[42]

One might expect close parallels between botanical and zoological literature. However, zoological knowledge had few applications in the medical realm and little practical value elsewhere; and consequently, there was no zoological counterpart of that repository of practical botanical knowledge—the herbal. As in the case of botany, there was an underlying Aristotelian textual tradition, for Aristotle had written a series of large and important zoological works. These were rendered into Latin (along with an influential commentary by Avicenna) and attracted considerable attention—not so much for the detailed zoological information they contained as for their bearing on more general issues in natural philosophy. Once again Albert the Great was one of the major figures, producing, in his *On Animals* and other works, a large body of descriptive and theoretical zoology. Of particular interest are his discussions of nutrition and embryology. His treatment of conception and embryological development, for example, was dependent not only on Aristotle's theories of conception, but to a very substantial degree on his own observations of the reproductive behavior of animals. The history of medieval zoology has yet to be written, but in Albert the Great we undoubtedly see the philosophical side near its zenith.[43]

Besides zoological works in the Aristotelian tradition, there were various other genres of literature on animals—two of which have attracted considerable attention. One of these consists of practical treatises on falconry. The most famous of the genre was written in Sicily by the Emperor Frederick II (about the middle of the thirteenth century) and entitled *On the Art of Hunting with Birds.* The most famous observation in this most famous treatise on birds is Frederick's experimental determination that vultures locate their food by sight rather than by smell—ascertained by observing their inability to find food when their eyes were covered.[44]

If Frederick's treatise on falconry seems remarkably practical and modern, devoid of the fanciful or metaphysical content that we have come to associate with the Middle Ages, our final example of medieval literature on animals goes to the other extreme. The medieval bestiary is often presented as an example of medieval inability to observe the world objectively and get zoological knowledge straight. Medieval bestiaries are all descended from an anonymous treatise entitled *Physiologus,* emanating from Alexandria and written in Greek (perhaps about the year 200 A.D.), subsequently translated into Latin and all of the major European vernacular languages. The *Physiologus* and the medieval books inspired by it are collections of animal lore arranged in short entries or chapters under the names of the respective animals—numbering from about forty in the *Physiologus* to more than a hundred in some of the later bestiaries.[45]

The typical entry in a bestiary begins with an etymological explanation of the animal's name. For example, the article on the horse in a twelfth-century bestiary maintains that it takes its name, *equus,* from the fact that when horses "are teamed in fours, they are 'levelled' (*equabantur*) and those which are pairs in shape and equals in pace are matched together."[46] If the animal has distinctive physical characteristics, these will be reported next, followed by an account of unusual or interesting behavior and a description of admirable and regrettable character traits. From this same twelfth-century bestiary, we learn that the hedgehog is covered with spikes and curls itself into a ball for protection; that the fox is a "fraudulent and ingenious animal" that plays dead in order to catch its prey; that cranes move about in military formation; that the serpent called "basilisk" can kill with the power of its glance; that the lynx's urine turns into a precious stone; that lions are compassionate and courageous, and that their eyebrows and manes offer a clue to their disposition. Finally, many (but not all) articles go on to draw a moral or make a theological point on the basis of the animal description. The hedgehog is an example of prudence, the crane of courtesy and responsibility. The fox is employed as a type of the devil, who entices carnal man through fraudulent behavior. And the male lion, breathing life into its stillborn offspring after three days, represents God the Father raising Christ from the dead.

How are we to judge such an odd mixture of fact, fancy, and parable? The bestiary certainly does not read like a modern zoology manual, and on this basis historians of science have sometimes portrayed the people who compiled the bestiaries as incompetent or unsuccessful zoologists. The assumption is that they were trying (or should have been trying) to write modern zoology manuals but could not figure out how to do it; and their most serious deficiency was their apparent inability to distinguish between fact and fancy. But it is, of course, ridiculous to insist that medieval people share our interests and priorities. That medieval scholars were capable of

Fig. 13.18. A page from a medieval bestiary, showing boar, ox, and bull. London, British Library, MS Harley 3244, fol. 47r (early 13th c.). By permission of the British Library.

writing something rather like a zoology manual is clear enough from the analogous case of the herbal or from the books on falconry that we have touched upon above. And their failure to make the bestiary into a zoological manual must, therefore, derive from the adoption of different aims.

What purpose, then, was the bestiary meant to serve? It was a collection of animal lore and mythology, rich in symbolism and associations, meant to instruct and entertain. And it surely did not occur, either to the compiler or to the reader, to inquire whether the stories were true in the sense that the claims of Aristotelian natural philosophy were expected to be true. A bestiary succeeded insofar as it effectively brought its reader into a world of traditional mythology, metaphor, and similitude.[47] We have similar mythologies of our own. Consider the lore surrounding the groundhog as a forecaster of the duration of winter, solemnly reported each February (at least in my part of the country) in newspapers and on radio and television. Does anybody believe in the truth of this forecast? Probably not; but to ask the question is to display a woeful misapprehension of the purpose of groundhog lore, which is not the "scientific" communication of meteorological truth, but participation in traditional community ritual, with all of the social and psychological benefits thus entailed.

Most of us become quite skillful at discriminating among different kinds of literary and artistic products in our own culture. We immediately know the difference between a scientific proposition, which must meet a variety of stiff epistemological tests in order to count as truly "scientific," and a Dr. Seuss story or a weather forecast offered to us by "Jimmy" the groundhog, which have quite different functions and must therefore be measured by different criteria. We need to become equally discerning in our study of medieval people and their achievements, including the various genres of art and literature they produced. Just as we have seen (chap. 11, above) that the medieval *mappamundi* generally had purposes quite different from those of a modern world atlas, so we must cease to presume that all medieval books that touch on natural phenomena were meant for philosophical or scientific purposes analogous to ours when we write a scientific textbook, and to understand that they may have been meant to please and inform their readers on a variety of other levels. As we acquire this kind of sophisticated discernment of the products of medieval culture, learning to judge the achievement in the light of the aim, we will be on our way to a fuller appreciation of the character, the achievements, and, yes, the charm of the Middle Ages.

FOURTEEN

The Legacy of Ancient and Medieval Science

THE CONTINUITY DEBATE

The historian's task is not to grade the past but to understand it. And my purpose in writing this book has been to describe the ancient and medieval scientific tradition, rather than to assess its merit or worth. But there seems little likelihood that the question of worth can be permanently avoided; it has figured prominently in previous accounts of early science, and it undoubtedly lurks in the minds of many readers of this one. In conclusion, therefore, let us steal cautiously into this dangerous, and hitherto forbidden, territory.

The question of worth has assumed many different forms. Critics and detractors have frequently inquired whether the intellectual activities and achievements about which historians of ancient and medieval science write, and to which this book has been devoted, were *really* science. That is, did they resemble or anticipate modern science? A subtler and more useful statement of the question might be: what difference did the ancient and medieval scientific tradition make in the long run? Did it have a permanent or continuing influence on the course or the shape of Western science, or was it an inconsequential cul-de-sac that ultimately led nowhere? Or to pose the question in its most common form, were medieval and early modern science continuous with each other, or discontinuous? The latter is the celebrated "continuity question," which has been the basis of a persistent, running feud between medievalists and "early modernists." Let us undertake a brief examination of the continuity debate—beginning with an account of its origins, in order to situate ourselves historically.[1]

The opinion that emerged from seventeenth-century appraisals of the past philosophical tradition typically acknowledged a measure of Greek achievement but judged the Middle Ages to have been a period of philosophical stagnation, if not desolation. Francis Bacon (1561–1626) set the tone when he wrote in his *New Organon* (1620) that the ages between antiquity and his own era were "unprosperous" for the sciences: "For neither the Arabians nor the Schoolmen need be mentioned, who in the intermediate times rather crushed the sciences with a multitude of treatises, than increased their weight." Voltaire (1694–1778) continued the attack, writing of the "general decay and degeneracy" that characterized the Middle Ages, and of the "cunning and simplicity. . . . brutality and artifice," of the medieval mind. Voltaire's younger contemporary Condorcet (1743–94) placed the blame for all of this squarely on the shoulders of the medieval church, arguing that "the triumph of Christianity was the signal for the complete decadence of philosophy and the sciences."[2]

The view of Bacon, Voltaire, and Condorcet was sharpened and widely disseminated in the second half of the nineteenth century by the distinguished Swiss historian Jacob Burckhardt (1818–97), to whom the concept of the "Renaissance" in its modern form is usually credited. Burckhardt conceived the Renaissance (roughly, in his view, the period from 1300 to 1500) as a rebirth of classical (that is, Greek) culture after the long dark period of the Middle Ages. He argued in his *The Civilization of the Renaissance in Italy* (1860) that "the Middle Ages . . . spared themselves the trouble of induction and free inquiry." Over against this failure of the human spirit, he maintained that during the Italian Renaissance, in each scientific discipline, "investigators of the period, chiefly through their rediscovery of the results attained by antiquity, mark a new epoch, with which the modern period of the science in question begins."[3] Burckhardt's enthusiasm for the Renaissance period proved contagious, as the overblown prose of one of his early (and extremely influential) followers, John A. Symonds, reveals:

> Beauty is a snare, pleasure a sin, the world a fleeting show, man fallen and lost, death the only certainty, judgment inevitable, hell everlasting, heaven hard to win; ignorance is acceptable to God as a proof of faith and submission; abstinence and mortification are the only safe rules of life: these were the fixed ideas of the ascetic medieval Church. The Renaissance shattered and destroyed them, rending the thick veil which they had drawn between the mind of man and the outer world and flashing the light of reality upon the darkened places of his own nature. For the mystic teaching of the Church was substituted culture in the classical humanities; a new ideal was established, whereby man strove to make himself the monarch of the globe. . . . The Renaissance was

the liberation of the reason from a dungeon, the double discovery of the outer and the inner world. An external event
determined the direction which this outburst of the spirit of
freedom should take. This was the contact of the modern with
the ancient mind. . . . The modern genius felt confidence
in its own energies when it learned what the ancients had
achieved.[4]

By this account, progress within the Western scientific tradition bypassed
the Middle Ages, following a route from classical antiquity to the Italian
Renaissance to the European science of the sixteenth and seventeenth centuries. In short, the "new science" of the early modern era owed a great
deal to antiquity, but little or nothing to the Middle Ages.

Quite a different vision of the course of science was articulated in the
early years of the twentieth century by the French physicist and philosopher Pierre Duhem (1861–1916). While exploring the origins of the science of statics, Duhem encountered the works of a series of medieval
mathematicians and natural philosophers who, in his judgment, had laid
the foundations for modern science, anticipating some of the most fundamental achievements of Galileo and his contemporaries. Duhem ultimately concluded that "the mechanics and physics of which modern times
are justifiably proud proceed, by an uninterrupted series of scarcely per-
ceptible improvements, from doctrines professed in the heart of the medieval schools."[5] If Duhem was right, the origins of modern science are to be
found not in the repudiation of medieval scholasticism and the return to
ancient ideas and sources by the "humanists" of the Renaissance, but in the
teachings of medieval natural philosophers and the interaction between
Christian theology and scholastic natural philosophy within the medieval
universities.

Duhem's claims set off the continuity debate, which has erupted with
a certain regularity throughout the twentieth century. Early support for
Duhem's campaign to rehabilitate the scientific tradition came from the
influential medievalists Charles Homer Haskins (1870–1937) and Lynn
Thorndike (1882–1965), writing in the twenties and thirties.[6] The decades
after World War II saw a dramatic expansion of historical research on medieval science; increased activity led to improved status and fresh claims
about the magnitude of the medieval scientific achievement. One of the
leading figures in the postwar movement was Marshall Clagett (1916–),
who made his mark primarily through the editing and translation of medieval scientific and mathematical texts. Another was Anneliese Maier (1905–
71), who produced a series of brilliant studies in which she demonstrated

by example how to read the sources more carefully and with much closer attention to their philosophical context. While challenging many of Duhem's more extreme claims and offering an analysis of medieval natural philosophy far subtler and more cautious than his, Maier reaffirmed the importance of the medieval contribution, both conceptual and methodological, to the forging of modern science.[7]

The continuity debate was a relatively quiet affair in the first half of the twentieth century. It heated up, however, after Alistair Crombie (1915–) issued a pair of manifestos concerning the relationship between medieval and early modern science, in the form of two books. In the first book, a survey of medieval and early modern science published in 1952, Crombie argued that "it was the growth of . . . 13th- and 14th-century experimental and mathematical methods that brought about the movement which by the 17th century had become so striking as to be called the Scientific Revolution."[8] Crombie elaborated on this theme in the second book, published a year later, maintaining (a) that the critical feature of early modern science was its possession of the proper methodology for the practice of science, the methodology of experimentation, and (b) that this methodology was a creation of the later Middle Ages:

> The thesis of this book is that a systematic theory of experimental science was understood and practised by enough philosophers [of the thirteenth and fourteenth centuries] for their work to produce the methodological revolution to which modern science owes its origin. . . . What seems to be the first appearance of a clear understanding of the principles of modern experimental science is found in the writings of the [thirteenth-century] English logician, natural philosopher, and scholar, Robert Grosseteste.[9]

Crombie's argument provoked a sharp response from early modern specialists, the most formidable of whom was the distinguished Continental scholar Alexandre Koyré (1892–1964). Koyré denied the importance of methodology in the abstract for the origins of modern science, suggesting that "*too much* methodology is dangerous," especially in the early stages of a scientific tradition. Nor was Koyré convinced that the methodological prescriptions of the medieval methodological tradition were the "right" ones—that is, the ones actually employed by Galileo and the others to whom we generally credit the founding of a new science in the sixteenth and seventeenth centuries.[10] In Koyré's opinion, the "Scientific Revolution" of the sixteenth and seventeenth centuries was not an outgrowth or extension of medieval science, but an intellectual "mutation" that entailed the "dissolution" of the medieval world view:

> What the founders of modern science . . . had to do, was not to criticize and combat certain faulty theories, and to correct or to replace them by better ones. They had to do something quite different. They had to destroy one world and to replace it by another. They had to reshape the framework of our intellect itself, to restate and to reform its concepts, to evolve a new approach to Being, a new concept of knowledge, a new concept of science.[11]

Koyré's scholarship powerfully influenced his generation of historians of science. A. R. Hall (1920–), an outspoken representative of the Koyré school of thought, wrote of the "totality of intellectual change during the late Renaissance" and portrayed the Scientific Revolution as "the phenomenon of the displacement . . . of one idea of nature by an alternative; one 'world-view' by another."[12]

In recent decades the terms of the debate have been refined. Ernan McMullin has replied to Crombie's more extreme methodological claims. While acknowledging a substantial measure of conceptual and linguistic continuity between medieval and early modern science, McMullin cannot detect methodological continuity. Indeed, he finds methodology to be precisely the area in which modern science broke most decisively with medieval thought.[13] Thomas Kuhn has developed an influential theory of scientific revolutions in general, portraying them as brief periods of radical change ("paradigm-shifts," he calls them) intervening between relatively static periods of puzzle-solving activity (which he labels "normal science"). As for *the* Scientific Revolution of the sixteenth and seventeenth centuries, Kuhn sees it as a collection of smaller, and (to a considerable degree) independent, revolutions within specific disciplines. He distinguishes between the "classical" mathematical sciences, such as optics and astronomy, and the new "Baconian" experimental sciences, such as electricity and chemistry. He denies that revolutionary change was possible in the newly emerging "Baconian" sciences, because of the absence within the medieval tradition of well-developed theoretical antecedents susceptible to radical transformation. He therefore locates revolutionary change almost exclusively in the "classical" sciences of astronomy, mechanics, and optics.[14]

The continuity debate has also been complicated by developments in Renaissance scholarship. There has been a tendency in the past three or four decades to endow Renaissance science with a character uniquely its own—that is, to redefine the Renaissance scientific achievement in ways that distinguish it from both the natural philosophy of the Middle Ages and that of the modern period. In the forefront of this movement has been Frances Yates (1899–1981), who identified the Renaissance contribution to the "genuine science" of the seventeenth century with its fascination for magic and the occult. Yates, in turn, has become the target of revisionist efforts, and at this point the question of the Renaissance and its scientific achievement remains something of a muddle.[15]

This is perhaps sufficient historical background to reveal the nature of the continuity debate and to pose the problem to which the remainder of this chapter will be devoted. But first a warning. If the debate were amenable to easy resolution, it would have ended long ago. It is unlikely, therefore, that we can definitively settle the issue here. Indeed, a definitive resolution may be forever out of reach on questions of this type, where the historian undertakes not simply to describe historical change, nor even to identify the causes of historical change, but to weigh the relative importance of episodes of historical change. Such judgments are at several removes from the historical data and emerge only as those data are viewed from the vantage point of larger interpretive schemes, which are not themselves amenable to easy, direct, or independent confirmation.[16] Inevitably, personal preference will figure heavily in the calculation. In the pages that follow, therefore, I do not expect to offer the final word on the continuity question. I propose, rather, to bring this book to a close by offering a few reflections (of necessity, somewhat personal) on the nature and significance of the medieval scientific achievement.

THE MEDIEVAL SCIENTIFIC ACHIEVEMENT

I would like to begin by revealing the colors under which I sail on the continuity question. It seems to me unquestionable that the more extreme claims made on behalf of medieval science and its anticipation of early modern developments are not merely exaggerated, but false. I *do* believe (as I will make clear below) that medieval natural philosophers made many important and enduring contributions to the Western scientific tradition—contributions that helped to shape this tradition and serve, in part, to explain it. But medieval natural philosophers did not anticipate the basic elements of early modern science; and the latter was far more than an extension, adaptation, and fuller articulation of the medieval worldview. In short, I accept the historical construct of the "Scientific Revolution."[17]

Much of the energy exended on the continuity debate has been focused on the question of scientific methodology. A version of the discontinuist view that flourished literally for centuries maintained that precisely what differentiated seventeenth-century science from that of the Middle Ages was the discovery and practice of a new experimental methodology. And at the heart of Crombie's defense of the continuity thesis lay the claim that this experimental methodology was a creation of the Middle Ages. Both opinions now seem considerably overdrawn. Recent studies of medieval and seventeenth-century scientific methodology have revealed the complexity of methodological theory and practice in the two periods and the total inadequacy of the simple generalizations on which past debates have depended. These studies make clear that medieval natural philosophers gave serious, critical attention to the details of the Aristotelian methodology, and that out of these efforts came interesting refinements and even departures from Aristotelian methodology. But it is also evident that the Aristotelian fundamentals were never relinquished: medieval philosophers continued to believe that the proper method of knowing involved syllogistic demonstration—deduction from universal first principles or premises taken to have self-evidential status.[18]

Natural philosophers of the seventeenth century departed from Aristotle far more radically, coming gradually as the century progressed to an appreciation of the hypothetical status of scientific claims, the potency of experiment as a technique of confirmation and disconfirmation, and the broad utility of mathematics as an instrument of measurement and analysis. I think that we must judge the gap between the methodologies of the two periods to be smaller than portrayed in the strong discontinuist view, but substantially larger than portrayed by Crombie and the continuists. If, methodologically speaking, the seventeenth century did not see a new world, it certainly saw a new day.[19]

An even stronger case for discontinuity can be made, I believe, if (following Alexandre Koyré's lead) we shift our focus from methodology to worldview or metaphysics. The specific metaphysical developments that I have in mind are the rejection, by the "new scientists" of the seventeenth century (Galileo, Descartes, Gassendi, Boyle, Newton, and others), of Aristotle's metaphysics of nature, form and matter, substance, actuality and potentiality, the four qualities, and the four causes; and the resuscitation and reformulation of the corpuscular philosophy of the ancient atomists. This produced a radical conceptual shift, which destroyed the foundations of natural philosophy as practiced for nearly two thousand years.[20]

Consider some of the consequences. In exchange for the purposeful, organized, and (in many ways) organic world of Aristotelian natural philosophy, the new metaphysics offered a mechanical world of lifeless matter, incessant local motion, and random collision. It stripped away the sensible qualities so central to Aristotelian natural philosophy, offering them second-class citizenship, as secondary qualities, or even reducing them to the status of sensory illusions. For the explanatory capabilities of form and matter, it offered the size, shape, and motion of invisible corpuscles—elevating local motion to a position of preeminence among the categories of change and reducing all causality to efficient and material causality. And for Aristotelian teleology, which discovered purpose *within* nature, it substituted the purposes of a creator God, imposed on nature from the outside.

Moreover, the new metaphysics had far-reaching implications for other aspects of natural philosophy, including methodology; and it can be plausibly (perhaps even persuasively) argued that many of the methodological innovations of the seventeenth century were rooted in the new metaphysics. It seems clear, for instance, that the abandonment of the essential natures of Aristotelian natural philosophy (which are to be discovered only by an examination of things in their natural, unfettered state) encouraged a more manipulative or experimental approach to natural phenomena.[21] Moreover, there can be no doubt that stress on invisible corpuscular mechanisms compelled serious thought about hypotheses and their epistemological status. And finally, the shift of emphasis from Aristotelian qualities to the geometrical properties of corpuscles (shape, size, and motion) surely encouraged the application of mathematics to nature.

Before leaving the case for discontinuity, I must call attention to various other circumstances that differentiate early modern science from its medieval predecessor. Although the institutionalization of natural philosophy in the medieval universities was a development of extraordinary importance, the size, scope, and organization of science continued to increase in the sixteenth and seventeenth centuries.[22] There can be no doubt, moreover, that in the early modern period science found itself in new social circumstances, which influenced its practice and altered its shape.[23] The sixteenth and seventeenth centuries also saw crucial innovations in instrumentation (invention of the telescope and the microscope, for example, which made it possible to explore the remote and the minute).[24] And finally, there were decisive theoretical developments within specific disciplines during the sixteenth and seventeenth centuries: the emergence of heliocentric cosmology in the sixteenth century and its triumph in the seventeenth; and a new theory of motion and inertia, with its far-reaching implications for both terrestrial and celestial dynamics. I will return below to the question of change within specific disciplines.

If it is granted that early modern science was discontinuous with medieval science in the ways described—that radical conceptual changes intervened to produce a Scientific Revolution of the sixteenth and seventeenth centuries—what does the medieval contribution then look like? If medieval natural philosophers failed to anticipate the science of the sixteenth and seventeenth centuries, do they as a consequence fall into disgrace or fade into insignificance? Did they contribute anything to the scientific movement that made a difference in the long run?

Before attempting to answer these questions, I would like to assure myself that one extremely elementary, but absolutely essential, point is understood. The ancient and medieval scholars whose intellectual efforts have been described in this book did not set out to solve sixteenth- and seventeenth-century scientific problems. They were preoccupied with a problem of their own—namely, the need to comprehend the world in which *they* lived, within the bounds of an inherited conceptual framework that defined the important questions and suggested useful ways of answering them. For the later Middle Ages, this conceptual framework was a rich composite of Aristotelian, Platonic, and Christian thought, adopted by medieval scholars because of its explanatory power. And as long as it successfully answered the questions they were asking, or promised future success, they had absolutely no reason to abandon it. Their aim was not to anticipate future worldviews, but to explore, articulate, employ, and criticize their own; and their competence as natural philosophers must be judged accordingly. In short, we must forgive medieval scholars for being medieval and cease to castigate them for not being modern. If we are lucky, future generations will do us a similar favor.[25]

If we can agree that medieval natural philosophers are not to be judged competent or incompetent by the degree to which they anticipated future developments, the question remains: did the Middle Ages make significant contributions to the science of the seventeenth century? The answer is unquestionably affirmative. In critical ways, medieval natural philosophers prepared the ground and paved the way for seventeenth-century achievement; and when a new structure for science was built in the seventeenth century, it contained a great many medieval materials. Let us briefly enumerate some of the more important medieval contributions.

First, scholars of the later Middle Ages created a broad intellectual tradition, in the absence of which subsequent progress in natural philosophy would have been inconceivable. During the early Middle Ages, as we have seen, Europe had a very limited intellectual life and possessed only a thin and fragmentary version of ancient philosophy. From these primitive beginnings, medieval Europeans managed, by the end of the fourteenth century, to create an advanced philosophical culture. They began by mastering the Latin sources already at their disposal. Having accomplished this, they undertook a massive translation effort, by which they gained possession of the fruits of Greek and Islamic philosophy—most notably, for our purposes, the works of Aristotle and his Islamic commentators, the medical philosophy of Hippocrates and Galen (as elaborated by Islamic physicians), and the works of a host of Greek and Islamic writers on mathematics and mathematical science.

Second, having gained possession of Greek and Islamic philosophy, medieval European philosophers plunged with relish into the task of grappling with its contents. The translated materials formed a heterogeneous collection of sources, which spoke with many different voices; and great ingenuity was required to sort through the differences, negotiate compromises, and arbitrate disputes. The dominant element was no doubt Aristotelian philosophy, but we must never simplify matters to the point of supposing that the latter was a seamless unity, comprehensive in its coverage, or without serious rival. Moreover, Aristotelian philosophy was a living tradition, in a state of continuous flux, as scholars endeavored to grasp its implications, correct its errors, resolve its inconsistencies, and apply it to new problems. And, of course, the new materials had to make their peace with Christian doctrine and vice versa. The medieval achievement, then, was no less than the formulation of a synthesis of classical and Christian thought, which would provide a framework for creative thought, including creative thought about nature, for several centuries.[26]

Third, this synthesis gained an institutional home in the medieval schools and universities. Natural philosophy had a precarious existence in the ancient world and in medieval Islam precisely because of its failure to secure more than sporadic institutional support. In the universities of medieval Europe, by contrast, the classical tradition in natural philosophy became one of the central elements of the curriculum, encountered (if not mastered) by everybody who embarked on higher studies. To be educated meant, by definition, to be educated in the philosophical tradition emanating from antiquity, including its natural philosophy. There may be a temptation to judge the mastery and institutionalization of an ancient philosophical tradition as too derivative an achievement to be of any interest, but in truth this was a critically important step. If, as we know by hindsight, ancient thought supplied the foundation on which the Western scientific tradition would build, it follows that the reception, assimilation, and institutionalization of ancient thought was a prerequisite to the further construction of that particular edifice.

Fourth, medieval natural philosophers were not content to merge Aristotelian philosophy with other intellectual traditions and oversee its absorption into medieval thought; they also submitted it to minute scrutiny and searching appraisal. The critical process began almost as soon as Aristotelian philosophy became available, and it continued through the late Middle Ages and into the early modern period. Some of the scrutiny was compelled or inspired by the encounter with theological doctrine. For example, the condemnation of 1277 provoked a reexamination of ideas of place and space, and this reexamination led a number of philosophers to accept the radically anti-Aristotelian notion that the universe is (or could be) surrounded by an infinite void space. And, of course, Aristotle's doctrine of the soul and the eternity of the world, as well as the deterministic tendencies in Aristotelian philosophy, were "corrected" with theological encouragement.[27]

But much of the appraisal had no theological roots, arising rather from tensions internal to Aristotelian philosophy, from its failure to account for the world as perceived by medieval natural philosophers, or from the need to give due consideration to non-Aristotelian alternatives. For example, Aristotle's theory of matter, form, and substance was sufficiently imprecise, incomplete, and even self-contradictory to provoke debate and criticism.[28] In the fourteenth century, the anti-Aristotelian idea that the earth might rotate on its axis, long acknowledged as an imaginary state of affairs with interesting implications, was submitted to meticulous and ingenious analysis by John Buridan and Nicole Oresme;[29] the latter took the analysis as far as it was to go before Galileo. And, for a final example, the later Middle Ages saw a complete overhaul of Aristotle's theory of motion, including new ideas about the nature of motion and the application of quantitative techniques to both kinematic and dynamic problems.[30]

Whatever the driving forces behind it, the medieval appraisal of Aristotle had vitally important implications for the course of natural philosophy. After all, one could not appraise Aristotelian philosophy without working out its implications and endeavoring to fill its gaps; and these were the necessary preliminaries to serious criticism. Insofar as criticism emerged during the Middle Ages, it tended to be piecemeal rather than wholesale, and it rarely led to the repudiation of a basic Aristotelian principle; indeed, a good bit of the criticism amounted to nibbling around the corners of Aristotelian philosophy, rather than a "feeding frenzy" of the kind that occurred in the sixteenth and seventeenth centuries. But its important contribution was to create a critical climate in which Aristotelian doctrine was regularly and carefully scrutinized, and in which its fate depended on its explanatory power rather than any authoritative status it might possess. This prepared the ground for a much broader and more destructive critique of Aristotle in the early modern period.[31]

Fifth and last, I would like to return to the question, touched on above, of developments at the disciplinary level. Many of the influential discontinuists have taken a holistic approach to the scientific revolution, focusing attention on broad metaphysical and methodological innovations, on the grounds that developments at this level will inevitably exercise a pervasive influence on the whole of the scientific enterprise. They are inclined, that is, to see the scientific revolution as an example of global change, which derived its energy from a new conception of nature or a new conception of the proper method of exploring nature's secrets (or both) and culminated as the implications of these innovations were felt in the various scientific disciplines. This is what Koyré had in mind when he claimed that man "lost his place in the world, or, more correctly perhaps, lost the very world in which he was living and about which he was thinking, and had to transform and replace not only his fundamental concepts and attributes, but even the very framework of his thought."[32] On this basis, the discontinuists have been disposed to overlook change in particular disciplines, or to argue that these were merely specific manifestations of more general tendencies. One of the most articulate defenders of discontinuity, A. R. Hall, has rebuked Thomas Kuhn and others for attempting to fracture the scientific revolution into a series of disciplinary events, arguing that it "refuses to dissolve into fragments," but comprises "an unbroken and interlocking series of new discoveries combined with changes in ideas, and it is quite arbitrary to resolve this into chapters concerned with discrete problems."[33]

It should be clear from the opening paragraphs of this section that I am sympathetic to the Koyré-Hall point of view. I agree that change at the metaphysical and methodological level was a critical feature of the scientific revolution, reverberating throughout the whole of the scientific enterprise. Moreover, I share Hall's opinion that scientific disciplines in the sixteenth and seventeenth centuries were often closely linked. Nonetheless, it seems to me a serious mistake to *restrict* attention to global change and *ignore* change at the disciplinary level. No doubt the individual discipline was influenced by general conceptions of nature and by broad methodological principles, but it should hardly be necessary to argue that the strength and character of the connection varied from one discipline to another and that metaphysical and methodological influences interacted differentially with specific features of the various disciplines. Surely no discipline was entirely self-sufficient, but neither were all disciplines locked into identical patterns of development.

But how does this discussion of global versus disciplinary change impinge on our assessment of the medieval scientific contribution? The connection is quite simple. A decision to concentrate on the global aspects of the scientific revolution, or to define the scientific revolution as a global event, powerfully tilts the continuity debate in the direction of discontinuity, because it focuses attention on precisely those aspects of scientific change within which discontinuity was most prominent. For what most clearly separated the natural philosophy of the Middle Ages from that of the early modern period was the appearance of a new conception of nature and new methodologies capable of influencing a wide range of scientific activity and belief; whereas it was *within* the various subjects or disciplines that medieval scholars made many of their most enduring contributions. A decision to concentrate on the global aspects of scientific change is therefore a decision to look for discontinuities. All interested parties would now grant (in principle) that there must have been elements of both continuity and discontinuity in the transition from medieval to early modern science; but to find them we must be willing to look for them in their customary habitats.[34]

If we shift our attention to developments within specific disciplines, I believe that a persuasive case can be made for a significant measure of linguistic, conceptual, and theoretical continuity between the Middle Ages and the early modern period. The questions asked by the "new scientists" of the seventeenth century were frequently those furnished by the medieval tradition. Much of the vocabulary of seventeenth-century science and many of the concepts denoted by that vocabulary were continuous with medieval usage. And on occasion medieval theories survived to be integrated into early modern science.

Examples are not difficult to find.[35] Galileo's analysis of the kinematics of falling bodies was, to a very considerable extent, an elaboration and application of kinematic principles developed at Oxford and Paris in the fourteenth century. The fact that Galileo saw the difference between kinematics and dynamics already reveals the influence of the tradition descending from Bradwardine and Oresme. As we probe Galileo's kinematics, it becomes apparent that the conceptual framework within which he was working—including conceptions of space, time, velocity, and acceleration—was that of medieval kinematics. His mathematical approach borrowed a great deal from the fourteenth century. And prominent in the finished Galilean theory were specific theorems of medieval origin, including the "mean-speed theorem" or "Merton rule." Indeed the mathematical relationships now considered the embodiment of Galileo's kinematic achievement ($v \propto t$ and $s \propto t^2$) are both simple elaborations of definitions or theorems articulated in the fourteenth century.[36]

Optics, especially in its more geometrical aspects, is another science that displays a high degree of continuity between the Middle Ages and the early modern period. For example, Kepler's theory of the retinal image (the claim that an inverted image of the visual field on the back of the eye is responsible for vision) was a brilliant achievement and an important innovation in visual theory. But it does not follow that the theory of the retinal image was revolutionary. It was the answer to an old question, worked out entirely within the medieval conceptual framework, obtained not by the repudiation of any of the fundamental principles of the discipline but by the determination to take those principles seriously. Likewise, Kepler's solution of the classic problem of radiation through small apertures (that is, his explanation of the puzzling fact that solar radiation projected through a square or triangular aperture gives rise, under suitable conditions, to a circular image of the sun) entailed no new geometrical principles, but merely the more rigorous application of the traditional axioms of the discipline.[37]

Additional examples could easily be produced. Copernican astronomy preserved the basic aims and principles of astronomy as it had been practiced since Ptolemy. Continuities were equally present in astrology, alchemy, anatomy, physiology, medicine, and natural history. As early modern science emerged in the sixteenth and seventeenth centuries, it maintained a complicated relationship to the past. Radically new in important aspects of its metaphysics and methodology, it nonetheless incorporated innumerable pieces of the medieval scientific achievement, sometimes unchanged, sometimes remolded to fit a new context. In order to demand respect for the medieval scientific achievement, we need not denigrate or diminish that of the sixteenth and seventeenth centuries. We need merely understand that the former shaped the latter and is therefore part of the ancestry of modern science. If we hope to understand what it means to inhabit the world of modern science, we cannot afford to be ignorant of the itinerary that brought us to it.

Notes

CHAPTER ONE

1. Bertrand Russell, *A History of Western Philosophy,* 2d ed., p. 514.

2. This point has been nicely made by David Pingree, "Hellenophilia versus the History of Science," manuscript of a lecture delivered at Harvard University, November 1990.

3. On ancient and medieval attitudes toward technology, see Elspeth Whitney, *Paradise Restored.*

4. The discussion of oral tradition in this section is heavily indebted to Jack Goody and Ian Watt, "The Consequences of Literacy" (p. 306 for the quoted phrase); Jack Goody, *The Domestication of the Savage Mind;* and Jan Vansina, *Oral Tradition as History.* See also Bronislaw Malinowski, *Myth in Primitive Psychology.*

5. This is certainly true of prehistoric cultures. Contemporary preliterate communities may have seen or heard about writing by contact with the literate world outside, but until they themselves have learned to write it is doubtful whether they can be said to have grasped the *idea* of writing.

6. Goody and Watt, "Consequences of Literacy," pp. 307–11. On oral tradition as "charter," see Malinowski, *Myth in Primitive Psychology,* pp. 42–44.

7. "Myth and Reality," in H. Frankfort, H. A. Frankfort, John A. Wilson, and Thorkild Jacobsen, *Before Philosophy,* pp. 24–25.

8. Jan Vansina, *The Children of Woot,* pp. 30–31, 198; Vansina, *Oral Tradition,* pp. 117, 125–29.

9. Vansina, *Oral Tradition,* pp. 130–33.

10. Vansina, *Children of Woot,* pp. 30–31. On origin myths and their relation to worldview, see also Vansina, *Oral Tradition,* pp. 133–37.

11. John A. Wilson, "The Nature of the Universe," in Frankfort et al., *Before Philosophy,* p. 63. For a recent and thorough discussion of Egyptian cosmology and cosmogony, see Marshall Clagett, *Ancient Egyptian Science,* vol. 1, pt. 1, pp. 263–372. On Egyptian religion, see James H. Breasted, *Development of Religion and Thought in Ancient Egypt.*

12. On Babylonian creation myths, see Thorkild Jacobsen, "Mesopotamia: The Cosmos as State," in Frankfort et al., *Before Philosophy,* chap. 5; S. G. F. Brandon, *Creation Legends of the Ancient Near East,* chap. 3.

13. On primitive or folk medicine, see Henry E. Sigerist, *A History of Medicine,* vol. 1: *Primitive and Archaic Medicine;* John Scarborough, ed., *Folklore and Folk Medicines.*

14. The idea of "primitive mentality" was developed by Lucien Lévy-Bruhl in his *How Natives Think;* for a critique, see Goody, *Domestication of the Savage Mind,* chap. 1; G. E. R. Lloyd, *Demystifying Mentalities,* introduction.

15. On "truth," especially "historical truth," see Vansina, *Oral Tradition,* pp. 21–24, 129–33.

16. Goody and Watt, "Consequences of Literacy," pp. 311–19. See also Barry Powell's reconstruction of the invention of Greek alphabetic writing: *Homer and the Origin of the Greek Alphabet.*

17. Goody, *Domestication of the Savage Mind,* p. 76.

18. Ibid., chap. 3.

19. Ibid., chap. 5.

20. Goody and Watt, "Consequences of Literacy," pp. 319–43; Lloyd, *Demystifying Mentalities,* chap. 1.

21. The answer is 14. On Egyptian mathematics, see Otto Neugebauer, *The Exact Sciences in Antiquity,* chap. 4; B. L. van der Waerden, *Science Awakening: Egyptian, Babylonian and Greek Mathematics,* chap. 1; G. J. Toomer, "Mathematics and Astronomy," in J. R. Harris, ed., *The Legacy of Egypt,* pp. 27–54; R. J. Gillings, "The Mathematics of Ancient Egypt"; and Carl B. Boyer, *A History of Mathematics,* chap. 2.

22. Richard Parker, "Egyptian Astronomy, Astrology, and Calendrical Reckoning."

23. On Babylonian mathematics, see Neugebauer, *Exact Sciences in Antiquity,* chaps. 2–3; van der Waerden, *Science Awakening,* chaps. 2–3; van der Waerden, "Mathematics and Astronomy in Mesopotamia"; and Boyer, *History of Mathematics,* chap. 3.

24. For an analysis of the question of ancient "algebra," see Sabetai Unguru, "History of Ancient Mathematics: Some Reflections on the State of the Art"; and Unguru, "On the Need to Rewrite the History of Greek Mathematics."

25. On Mesopotamian or Babylonian astronomy, see Neugebauer, *Exact Sciences in Antiquity,* chap. 5; B. L. van der Waerden, with Peter Huber, *Science Awakening 2: The Birth of Astronomy,* chaps. 2–8; van der Waerden, "Mathematics and Astronomy in Mesopotamia"; Asger Aaboe, "On Babylonian Planetary Theories"; and the essays collected in Neugebauer, *Astronomy and History.* For a highly technical account, see Otto Neugebauer, *A History of Ancient Mathematical Astronomy,* 1:347–555.

26. On Babylonian astrology and its relationship to astronomy, see van der Waerden and Huber, *Science Awakening 2,* chap. 5.

27. Neugebauer, *Exact Sciences in Antiquity,* pp. 104–9; van der Waerden and Huber, *Science Awakening 2,* chap. 6. For a more popular account, see Stephen Toulmin and June Goodfield, *The Fabric of the Heavens,* chap. 1.

28. Sigerist, *History of Medicine,* 1:276. On Egyptian medicine, besides Sigerist, see Paul Ghalioungui, *The House of Life, Per Ankh: Magic and Medical Science in Ancient Egypt;* Ghalioungui, *The Physicians of Pharaonic Egypt;* and John R. Harris, "Medicine," in Harris, ed., *The Legacy of Egypt.* On surgery, see Guido Majno, *The Healing Hand,* chap. 3.

29. B. Ebbell, *The Papyrus Ebers.*

30. James Henry Breasted, *The Edwin Smith Surgical Papyrus.*

31. On Mesopotamian medicine, see Sigerist, *History of Medicine,* 1: pt. 4; Robert Biggs, "Medicine in Ancient Mesopotamia"; Majno, *Healing Hand,* chap. 2.

CHAPTER TWO

1. For a convenient introduction to Homer, see Jasper Griffin, *Homer;* or M. I. Finley, *The World of Odysseus.*

2. Hesiod, *Theogony and Works and Days,* trans., with introduction and notes, by M. L. West.

3. On Greek mythology, see Edith Hamilton, *Mythology.* The quoted passage is from Homer's *Odyssey,* bk. 5, trans. S. H. Butcher and Andrew Lang, in *The Complete Works of Homer* (New York: Modern Library, [1935]), pp. 79–82.

4. *The Poems of Hesiod,* trans. R. M. Frazer, p. 32. On Hesiod see also Friedrich Solmsen, *Hesiod and Aeschylus.*

5. See also the interesting analysis of this problem by Paul Veyne, *Did the Greeks Believe in Their Myths?*

6. I owe my outlook on these matters especially to G. E. R. Lloyd, *Early Greek Science: Thales to Aristotle,* chap. 1; Lloyd, *Magic, Reason and Experience;* Lloyd, *The Revolutions of Wisdom;* and Gregory Vlastos, *Plato's Universe,* chap. 1. See these works for additional bibliography.

7. On the Milesians, see Lloyd, *Early Greek Science,* chap. 2; David Furley, *The Greek Cosmologists,* vol. 1: *The Formation of the Atomic Theory and its Earliest Critics;* G. S. Kirk and J. E. Raven, *The Presocratic Philosophers,* chaps. 2–4; and Jonathan Barnes, *The Presocratic Philosophers,* 1: chaps. 2–3. On Thales and astronomy, see D. R. Dicks, *Early Greek Astronomy to Aristotle,* pp. 42–44.

8. Kirk and Raven, *Presocratic Philosophers,* p. 87.

9. Charles H. Kahn, *Anaximander and the Origins of Greek Cosmology,* p. 233.

10. G. E. R. Lloyd, *Demystifying Mentalities,* esp. chap. 1; Lloyd, *Early Greek Science,* pp. 10–15.

11. Kirk and Raven, *Presocratic Philosophers,* p. 199. On the interpretation of this passage, see also Furley, *Greek Cosmologists,* pp. 35–36; Barnes, *Presocratic Philosophers,* 1:60–64.

12. On the atomists, see Furley, *Greek Cosmologists,* chaps. 9–11; Kirk and Raven, *Presocratic Philosophers,* chap. 17; Barnes, *Presocratic Philosophers,* 2:40–75; Cyril Bailey, *The Greek Atomists and Epicurus.*

13. Kirk and Raven, *Presocratic Philosophers,* pp. 328–29. See also Furley, *Greek Cosmologists,* chap. 7.

14. Aristotle, *Metaphysics,* I.5.985^{b}33–986^{a}2, in *The Complete Works of Aristotle,* ed. Jonathan Barnes, 2:1559. On the Pythagoreans, see also Kirk and Raven, *Presocratic Philosophers,* chap. 9; Furley, *Greek Cosmologists,* chap. 5; Barnes, *Presocratic Philosophers,* 2:76–94; and Lloyd, *Early Greek Science,* chap. 3.

15. In the nineteenth century William Stanley Jevons nicely captured this version of the Pythagorean vision: "Not without reason did Pythagoras represent the world as ruled by number. Into almost all our acts of thought number enters, and in proportion as we can define numerically we enjoy exact and useful knowledge of the

universe." From Jevons's *Principles of Science,* employed as an epigraph by Margaret Schabas, *A World Ruled by Number: William Stanley Jevons and the Rise of Mathematical Economics* (Princeton: Princeton University Press, 1990).

16. Lloyd, *Early Greek Science,* pp. 36–37; Furley, *Greek Cosmologists,* pp. 33–36.

17. Kirk and Raven, *Presocratic Philosophers,* p. 271. Also, Furley, *Greek Cosmologists,* pp. 36–42; Lloyd, *Early Greek Science,* pp. 37–39; and Barnes, *Presocratic Philosophers,* 1: chaps. 10–11.

18. Kirk and Raven, *Presocratic Philosophers,* chap. 11; Barnes, *Presocratic Philosophers,* 1: chaps. 12–13. Aristotle's comment is found in his *Physics,* VI.2.233ᵃ 22–23. In a second paradox, Zeno describes a race between Achilles (noted for his swiftness) and a tortoise (noted for its slowness): if the tortoise is given a head start, however small, Achilles will never be able to catch it, since by the time Achilles reaches the tortoise's starting point, the tortoise will have moved beyond it to a new position; by the time Achilles reaches this new position, the tortoise will have moved beyond that; and so on, *ad infinitum.*

19. Kirk and Raven, *Presocratic Philosophers,* p. 271.

20. Lloyd, *Early Greek Science,* chap. 4.

21. Kirk and Raven, *Presocratic Philosophers,* p. 422. See also Lloyd, *Early Greek Science,* chap. 4.

22. Kirk and Raven, *Presocratic Philosophers,* pp. 325, 394.

23. The scholarship on Plato is enormous. For brief, recent introductions, see R. M. Hare, *Plato;* and David J. Melling, *Understanding Plato.* I have been influenced especially by Vlastos, *Plato's Universe,* and by the translation-commentaries of the various Platonic dialogues by Francis M. Cornford.

24. Plato, *Republic,* bk. VII, 514a–521b.

25. Lloyd, *Early Greek Science,* pp. 68–72. Plato, *Phaedo,* 65b; Plato, *Republic,* bk. VII, 532, trans. Francis M. Cornford, p. 252.

26. Vlastos, *Plato's Universe,* chap. 2. On Plato's cosmology, see also *Plato's Cosmology: The "Timaeus" of Plato,* trans. and commentary by Francis M. Cornford; and Richard D. Mohr, *The Platonic Cosmology.*

27. Vlastos, *Plato's Universe,* chap. 3.

28. Ibid., chap. 2.

29. The quoted passages are from Plato, *Timaeus,* trans. Cornford, 30d, p. 40; 34b, p. 58.

30. Vlastos, *Plato's Universe,* pp. 61–65; Friedrich Solmsen, *Plato's Theology.*

CHAPTER THREE

1. For more on the Lyceum, see below, chap. 4.

2. There is a great deal of excellent introductory literature on Aristotle; see especially Jonathan Barnes, *Aristotle;* Abraham Edel, *Aristotle and His Philosophy;* and G. E. R. Lloyd, *Aristotle: The Growth and Structure of His Thought.*

3. Barnes, *Aristotle,* pp. 32–51; Edel, *Aristotle,* chaps. 3–4; Lloyd, *Aristotle,* chap. 3.

4. The technical term used to denote this doctrine of Aristotle is "hylomorphism," from *hylē,* the Greek term for matter, and *morphē,* the Greek term for form.

5. On Aristotle's epistemology, see especially Edel, *Aristotle*, chaps. 12–15; Lloyd, *Aristotle*, chap. 6; Jonathan Lear, *Aristotle: The Desire to Understand*, chap. 4; and Marjorie Grene, *A Portrait of Aristotle*, chap. 3.

6. On this subject, see Jonathan Barnes, "Aristotle's Theory of Demonstration"; G. E. R. Lloyd, *Magic, Reason and Experience*, pp. 200–220.

7. On change, see especially Edel, *Aristotle*, pp. 54–60; and Sarah Waterlow, *Nature, Change, and Agency in Aristotle's "Physics,"* chaps. 1, 3.

8. On Aristotle's conception of "nature," see Waterlow, *Nature, Change, and Agency*, chaps. 1–2; James A. Weisheipl, "The Concept of Nature."

9. Waterlow, *Nature, Change, and Agency*, pp. 33–34; Ernan McMullin, "Medieval and Modern Science: Continuity or Discontinuity?" pp. 103–29, esp. 118–19.

10. Edel, *Aristotle*, chap. 5.

11. See especially Friedrich Solmsen, *Aristotle's System of the Physical World*; and Lloyd, *Aristotle*, chaps. 7–8.

12. *On the Heavens*, I.4.270^{b}13–16, quoted from *The Complete Works of Aristotle*, ed. Jonathan Barnes, 1:451.

13. Lloyd, *Aristotle*, chap. 7.

14. Ibid., chap. 8. On alchemy, see below, chap. 12.

15. For Aristotle on the void, see Solmsen, *Aristotle's System of the Physical World*, pp. 135–43; David Furley, *Cosmic Problems*, pp. 77–90.

16. Furley, *Cosmic Problems*, chaps. 12–13.

17. Aristotle deals with the shape of the earth in *On the Heavens*, II.13. See also D. R. Dicks, *Early Greek Astronomy to Aristotle*, pp. 196–98. On the myth that ancient and medieval people believed in a flat earth, see Jeffrey B. Russell, *Inventing the Flat Earth: Columbus and Modern Historians*.

18. Waterlow, *Nature, Change, and Agency*, pp. 103–4.

19. For a careful analysis of the fine points, see James A. Weisheipl, "The Principle *Omne quod movetur ab alio movetur* in Medieval Physics." Reprinted in Weisheipl, *Nature and Motion in the Middle Ages*, pp. 75–97.

20. On natural motion, see Aristotle's *On the Heavens*, I.6, and *Physics*, IV.8. On forced motion, see *Physics*, VIII.5. For discussion, see Marshall Clagett, *The Science of Mechanics in the Middle Ages*, pp. 421–33; Clagett, *Greek Science in Antiquity*, pp. 64–68.

21. Lloyd, *Aristotle*, pp. 139–58.

22. There has been a recent burst of interest in Aristotle's biology. See especially Lloyd, *Aristotle*, chap. 4; Lloyd, *Early Greek Science*, pp. 115–24; Anthony Preus, *Science and Philosophy in Aristotle's Biological Works*; Martha Craven Nussbaum, *Aristotle's "De motu animalium"*; Pierre Pellegrin, *Aristotle's Classification of Animals*; and Allan Gotthelf and James G. Lennox, eds., *Philosophical Issues in Aristotle's Biology*. Older sources, still useful, are W. D. Ross, *Aristotle: A Complete Exposition of His Works and Thought*, 5th ed., chap. 4; and Thomas E. Lones, *Aristotle's Researches in Natural Science*.

23. Aristotle, *On the Parts of Animals*, I.5. See also Lloyd, *Aristotle*, pp. 69–73.

24. Lloyd, *Aristotle*, pp. 76–81, 86–90; Lloyd, *Early Greek Science*, pp. 116–18; Pellegrin, *Aristotle's Classification of Animals*.

25. *History of Animals*, VI.3.561^{a}3–19, in *Complete Works*, ed. Barnes, 1:883.

26. Lloyd, *Aristotle,* pp. 90–93; D. M. Balme, "The Place of Biology in Aristotle's Philosophy."

27. Aristotle, *De generatione animalium,* II.1.733^{b}25–27, in *Complete Works,* ed. Barnes, 1:1138.

28. On Aristotle's doctrine of the soul and its faculties, see Lloyd, *Aristotle,* chap. 9; Ross, *Aristotle,* chap. 5; Ackrill, *Aristotle,* pp. 68–78.

29. Aristotle, *De generatione animalium,* II.1.733^{a}34–733^{b}14, in *Complete Works,* ed. Barnes, 1:1138. On biological reproduction, see also Ross, *Aristotle,* pp. 117–22; Preus, *Science and Philosophy in Aristotle's Biological Works,* pp. 48–107.

30. Aristotle, *On the Parts of Animals,* III.6.668^{b}33–669^{a}7. On teleology in Aristotle's biology, see also Ross, *Aristotle,* pp. 122–27; Nussbaum, *Aristotle's "De motu animalium,"* pp. 59–106.

31. On method in Aristotle's biology, see Lloyd, *Aristotle,* pp. 76–81; Lloyd, *Magic, Reason and Experience,* pp. 211–20; Nussbaum, *Aristotle's "De motu animalium,"* pp. 107–42.

CHAPTER FOUR

1. The classic source on ancient education, to be used with caution, is H. I. Marrou, *A History of Education in Antiquity.* More reliable is John Patrick Lynch, *Aristotle's School.* See also Robin Barrow, *Greek and Roman Education.*

2. Lynch, *Aristotle's School,* pp. 65–66. On sophistic teaching in general, see pp. 38–54.

3. On Plato's Academy, see Lynch, *Aristotle's School,* pp. 54–63; Harold Cherniss, *The Riddle of the Early Academy.*

4. Cherniss, *Riddle of the Early Academy,* p. 65.

5. On the Lyceum, see Lynch, *Aristotle's School,* chaps. 1, 3; also Felix Grayeff, *Aristotle and His School.*

6. Lynch, *Aristotle's School,* chap. 6.

7. The best source on the Alexandrian Museum and Library, including their social context, is P. M. Fraser, *Ptolemaic Alexandria,* esp. 1:305–35. See also Lynch, *Aristotle's School,* pp. 121–23, 194.

8. On Theophrastus as a natural philosopher, see G. E. R. Lloyd, *Greek Science after Aristotle;* J. B. McDiarmid, "Theophrastus," *Dictionary of Scientific Biography,* 13:328–34.

9. On Theophrastus and the Lyceum, see Lynch, *Aristotle's School,* pp. 97–108. The quotation comes from p. 101, with one minor alteration.

10. Ibid., pp. 101–3, 193.

11. On Strato, see Lloyd, *Greek Science after Aristotle,* pp. 15–20; Marshall Clagett, *Greek Science in Antiquity,* pp. 68–71; H. B. Gottschalk, "Strato of Lampsacus," *Dictionary of Scientific Biography,* 13:91–95; and David Furley, *Cosmic Problems,* pp. 149–60. On his relationship to the Lyceum, see Lynch, *Aristotle's School,* passim.

12. On the ancient Aristotelian commentators, see the articles collected by Richard Sorabji, ed., *Aristotle Transformed.*

13. For the quotations, see Diogenes Laertius, *Lives of Eminent Philosophers,*

trans. R. D. Hicks, 2:649, 667 (substituting "philosophy" for "science" in the latter passage). On Epicurean philosophy, see A. A. Long, *Hellenistic Philosophy,* 2d ed.; David J. Furley, *Two Studies in the Greek Atomists;* Elizabeth Asmis, *Epicurus' Scientific Method;* Lloyd, *Greek Science after Aristotle,* chap. 3; Cyril Bailey, *The Greek Atomists and Epicurus;* and the sources edited and translated by A. A. Long and D. N. Sedley, *The Hellenistic Philosophers,* 2 vols.

14. Quotations (from Lucretius's *De rerum natura,* II.15 and IV.840) are borrowed from Long and Sedley, *Hellenistic Philosophers,* 1:47; and Lucretius, *De rerum natura,* trans. W. H. D. Rouse and M. F. Smith, rev. 2d ed., p. 343.

15. Asmis, *Epicurus' Scientific Method,* chap. 8.

16. Lloyd, *Greek Science after Aristotle,* pp. 23–24.

17. On Stoic philosophy in general, see Lloyd, *Greek Science after Aristotle,* chap. 3; F. H. Sandbach, *The Stoics;* Long, *Hellenistic Philosophy;* Marcia L. Colish, *The Stoic Tradition from Antiquity to the Early Middle Ages;* the articles collected in Ronald H. Epp, ed., *Recovering the Stoics;* and the sources gathered in Long and Sedley, *The Hellenistic Philosophers.*

18. On Stoic natural philosophy, in addition to the sources cited above, see David E. Hahm, *The Origins of Stoic Cosmology;* and S. Sambursky, *Physics of the Stoics.*

19. A. A. Long, "The Stoics on World-Conflagration and Everlasting Recurrence," in Epp, *Recovering the Stoics,* pp. 13–37; Hahm, *Origins of Stoic Cosmology,* chap. 6.

20. Cicero, *On Divination,* I.125–26; quoted by Long and Sedley, *Hellenistic Philosophers,* 1:337.

CHAPTER FIVE

1. On this question, see Friedrich Solmsen, *Aristotle's System of the Physical World,* pp. 46–48, 259–62; David C. Lindberg, "On the Applicability of Mathematics to Nature"; James A. Weisheipl, *The Development of Physical Theory in the Middle Ages,* pp. 13–17, 48–62.

2. Aristotle, *Metaphysics,* XI.3.1061ᵃ30–35, trans. Hugh Tredennick, 2:67–69.

3. On Greek mathematics, see B. L. van der Waerden, *Science Awakening: Egyptian, Babylonian and Greek Mathematics,* chaps. 4–8; Carl B. Boyer, *A History of Mathematics,* chaps. 4–11; Thomas Heath, *A History of Greek Mathematics.* For an overview of recent research, see J. L. Berggren, "History of Greek Mathematics: A Survey of Recent Research."

4. Wilbur Knorr. *The Evolution of the Euclidean Elements.*

5. On Euclid, see Heath, *Greek Mathematics,* chap. 11; Boyer, *History of Mathematics,* chap. 7; also Thomas Heath's translation of the *Elements,* with lengthy and detailed commentary.

6. E. J. Dijksterhuis, *Archimedes;* T. L. Heath, ed., *The Works of Archimedes.* On Archimedes' medieval influence, see Marshall Clagett, ed. and trans., *Archimedes in the Middle Ages.*

7. On early Greek astronomy, see especially Bernard R. Goldstein and Alan C. Bowen, "A New View of Early Greek Astronomy"; D. R. Dicks, *Early Greek Astronomy to Aristotle;* Lloyd, *Early Greek Science,* chap. 7; Thomas Heath, *Aristarchus of*

Samos, The Ancient Copernicus. For a highly technical account, see Otto Neugebauer, *A History of Ancient Mathematical Astronomy,* 2:571–776.

8. For a useful discussion of the basic planetary phenomena and the two-sphere model, see Thomas S. Kuhn, *The Copernican Revolution,* chap. 1; Michael J. Crowe, *Theories of the World from Antiquity to the Copernican Revolution,* chap. 1.

9. On Plato's astronomical knowledge, see Dicks, *Early Greek Astronomy,* chap. 5.

10. Ibid., chap. 6.

11. Otto Neugebauer, "On the 'Hippopede' of Eudoxus"; and David Hargreave, "Reconstructing the Planetary Motions of the Eudoxean System."

12. Dicks, *Early Greek Astronomy,* chap. 7.

13. On Aristotle, see ibid., chap. 7; G. E. R. Lloyd, *Aristotle,* pp. 147–53. Aristotle dealt with the planetary spheres in his *Metaphysics,* XII.8. For further discussion of the debate over the aims of astronomy, see chap. 11, below.

14. Heath, *Aristarchus of Samos,* pt. 1, chap. 18; Otto Neugebauer, "On the Allegedly Heliocentric Theory of Venus by Heraclides Ponticus"; G. J. Toomer, "Heraclides Ponticus," *Dictionary of Scientific Biography,* 15:202–5; and especially Bruce S. Eastwood, "Heraclides and Heliocentrism: An Analysis of the Text and Manuscript Diagrams," an unpublished chapter (generously made available to me by Eastwood), scheduled for publication in his book, provisionally entitled *Before Copernicus: Planetary Theory and the Circumsolar Idea from Antiquity to the Twelfth Century.* On the subsequent history of the idea of a sun-centered motion for Mercury and Venus, see Eastwood, "Kepler as Historian of Science: Precursors of Copernican Heliocentrism according to *De revolutionibus,* I, 10."

15. Heath, *Aristarchus of Samos,* pt. 2; G. E. R. Lloyd, *Greek Science after Aristotle,* pp. 53–61. We have almost no details about Aristarchus's life. Since the island of Samos was under Ptolemaic rule during his lifetime, it is possible that he undertook astronomical and cosmological studies in Alexandria; see P. M. Fraser, *Ptolemaic Alexandria,* 1:397; William H. Stahl, "Aristarchus of Samos," *Dictionary of Scientific Biography,* 1:246.

16. In general, parallax (or "geometrical parallax") is the shift in apparent position of a celestial body against the stellar background, caused by a change of viewpoint on the part of the observer. In the present case, the absence of solar parallax means the absence of any discernible effect on the sun's position owing to different points of observation on the surface of the earth.

17. Heath, *Aristarchus of Samos,* pt. 2, chap. 3; G. J. Toomer, "Hipparchus," *Dictionary of Scientific Biography,* 15:205–24; D. R. Dicks, "Eratosthenes," *Dictionary of Scientific Biography,* 4:388–93; Albert Van Helden, *Measuring the Universe,* chap. 2.

18. Otto Neugebauer, "Apollonius' Planetary Theory"; Toomer, "Hipparchus." On Hellenistic astronomy more generally, see Neugebauer, *Ancient Mathematical Astronomy,* 2:779–1058.

19. For introductions to Ptolemy, see Lloyd, *Greek Science after Aristotle,* chap. 8; Crowe, *Theories of the World,* chaps. 3–4. For more technical discussions, see G. J. Toomer, "Ptolemy," *Dictionary of Scientific Biography,* 11:186–206; Neugebauer, *Ancient Mathematical Astronomy,* 1:21–343; Olaf Pedersen, *A Survey of the Almagest;* Ptolemy, *Almagest,* ed. and trans. G. J. Toomer.

20. Toomer, "Ptolemy," pp. 192–94.

21. Bernard R. Goldstein, *The Arabic Version of Ptolemy's "Planetary Hypotheses"*; G. E. R. Lloyd, "Saving the Appearances." See also below, chap. 11, pp. 264–67.

22. On ancient theories of vision, see David C. Lindberg, *Theories of Vision from al-Kindi to Kepler,* chap. 1.

23. On the geometrical approach to vision, see especially A. Mark Smith, "Saving the Appearances of the Appearances"; also Albert Lejeune, *Euclide et Ptolémée;* Lindberg, *Theories of Vision,* pp. 11–17.

24. On Ptolemy, see Albert Lejeune, *Recherches sur la catoptrique grecque;* Lejeune, *Euclide et Ptolémée;* A. Mark Smith, "Ptolemy's Search for a Law of Refraction." For a French translation of Ptolemy's *Optics,* see *L'Optique de Claude Ptolémée,* ed. and trans. Albert Lejeune.

25. For example, Ptolemy constructed the following table, comparing angles of incidence with their corresponding angles of refraction for light (or visual rays) passing from air to water:

Angle of incidence	10°	20°	30°	40°	50°	60°	70°	80°
Angle of refraction	8°	$15\frac{1}{2}$°	$22\frac{1}{2}$°	29°	35°	$40\frac{1}{2}$°	$45\frac{1}{2}$°	50°

Note that the differences between successive angles of refraction from an arithmetic series: $7\frac{1}{2}$, 7, $6\frac{1}{2}$, 6, $5\frac{1}{2}$, 5, $4\frac{1}{2}$. For an analysis of these results, see Lejeune, *Recherches,* pp. 152–66; Smith, "Ptolemy's Search for a Law of Refraction."

26. See Marshall Clagett, *The Science of Mechanics in the Middle Ages,* chap. 1.

27. For analyses of Archimedes' work, see the sources cited above, n. 6.

CHAPTER SIX

1. On "primitive" Greek medicine, see Fridolf Kudlien, "Early Greek Primitive Medicine." On varieties of Greek medical practitioners, see Owsei Temkin, "Greek Medicine as Science and Craft"; Lloyd, *Magic, Reason and Experience,* pp. 37–49. On Greek and Roman medicine in general, see the useful bibliography of recent studies by John Scarborough, "Classical Antiquity: Medicine and Allied Sciences, An Update."

2. Ludwig Edelstein, "The Distinctive Hellenism of Greek Medicine," reprinted in Edelstein, *Ancient Medicine,* ed. Owsei Temkin and C. Lilian Temkin, pp. 367–97; for the attitudes of Homer and Hesiod, see pp. 376–78. Also James Longrigg, "Presocratic Philosophy and Hippocratic Medicine." On magic and religion in Greek medicine, see Ludwig Edelstein, "Greek Medicine and Its Relation to Religion and Magic"; G. E. R. Lloyd, *Magic, Reason and Experience,* chap. 1; and Lloyd, *The Revolutions of Wisdom,* chap. 1.

3. Emma J. Edelstein and Ludwig Edelstein, *Asclepius: A Collection and Interpretation of the Testimonies,* 1:235.

4. There is an enormous literature on Hippocratic medicine. For recent interpretations, see Wesley D. Smith, *The Hippocratic Tradition;* and G. E. R. Lloyd's introduction to his edition of the *Hippocratic Writings.* See also Lloyd, *Early Greek Science,* chap. 5; Lloyd, *Magic, Reason and Experience,* passim; Longrigg, "Presocratic Philosophy and Hippocratic Medicine"; and the first three articles in Edelstein's *Ancient Medicine.*

5. On the relationship of medicine to philosophy, see Longrigg, "Presocratic Philosophy and Hippocratic Medicine"; Ludwig Edelstein, "The Relation of Ancient Philosophy to Medicine"; Lloyd, *Magic, Reason and Experience,* pp. 86–98.

6. Quoted from the translation of J. Chadwick and W. N. Mann, in *Hippocratic Writings,* ed. Lloyd, pp. 237–38. On medicine and the supernatural in the Hippocratic corpus, see especially G. E. R. Lloyd, *Revolutions of Wisdom,* chap. 1; Lloyd *Magic, Reason and Experience,* chap. 1; and Longrigg, "Presocratic Philosophy and Hippocratic Medicine."

7. *The Nature of Man,* trans. J. Chadwick and W. N. Mann, in *Hippocratic Writings,* ed. Lloyd, p. 262. The theory of four humors did not dominate Hippocratic physiology, as it would Galenic and subsequent physiology. Some Hippocratic writers accepted only two humors (usually bile and phlegm), and of course many discussed no humoral theory at all.

8. *Epidemics,* I.i, trans. J. Chadwick and W. N. Mann, in *Hippocratic Writings,* ed. Lloyd, pp. 87–88, with minor changes of punctuation.

9. Trans. J. Chadwick and W. N. Mann, in *Hippocratic Writings,* ed. Lloyd, p. 79. Despite his skepticism, the author of this treatise advances hypotheses of his own.

10. Ibid., p. 247.

11. Edelstein, "Greek Medicine and Its Relation to Religion and Magic," in *Ancient Medicine,* pp. 241–43.

12. See *Hippocrates with an English Translation,* 4:423, 437.

13. On Hellenistic medicine, in addition to sources cited below, see John Scarborough, *Roman Medicine;* Ralph Jackson, *Doctors and Diseases in the Roman Empire.*

14. For an excellent survey of these developments, see James Longrigg, "Anatomy in Alexandria in the Third Century B.C."

15. On Herophilus, see the authoritative volume by Heinrich von Staden, *Herophilus;* also Longrigg, "Superlative Achievement," pp. 164–77.

16. On Erasistratus, see James Longrigg, "Erasistratus," *Dictionary of Scientific Biography,* 4:382–86; Longrigg, "Superlative Achievement," pp. 177–84; G. E. R. Lloyd, *Greek Science after Aristotle,* pp. 80–85.

17. On the medical sects, see Heinrich von Staden, "Hairesis and Heresy: The Case of the *haireseis iatrikai*"; also Michael Frede, "The Method of the So-Called Methodical School of Medicine"; Ludwig Edelstein, "The Methodists"; Edelstein, "Empiricism and Skepticism in the Teaching of the Greek Empiricist School"; and P. M. Fraser, *Ptolemaic Alexandria,* 1:338–76.

18. On Galen's life and times, see Vivian Nutton, "The Chronology of Galen's Early Career"; Nutton, "Galen in the Eyes of His Contemporaries"; and John Scarborough, "The Galenic Question."

19. Fraser, *Ptolemaic Alexandria,* 1:339. On Galen's thought, see Owsei Temkin, *Galenism;* Luis García Ballester, "Galen as a Medical Practitioner: Problems in Diagnosis"; Smith, *Hippocratic Tradition,* chap. 2; John Scarborough, "Galen Redivivus: An Essay Review"; Phillip De Lacy, "Galen's Platonism"; the essays contained in Fridolf Kudlien and Richard J. Durling, eds., *Galen's Method of Healing;* and Lloyd, *Greek Science after Aristotle,* chap. 9. Also the introductions to Galen's *On the Usefulness of the Parts of the Body,* ed. and trans. Margaret T. May; and Peter Brain, *Galen on Bloodletting.*

20. I.2, quoted from García Ballester, "Galen as a Medical Practitioner," with minor improvements in punctuation.

21. I have been strongly tempted to prepare a schematic diagram of Galen's physiological system. In the end, however, I have reluctantly concluded that this cannot be done without foisting an unacceptable amount of modern anatomical and physiological knowledge onto Galen. For previous attempts to diagram Galen's physiology, see Charles Singer, *A Short History of Anatomy and Physiology from the Greeks to Harvey,* p. 61; Karl E. Rothschuh, *History of Physiology,* p. 19.

22. On (at least) one occasion Galen referred to the possibility of a "natural spirit" or "natural pneuma" in the venous blood; this suggestion was taken up by his followers, who made it a canonical part of the Galenic system; see Owsei Temkin, "On Galen's Pneumatology."

23. *On the Natural Faculties,* trans. A. J. Brock, III.15, p. 321, with minor editing, following Lloyd, *Greek Science after Aristotle,* p. 149.

24. In addition to works on Galen cited above, see *Galen, On Respiration and the Arteries,* ed. and trans. David J. Furley and J. S. Wilkie.

25. On Galen's methodology, in addition to works cited above, see Galen, *Three Treatises on the Nature of Science.*

26. III.10, trans. May, 1:189.

27. VII.1, trans. May, 2:729–31.

28. For a good example, see George Sarton, *Galen of Pergamon.*

29. Temkin, *Galenism,* p. 24.

CHAPTER SEVEN

1. Horace, *Epistles,* II.1.156.

2. On these developments, see especially Elizabeth Rawson, *Intellectual Life in the Late Roman Republic.*

3. On Cicero, see below.

4. See especially William H. Stahl, *Roman Science,* pp. 50 (where Stahl refers to the "curse of the popularizer") and 55 (where he labels popularizers as "hacks").

5. On Aratus, see ibid., pp. 36–38. Stahl is also one of the best sources on the Roman popularization of Greek science.

6. See, for example, Arnold Reymond, *History of the Sciences in Greco-Roman Antiquity,* trans. Ruth Gheury de Bray, p. 92.

7. Quoted by Cicero, *De re publica,* I.xviii.30, trans. Clinton Walker Keyes (London: Heinemann, 1928), p. 55.

8. For attempts to reconstruct the scientific content of Varro's *Disciplines,* see Rawson, pp. 158–64; Stephen Gersh, *Middle Platonism and Neoplatonism: The Latin Tradition,* 2:825–40; William H. Stahl, Richard Johnson, and E. L. Burge, *Martianus Capella and the Seven Liberal Arts,* 1:44–53.

9. When we speak of "Platonists" in this period, we always mean members of one or another of the philosophical traditions emanating from Plato and the Academy. Many of these "Platonists" defended doctrines that Plato would have repudiated. For a useful analysis of Cicero's philosophy and its relationship to the Platonic tradition, see Gersh, *Middle Platonism and Neoplatonism,* 1:53–154.

10. On Lucretius, see Stahl, *Roman Science,* pp. 80–83.

11. On these authors, see Stahl, *Roman Science,* chap. 6; Gersh, *Middle Platonism and Neoplatonism,* chap. 3; and the relevant articles in the *Dictionary of Scientific Biography.* On Seneca, see also *Physical Science in the Time of Nero: Being a Translation of the "Quaestiones naturales" of Seneca,* trans. John Clarke. For Celsus, see his *On Medicine,* trans. W. G. Spencer.

12. See the articles in Roger French and Frank Greenaway, eds., *Science in the Early Roman Empire: Pliny the Elder, His Sources and Influence;* also Stahl, *Roman Science,* chap. 7; for an older, but fuller, analysis see Lynn Thorndike, *A History of Magic and Experimental Science,* 1:41–99.

13. On Pliny's method, see A. Locher, "The Structure of Pliny the Elder's Natural History."

14. Pliny (the Younger), *Letters,* trans. William Melmoth, revised by W. M. L. Hutchinson, III.5, 1:198.

15. *Natural History,* II.25–37, II.54, II.86, VII.2, IX.8.

16. *Natural History,* II.6–22; Olaf Pedersen, "Some Astronomical Topics in Pliny"; and Bruce S. Eastwood, "Plinian Astronomy in the Middle Ages and Renaissance."

17. Gersh, *Middle Platonism and Neoplatonism,* chap. 7; Macrobius, *Commentary on the Dream of Scipio,* trans. with introduction and notes by William H. Stahl; Stahl, *Roman Science,* pp. 153–69.

18. On Martianus, see Stahl et al., *Martianus Capella,* 1:9–20. This work also contains a full translation of *The Marriage of Philology and Mercury,* with accompanying commentary.

19. Ibid., 2:278.

20. There has been considerable discussion of the sources of Martianus's astronomical knowledge. See ibid., 1:50–53; Eastwood, "Plinian Astronomy in the Middle Ages and Renaissance," pp. 198–99.

21. On Martianus's theory of the inferior planets and its subsequent history, see Bruce S. Eastwood, "Kepler as Historian of Science: Precursors of Copernican Heliocentrism according to *De revolutionibus* I, 10."

22. For a list of other translators and their translations, see Marshall Clagett, *Greek Science in Antiquity,* pp. 154–56.

23. On this question, see Gersh, *Middle Platonism and Neoplatonism,* pp. 421–34. Gersh also discusses Calcidius's philosophical stance.

24. On Boethius, see Lorenzo Minio-Paluello, "Boethius, Anicius Manlius Severinus," *Dictionary of Scientific Biography,* 2:228–36; Gersh, *Middle Platonism and Neoplatonism,* chap. 9; Clagett, *Greek Science in Antiquity,* pp. 150–53.

25. On this subject, see David C. Lindberg, "Science and the Early Church," which contains additional bibliography; also Lindberg, "Science as Handmaiden: Roger Bacon and the Patristic Tradition." For a brief history of the early church, see Henry Chadwick, *The Early Church.*

26. See especially Henry Chadwick, *Early Christian Thought and the Classical Tradition;* Charles N. Cochrane, *Christianity and Classical Culture;* A. H. Armstrong and R. A. Markus, *Christian Faith and Greek Philosophy.*

27. The use of the feminine gender (handmaiden instead of manservant) in Augustine's metaphor may interest some. Augustine's choice has nothing to do with notions of female inferiority, but derives simply from the gender (gram-

matically speaking) of the Latin noun *philosophia*. The mistress, *theologia*, is also feminine.

28. On Roman education, see especially Stanley F. Bonner, *Education in Ancient Rome;* also H. I. Marrou, *A History of Education in Antiquity;* N. G. Wilson, *Scholars of Byzantium,* esp. pp. 8–27; and Robin Barrow, *Greek and Roman Education.* On early medieval education, see Pierre Riché, *Education and Culture in the Barbarian West, Sixth through Eighth Centuries;* M. L. W. Laistner, *Thought and Letters in Western Europe, A.D. 500–900,* new ed., chaps. 2–3.

29. On monasticism and monastic schools, see Jean Leclercq, O.S.B., *The Love of Learning and the Desire for God: A Study of Monastic Culture;* Riché, *Education and Culture,* chap. 4.

30. The evidence is persuasively presented in M. M. Hildebrandt, *The External School in Carolingian Society.*

31. Laistner, *Thought and Letters,* chap. 5.

32. On Cassiodorus and Vivarium, see James J. O'Donnell, *Cassiodorus.*

33. On Isidore, see Stahl, *Roman Science,* pp. 213–23; J. N. Hillgarth, "Isidore of Seville, St.," *Dictionary of the Middle Ages,* 6:563–66; H. Liebeschütz, "Boethius and the Legacy of Antiquity," in A. H. Armstrong, ed., *The Cambridge History of Later Greek and Early Medieval Philosophy,* pp. 555–64; Jacques Fontaine, *Isidore de Séville et la culture classique dans l'Espagne wisigothique;* and Ernest Brehaut, *An Encyclopedist of the Dark Ages: Isidore of Seville.*

34. On Bede, see Stahl, *Roman Science,* pp. 223–32; Charles W. Jones, "Bede," *Dictionary of the Middle Ages,* 2:153–56; Wesley M. Stevens, *Bede's Scientific Achievement;* Peter Hunter Blair, *The World of Bede,* esp. chap. 24; and Clagett, *Greek Science in Antiquity,* pp. 160–65.

CHAPTER EIGHT

1. On learning in the Byzantine Empire, see N. G. Wilson, *Scholars of Byzantium;* F. E. Peters, *The Harvest of Hellenism.*

2. A convenient discussion of the ancient Greek commentators on Aristotle is to be found in Richard Sorabji's General Introduction to Christian Wildberg's translation of John Philoponus's *Against Aristotle on the Eternity of the World,* pp. 1–17. On Themistius and Simplicius, see G. Verbeke's articles in the *Dictionary of Scientific Biography,* 12:440–43; 13:307–9; Ilsetraut Hadot, ed., *Simplicius: sa vie, son oeuvre, sa survie.* On Philoponus, see Richard Sorabji, ed., *Philoponus and the Rejection of Aristotelian Science.*

3. For an excellent analysis of the process of cultural diffusion in general, see F. E. Peters, *Allah's Commonwealth;* also Peters, *Aristotle and the Arabs: The Aristotelian Tradition in Islam;* and *Harvest of Hellenism.* A great deal of useful information is contained in De Lacy O'Leary, *How Greek Science Passed to the Arabs.*

4. On these developments, see W. H. C. Frend, *The Rise of the Monophysite Movement.*

5. The best sources are Peters, *Aristotle and the Arabs,* chap. 2; and Peters, *Allah's Commonwealth,* introduction and chap. 5. See also Arthur Vööbus, *History of the School of Nisibis.*

6. O'Leary, *How Greek Science Passed to the Arabs,* pp. 150–53; Peters, *Allah's*

Commonwealth, pp. 318, 377–78, 383, 529; Peters, *Aristotle and the Arabs,* pp. 44–45, 53, 59; and Majid Fakhry, *A History of Islamic Philosophy,* pp. 15–16.

7. For a reappraisal of the Jundishapur legend, see Michael W. Dols, "The Origins of the Islamic Hospital: Myth and Reality." I am grateful also for a discussion of the problem with Vivian Nutton, whose work on the subject is forthcoming.

8. Among the innumerable books on the early history of Islam, the following are particularly useful: Peters, *Allah's Commonwealth;* G. E. von Grunebaum, *Classical Islam;* and Philip K. Hitti, *History of the Arabs from the Earliest Times to the Present.* For lavish illustrations accompanied by an excellent text, see Bernard Lewis, ed., *Islam and the Arab World.*

9. On Ḥunayn, see Lufti M. Sa'di, "A Bio-Bibliographical Study of Hunayn ibn Is-haq al-Ibadi (Johannitius)"; and the two articles on Ḥunayn by G. C. Anawati and Albert Z. Iskandar, in the *Dictionary of Scientific Biography,* 15:230–49. On the translations more generally, see Peters, *Allah's Commonwealth* and *Aristotle and the Arabs;* also O'Leary, *How Greek Science Passed to the Arabs;* and Fakhry, *History of Islamic Philosophy,* pp. 16–31.

10. This point is emphasized by G. E. von Grunebaum, *Islam: Essays in the Nature and Growth of a Cultural Tradition,* chap. 6. It is applied specifically to the mathematical sciences by George Saliba, "The Development of Astronomy in Medieval Islamic Society," esp. pp. 217–21.

11. For a survey of Islamic medicine, see the introductory essay in *Medieval Islamic Medicine: Ibn Riḍwān's Treatise "On the Prevention of Bodily Ills in Egypt,"* trans. with an introduction by Michael W. Dols.

12. See especially, von Grunebaum, *Islam: Essays in the Nature and Growth of a Cultural Tradition,* chap. 6. For a somewhat less extreme statement of the same view, see Peters, *Aristotle and the Arabs,* chap. 4.

13. The most eloquent statement of the "appropriation thesis" is by A. I. Sabra, "The Appropriation and Subsequent Naturalization of Greek Science in Medieval Islam"; see also Sabra, "The Scientific Enterprise."

14. On Muslim education, see Bayard Dodge, *Muslim Education in Medieval Times;* George Makdisi, *The Rise of Colleges: Institutions of Learning in Islam and the West;* Peters, *Aristotle and the Arabs,* chap. 4; Fazlur Rahman, *Islam,* 2d ed., chap. 11; and Mehdi Nakosteen, *History of Islamic Origins of Western Education.*

15. "Physics, History of," *The Catholic Encyclopedia* (1911), 11:48.

16. There is no acceptable general history of Islamic science. For an excellent sketch, see A. I. Sabra, "Science, Islamic," *Dictionary of the Middle Ages,* 11:81–88; and "The Scientific Enterprise." See also Max Meyerhof, "Science and Medicine," and Carra de Vaux, "Astronomy and Mathematics," both in Thomas Arnold and Alfred Guillaume, eds., *The Legacy of Islam;* E. S. Kennedy, "The Exact Sciences," in R. N. Frye, ed., *The Cambridge History of Iran,* 4:378–95; and Kennedy, "The Arabic Heritage in the Exact Sciences." There is also an excellent, growing specialist literature on particular scientific disciplines.

17. Al-Kindī is quoted by Richard Walzer, "Arabic Transmission of Greek Thought to Medieval Europe," pp. 172–73, 175 (with minor changes). For al-Bīrūnī, see de Vaux, "Astronomy and Mathematics," p. 376.

18. See A. I. Sabra, "Al-Farghānī," *Dictionary of Scientific Biography,* 4:541–45; B. A. Rosenfeld and A. T. Grigorian, "Thābit ibn Qurra," *Dictionary of Scientific Bi-*

ography, 13:288–95; and Willy Hartner, "Al-Battānī," *Dictionary of Scientific Biography,* 1:507–16. On Islamic astronomy, see also below, chap. 11, and the sources cited there.

19. See David C. Lindberg, *Theories of Vision from al-Kindi to Kepler,* esp. chap. 4; A. I. Sabra, "Ibn al-Haytham," *Dictionary of Scientific Biography,* 6:189–210.

CHAPTER NINE

1. John Marenbon, *Early Medieval Philosophy (480–1150),* chaps. 4–5; M. L. W. Laistner, *Thought and Letters in Western Europe,* chaps. 3–4; G. R. Evans, *The Thought of Gregory the Great,* pp. 55–68.

2. For a careful discussion of the exact meaning and significance of the decree to establish monastery schools, see M. M. Hildebrandt, *The External School in Carolingian Society.* On Alcuin and the Carolingian educational reforms more generally, see Heinrich Fichtenau, *The Carolingian Empire,* chap. 4; John Marenbon, *From the Circle of Alcuin to the School of Auxerre,* chap. 2; Laistner, *Thought and Letters,* chap. 7.

3. On Eriugena and his circle, see John J. O'Meara, *Eriugena;* Marenbon, *Early Medieval Philosophy,* chap. 6; Marenbon, *Circle of Alcuin,* chaps. 3–4.

4. On Gerbert, see Harriet Pratt Lattin, ed. and trans., *The Letters of Gerbert with His Papal Privileges as Sylvester II;* Cora E. Lutz, *Schoolmasters of the Tenth Century,* chap. 12; Uta Lindgren, *Gerbert von Aurillac und das Quadrivium: Untersuchungen zur Bildung im Zeitalter der Ottonen.* On the Ripoll manuscript, see J. M. Millas-Vallicrosa, "Translations of Oriental Scientific Works."

5. On the technology of this period, see especially Lynn White, Jr., *Medieval Technology and Social Change;* and Jean Gimpel, *The Medieval Machine: The Industrial Revolution of the Middle Ages.* On the water wheel, see Terry S. Reynolds, *Stronger than a Hundred Men: A History of the Vertical Water Wheel,* chap. 2.

6. See David Herlihy, "Demography," *Dictionary of the Middle Ages,* 4:136–48.

7. On the medieval schools, see Nicholas Orme, *English Schools of the Middle Ages;* John J. Contreni, "Schools, Cathedral," *Dictionary of the Middle Ages,* 11:59–63; Contreni, *The Cathedral School of Laon from 850 to 930;* Marenbon, *Early Medieval Philosophy,* chap. 10; John W. Baldwin, *The Scholastic Culture of the Middle Ages,* chap. 3; Richard W. Southern, "The Schools of Paris and the School of Chartres"; Southern, "From Schools to University"; and Paul F. Grendler, *Schooling in Renaissance Italy,* esp. chap. 1.

8. See Southern, "The Schools of Paris and the School of Chartres," pp. 114–18; Jean Leclercq, "The Renewal of Theology," pp. 72–73.

9. See Richard W. Southern, "Humanism and the School of Chartres"; the vehement reply by Nikolaus Häring, "Chartres and Paris Revisited"; and Southern's rejoinder in "The Schools of Paris and the School of Chartres."

10. Charles Homer Haskins, *The Renaissance of the Twelfth Century,* chaps. 4, 7.

11. Colin Morris, *The Discovery of the Individual, 1050–1200,* p. 46. On the rationalistic turn of the eleventh and twelfth centuries, see also the ambitious book by Alexander Murray, *Reason and Society in the Middle Ages.*

12. Although his intellectual formation occurred, at least in part, within the monastic tradition—in his late twenties he studied at the monastery of Bec in north-

ern France—Anselm faithfully represents the broader intellectual currents of his day and did much to shape the theological traditions of the twelfth-century schools.

13. Jasper Hopkins, *A Companion to the Study of St. Anselm;* G. R. Evans, *Anselm and a New Generation;* Richard W. Southern, *Saint Anselm,* esp. pp. 123–37; and Southern, *Medieval Humanism,* chap. 2. On the distinction between monastic and "scholastic" theology in the twelfth century, see Jean Leclercq, "The Renewal of Theology."

14. Abelard's *Epistolae,* no. 17, in *Patrologia latina,* ed. J.-P. Migne, vol. 178 (Paris: J.-P. Migne, 1855), col. 375. For a brief account of Abelard's life and thought, see David E. Luscombe, *Peter Abelard;* Luscombe, "Peter Abelard."

15. On twelfth-century Platonism, see M.-D. Chenu, *Nature, Man, and Society in the Twelfth Century,* chap. 2; and Tullio Gregory, "The Platonic Inheritance." On other specific aspects of twelfth-century philosophy, see the citations below. On twelfth-century natural philosophy in general, see chap. 1 of Chenu and the essays in Dronke, *History of Twelfth-Century Western Philosophy,* especially chap. 1: Winthrop Wetherbee, "Philosophy, Cosmology, and the Twelfth-Century Renaissance," pp. 21–53. Older, but still useful, are Charles Homer Haskins, *Studies in the History of Mediaeval Science;* and Lynn Thorndike, *A History of Magic and Experimental Science,* vol. 2, chaps. 35–50.

16. Nikolaus M. Häring, "The Creation and Creator of the World according to Thierry of Chartres and Clarenbaldus of Arras"; Peter Dronke, "Thierry of Chartres"; J. M. Parent, *La doctrine de la création dans l'école de Chartres.*

17. On the idea of nature, see Tullio Gregory, "La nouvelle idée de nature et de savoir scientifique au XIIe siècle"; and a number of the essays contained in *La filosofia della natura nel medioevo.*

18. On William of Conches, see Tullio Gregory, *Anima mundi: La filosofia di Guglielmo di Conches e la scuola di Chartres;* Dorothy Elford, "William of Conches"; Thorndike, *History of Magic,* vol. 2, chap. 37. On Adelard of Bath, see Charles Burnett, ed., *Adelard of Bath.* For the quoted passages, see William of Conches, *Philosophia mundi,* ed. Gregor Maurach (Pretoria: University of South Africa, 1974), I.22, pp. 32–33 (a text slightly different from, and better than, that in Migne's *Patrologia latina*); Adelard of Bath, *Quaestiones naturales,* ed. M. Müller (*Beiträge zur Geschichte der Philosophie des Mittelalters,* vol. 31, pt. 2) (Münster: Aschendorff, 1934), p. 8, quoted by William J. Courtenay, "Nature and the Natural in Twelfth-Century Thought," p. 10; and Beryl Smalley, *The Study of the Bible in the Middle Ages,* p. 144. Chenu (*Nature, Man, and Society*) and Courtenay offer useful summaries and analyses of the problem.

19. The quoted passages are taken from Tullio Gregory, "The Platonic Inheritance," pp. 65, 57. Cf. similar remarks by Adelard of Bath, *Quaestiones naturales,* 4, p. 8; quoted by Courtenay, "Nature and the Natural," p. 10.

20. William J. Courtenay, "Nature and the Natural in Twelfth-Century Thought" and "The Dialectic of Divine Omnipotence," both in Courtenay's *Covenant and Causality in Medieval Thought,* chaps. 3–4.

21. On humanism, see Morris, *Discovery of the Individual;* Southern, *Medieval Humanism,* chap. 4. For a significant qualification, see Caroline Walker Bynum, "Did the Twelfth Century Discover the Individual?"

22. The best sketch of the history of medieval astrology is to be found in Olaf

Pedersen, "Astrology," *Dictionary of the Middle Ages,* 1:604–10. For further discussion and additional bibliography, see the final section of chap. 11, below.

23. On mathematics in the twelfth century, see Charles Burnett, "Scientific Speculations"; Gillian R. Evans, *Old Arts and New Theology,* pp. 119–136; Evans, "The Influence of Quadrivium Studies in the Eleventh- and Twelfth-Century Schools"; and Guy Beaujouan, "The Transformation of the Quadrivium." For the quoted passage, see Häring, "The Creation and Creator of the World according to Thierry of Chartres," p. 196.

24. For a general discussion of the translations, see David C. Lindberg, "The Transmission of Greek and Arabic Learning to the West"; Marie-Thérèse d'Alverny, "Translations and Translators"; Millas-Vallicrosa, "Translations of Oriental Scientific Works"; Charles S. F. Burnett, "Translation and Translators, Western European," *Dictionary of the Middle Ages,* 12:136–42; Jean Jolivet, "The Arabic Inheritance"; and Haskins, *Studies in the History of Mediaeval Science,* passim.

25. Michael McVaugh, "Constantine the African," *Dictionary of Scientific Biography,* 3:393–95.

26. Richard Lemay, "Gerard of Cremona," *Dictionary of Scientific Biography,* 15:173–92. For a list of Gerard's translations, see the document translated by Michael McVaugh, in Edward Grant, ed., *A Source Book in Medieval Science,* pp. 35–38.

27. For two different opinions, see Lemay, "Gerard of Cremona," pp. 174–75; d'Alverny, "Translations and Translators," pp. 453–54.

28. Lorenzo Minio-Paluello, "Moerbeke, William of," *Dictionary of Scientific Biography,* 9:434–40.

29. On the importance of astrology in the revival of Aristotle, see Richard Lemay, *Abu Ma'shar and Latin Aristotelianism in the Twelfth Century.*

30. M. B. Hackett, "The University as a Corporate Body," p. 37.

31. Excellent introductions to the history of the universities are to be found in John W. Baldwin, *The Scholastic Culture of the Middle Ages;* Astrik L. Gabriel, "Universities," in *Dictionary of the Middle Ages,* 12:282–300; and Alan B. Cobban, *The Medieval Universities: Their Development and Organization.* Older classics, still useful, are Charles H. Haskins, *The Rise of Universities;* and Hastings Rashdall, *The Universities of Europe in the Middle Ages,* ed. F. M. Powicke and A. B. Emden, 3 vols. For excellent recent work on the English universities, see Catto, *History of the University of Oxford,* vol. 1; William J. Courtenay, *Schools and Scholars in Fourteenth-Century England;* and Alan B. Cobban, *The Medieval English Universities: Oxford and Cambridge to c. 1500.* On Paris, see Stephen C. Ferruolo, *The Origins of the University: The Schools of Paris and Their Critics, 1100–1215.*

32. On patronage and privileges, see Pearl Kibre, *Scholarly Privileges in the Middle Ages;* and Guy Fitch Lytle, "Patronage Patterns and Oxford Colleges, c. 1300–c. 1530."

33. I owe these estimates to my colleague, William J. Courtenay.

34. For the data, see James H. Overfield, "University Studies and the Clergy in Pre-Reformation Germany," pp. 277–86.

35. For actual data on student mortality, see Guy Fitch Lytle, "The Careers of Oxford Students in the Later Middle Ages," p. 221.

36. There is a great deal of useful literature on the curriculum of the medieval

universities. In general, see Baldwin, *Scholastic Culture;* James A. Weisheipl, "Curriculum of the Faculty of Arts at Oxford in the Fourteenth Century"; Weisheipl, "Developments in the Arts Curriculum at Oxford in the Early Fourteenth Century"; and the relevant articles in Catto, *The Early Oxford Schools,* vol. 1 of *The History of the University of Oxford.*

37. On science in the medieval curriculum, in addition to the works of Baldwin and Weisheipl cited above, see Pearl Kibre, "The *Quadrivium* in the Thirteenth Century Universities (with Special Reference to Paris)." Also Guy Beaujouan, "Motives and Opportunities for Science in the Medieval Universities"; Edward Grant, "Science and the Medieval University"; James A. Weisheipl, "Science in the Thirteenth Century"; and Edith Dudley Sylla, "Science for Undergraduates in Medieval Universities."

38. It needs to be stressed that learning in the Middle Ages was conceived as the mastery of a set of standard texts. This is in contrast to the modern view, which views education as the mastery of certain subjects and consider the choice of specific texts to be incidental.

CHAPTER TEN

1. On the earliest dissemination of Aristotle's works in the West, see Aleksander Birkenmajer, "Le rôle joué par les médecins et les naturalistes dans la réception d'Aristote au XIIe et XIIIe siècles"; Richard Lemay, *Abu Ma'Shar and Latin Aristotelianism in the Twelfth Century.* On the reception of Aristotle in the universities, see the excellent discussion by Fernand Van Steenberghen, *Aristotle in the West;* a parallel analysis can be found in Van Steenberghen's *The Philosophical Movement in the Thirteenth Century.* The useful discussion by David Knowles, *The Evolution of Medieval Thought,* is based largely on Van Steenberghen. For an excellent survey of Aristotelianism in the West, see William A. Wallace, "Aristotle in the Middle Ages," *Dictionary of the Middle Ages,* 1:456–69. On Oxford, see Van Steenberghen, *Aristotle in the West,* chap. 6; D. A. Callus, "Introduction of Aristotelian Learning to Oxford."

2. On Aristotelianism at Paris, see Van Steenberghen, *Aristotle in the West,* chaps. 4–5. See also John W. Baldwin, *Masters, Princes, and Merchants: The Social Views of Peter the Chanter and His Circle,* 1:104–7; and Richard C. Dales, *The Intellectual Life of Western Europe in the Middle Ages,* pp. 243–46. For a translation of documents bearing on the events in Paris, see Lynn Thorndike, *University Records and Life in the Middle Ages,* pp. 26–40; reprinted, with additional notes, in Edward Grant, ed., *A Source Book in Medieval Science,* pp. 42–44.

3. For the Latin text, see Henricus Denifle and Aemilio Chatelain, *Chartularium Universitatis Parisiensis,* 4 vols. (Paris: Delalain, 1889–97), 1:138, 143. For another English translation, which includes more of the text, see Thorndike, *University Records,* p. 40.

4. Van Steenberghen, *Aristotle in the West,* pp. 89–110; David C. Lindberg, ed. and trans., *Roger Bacon's Philosophy of Nature,* pp. xvi–xvii.

5. Van Steenberghen, *Aristotle in the West,* pp. 17–18, 64–66, 127–28. A brief account of Avicenna's philosophy can be found in Majid Fakhry, *A History of Islamic Philosophy,* pp. 147–83; and G. C. Anawati and Albert Z. Iskandar, "Ibn Sīnā," *Dictionary of Scientific Biography,* 15:494–501.

6. Van Steenberghen, *Aristotle in the West,* pp. 18–20, 89–93. The most important translator of Averroes was Michael Scot (d. ca. 1235), beginning in 1217 and continuing into the 1230s, but there is no evidence that his translations were used at Paris until after 1230; see ibid., pp. 89–94; Lorenzo Minio-Paluello, "Michael Scot," *Dictionary of Scientific Biography,* 9:361–65. On Averroes' philosophy, see Fakhry, *History of Islamic Philosophy,* pp. 302–25; Roger Arnaldez and Albert Z. Iskandar, "Ibn Rushd," *Dictionary of Scientific Biography,* 12:1–9.

7. See, for example, Aristotle, *On the Heavens,* I.10–11. For a discussion of Aristotle's doctrine, see Friedrich Solmsen, *Aristotle's System of the Physical World,* pp. 51, 266–74, 288, 422–24.

8. See, for example, St. Thomas Aquinas, Siger of Brabant, and St. Bonaventure, *On the Eternity of The World,* trans. Cyril Vollert et al.; Boethius of Dacia, *On the Supreme Good, On the Eternity of the World, On Dreams;* Richard C. Dales, "Time and Eternity in the Thirteenth Century," *Journal of the History of Ideas,* 49 (1988): 27–45. For a full account of medieval discussions, see Dales, *Medieval Discussions of the Eternity of the World.*

9. For a short account of determinism and indeterminism in Aristotle, see Abraham Edel, *Aristotle and His Philosophy,* pp. 95, 389–401. For a full analysis, see Richard Sorabji, *Necessity, Cause, and Blame.* For an excellent analysis of the Islamic attack on this problem, see Barry S. Kogan, *Averroes and the Metaphysics of Causation.*

10. For the biblical doctrine, see Matthew 10:29–31.

11. On Aristotle's theory of the soul, see G. E. R. Lloyd, *Aristotle,* chap. 9. On the Christian response, see Fernand Van Steenberghen, *Thomas Aquinas and Radical Aristotelianism,* pp. 29–70; Knowles, *Evolution of Medieval Thought,* pp. 206–18, 292–96.

12. For an extended account of Averroistic monopsychism and the Western response, see Van Steenberghen, *Thomas Aquinas and Radical Aristotelianism,* pp. 29–74.

13. For a full account, see William J. Courtenay, *Teaching Careers at the University of Paris in the Thirteenth and Fourteenth Centuries.*

14. On Grosseteste and his scholarly career, see the excellent study by James McEvoy, *The Philosophy of Robert Grosseteste;* for the dating of Grosseteste's commentary on the *Posterior Analytics,* see pp. 512–14. On Grosseteste's life and work, see also D. A. Callus, ed., *Robert Grosseteste, Scholar and Bishop;* and Richard W. Southern, *Robert Grosseteste.* On Grosseteste's investigation of Aristotle's logic and its effect on his scientific methodology, see the somewhat overstated analysis by A. C. Crombie, *Robert Grosseteste and the Origins of Experimental Science, 1100–1700,* chaps. 3–4; also William A. Wallace, *Causality and Scientific Explanation,* 1:28–47.

15. On Grosseteste's cosmogony, see below, chap. 11, and the accompanying notes.

16. On Bacon's scientific career, see Stewart C. Easton, *Roger Bacon and His Search for a Universal Science;* Theodore Crowley, *Roger Bacon: The Problem of the Soul in His Philosophical Commentaries.* For a convenient biographical sketch, see Lindberg, *Bacon's Philosophy of Nature,* pp. xv–xxvi.

17. On the term "handmaiden" and its gender implications, see above, chap. 7, n. 27.

18. *The Opus majus of Roger Bacon,* ed. John H. Bridges, 3 vols. (London: Williams and Norgate, 1900), 3:36. On Bacon's defense of the new philosophy, see David C. Lindberg, "Science as Handmaiden: Roger Bacon and the Patristic Tradition."

19. Bonaventure's position in relation to the various philosophical traditions of the thirteenth century has been much disputed. For an account of the alternatives and an attempt to adjudicate among them, see Van Steenberghen, *Aristotle in the West,* pp. 147–62; Knowles, *Evolution of Medieval Thought,* pp. 236–48; and John Francis Quinn, *The Historical Constitution of St. Bonaventure's Philosophy,* esp. pp. 841–96. These works will lead the reader to additional sources.

20. On Albert's life and works, see James A. Weisheipl, "The Life and Works of St. Albert the Great," in Weisheipl, ed., *Albertus Magnus and the Sciences,* pp. 13–51; also Appendix I to the same volume, pp. 565–77.

21. Quoted by Benedict M. Ashley, "St. Albert and the Nature of Natural Science," in Weisheipl, *Albertus Magnus and the Sciences,* p. 78. On Albert's thought, see the essays contained in this volume; also Van Steenberghen, *Aristotle in the West,* pp. 167–81; and Francis J. Kovach and Robert W. Shahan, eds., *Albert the Great: Commemorative Essays.*

22. On Albert's sources, see the various essays in Weisheipl, *Albertus Magnus and the Sciences.*

23. Karen Reeds, "Albert on the Natural Philosophy of Plant Life," in Weisheipl, *Albertus Magnus and the Sciences,* p. 343. On Albert as an observer of flora, fauna, and minerals, see the excellent essays in this volume.

24. Albert's theory of the soul is discussed by Anton C. Pegis, *St. Thomas and the Problem of the Soul in the Thirteenth Century,* chap. 3; and by Katharine Park, "Albert's Influence on Medieval Psychology." For Albert's views on the eternity of the world, see the introduction to Thomas Aquinas, Siger of Brabant, and Bonaventure, *On the Eternity of the World,* trans. Vollert et al., p. 13.

25. On Albert's naturalist program and the question of Noah's flood, see Albert's *De causis proprietatibus elementorum,* I.2.9, in Albert the Great, *Opera omnia,* ed. Augustus Borgnet, 38 vols. (Paris, Vivès, 1890–99), 9:618–19. Cf. Lynn Thorndike, *History of Magic and Experimental Science,* 2:535.

26. The literature on Thomas Aquinas is enormous. On his life, see James A. Weisheipl, *Friar Thomas d'Aquino: His Life, Thought and Works.* Useful summaries of his scholarly achievement (here arranged in order of ascending length) are Knowles, *Evolution of Medieval Thought,* chap. 21; Ralph McInerny, *St. Thomas Aquinas;* M.-D. Chenu, *Toward Understanding St. Thomas;* and Etienne Gilson, *The Christian Philosophy of St. Thomas Aquinas.* Most discussions of Thomas's philosophy (including all of the above) have been written by modern-day Thomists, committed to Thomas's philosophy and not averse to extolling its virtues. These works are marred, therefore, by a tendency to see Thomas (because he was "right") as the glorious culmination of medieval thought. For a brief account that manages to capture the essentials of Thomas's achievement while avoiding value judgments, see Julius Weinberg, *A Short History of Medieval Philosophy,* chap. 9.

27. Thomas Aquinas, *Faith, Reason and Theology: Questions I–IV of His Commentary on the De Trinitate of Boethius,* trans. Armand Maurer, p. 48. The first two of these four questions are devoted to the legitimacy of employing philosophy in matters of faith.

28. Ibid., pp. 48–49.

29. For an excellent analysis of Thomas's position on the eternity of the world and the nature of the soul, see Van Steenberghen, *Aquinas and Radical Aristotelianism,* chaps. 1–2.

30. For a discussion of radical Aristotelianism and its consequences, see the excellent survey by Edward Grant, "Science and Theology in the Middle Ages."

31. In the judgment of Siger's foremost modern interpreter, this was not a matter of caving in to theology, but of being led by the force of Thomas's *philosophical* arguments to rethink and correct his own philosophical position. See Fernand Van Steenberghen, *Les oeuvres et la doctrine de Siger de Brabant; Aristotle in the West,* pp. 209–29; and *Aquinas and Radical Aristotelianism,* pp. 6–8, 35–43, 89–95. It is inconceivable to me that Siger's philosophical purity would not have been compromised in some measure by the need to arrive at a theologically orthodox conclusion.

32. Boethius of Dacia, *On the Supreme Good, On the Eternity of the World, On Dreams,* pp. 36–67, quoting from p. 47.

33. For a short account of the condemnations, see Van Steenberghen, *Aristotle in the West,* chap. 9; John F. Wippel, "The Condemnations of 1270 and 1277 at Paris"; Edward Grant, "The Condemnation of 1277, God's Absolute Power, and Physical Thought in the Late Middle Ages." For an extended analysis of the condemnations in relation to natural philosophy, see Pierre Duhem, *Le système du monde,* vol. 6; Roland Hissette, *Enquête sur les 219 articles condamnés à Paris le 7 mars 1277.* For a translation of the decree of 1277 and the condemned propositions, see Ralph Lerner and Muhsin Mahdi, eds., *Medieval Political Philosophy: A Sourcebook* (New York: Free Press of Glencoe, 1963), pp. 335–54; a selection of propositions relevant to natural philosophy appears, with introduction and running commentary, in Edward Grant, *A Source Book in Medieval Science,* pp. 45–50.

34. There are at least two possible interpretations of this straight-line motion, which the radical Aristotelians denied God the ability to bestow on the heaven: (a) a motion of translation, which would have moved the heaven as a whole and its contents (that is, the entire cosmos) in one direction or another; and (b) rectilinear descent of the heaven, or a portion of the heaven, toward the center of the cosmos. The former interpretation was certainly current by the middle of the fourteenth century, when it was expressed by John Buridan in his commentary on Aristotle's *Physics;* through the influence of Pierre Duhem it has since become standard. The latter interpretation, Roland Hissette has recently shown, is probably what the framers of this article of the condemnation had in mind. Fortunately, for our purposes it does not matter which of the two interpretations was held by any particular historical actor, since the essential point is the same in either case—namely, that God cannot cause rectilinear motions that threaten to leave a vacuum in their wake. See Pierre Duhem, *Etudes sur Léonard de Vinci,* 2:412; Anneliese Maier, *Zwischen Philosophie und Mechanik,* pp. 122–24; Edward Grant, "The Condemnation of 1277, God's Absolute Power, and Physical Thought in the Late Middle Ages," pp. 226–31; Hissette, *Enquête sur les 219 articles,* pp. 118–20.

35. Duhem, *Etudes sur Léonard de Vinci,* 2:412; Duhem, *Système du monde,* 6:66. For survival of the Duhem claim, qualified and weakened but still recognizable, see Edward Grant, "Late Medieval Thought, Copernicus, and the Scientific Revolution"; and "Condemnation of 1277."

36. On the question of void space, see Edward Grant, *Much Ado about Nothing: Theories of Space and Vacuum from the Middle Ages to the Scientific Revolution;* also Grant, "Condemnation of 1277," pp. 232–34.

37. For an excellent historical analysis of the question of divine omnipotence, see Francis Oakley, *Omnipotence, Covenant, and Order.*

38. Transubstantiation is the process by which, according to Catholic doctrine, the eucharistic bread and wine are transformed into the body and blood of Christ.

39. See Grant, "Science and Theology in the Middle Ages," pp. 54–70; Grant, *Nicole Oresme and the Kinematics of Circular Motion.*

40. For the impact of the condemnations on natural philosophy, see Grant, "Condemnation of 1277."

41. William A. Wallace, "Thomism and Its Opponents," *Dictionary of the Middle Ages,* 12:38–45; Knowles, *Evolution of Medieval Thought,* chap. 24; Grant, "Condemnation of 1277." The quoted passages come, respectively, from Marshall Clagett, *The Science of Mechanics in the Middle Ages,* p. 536 (with minor modifications); and Nicole Oresme, *Le livre du ciel et du monde,* ed. and trans. A. D. Menut and A. J. Denomy, p. 369.

42. On late medieval and Renaissance Aristotelianism, see John Herman Randall, Jr., *The School of Padua and the Emergence of Modern Science;* Charles B. Schmitt, *Aristotle and the Renaissance.* For the quotation (both English translation and Latin text), see William J. Courtenay and Katherine H. Tachau, "Ockham, Ockhamists, and the English-German Nation at Paris, 1339–1341," pp. 61, 86.

43. On the epistemological discussions of the late thirteenth and fourteenth centuries, see Marilyn McCord Adams, *William Ockham,* 1:551–629; Eileen Serene, "Demonstrative Science," in Norman Kretzmann, Anthony Kenny, and Jan Pinborg, eds., *The Cambridge History of Later Medieval Philosophy,* pp. 496–517. On Ockham, see also William J. Courtenay, "Ockham, William of," *Dictionary of the Middle Ages,* 9:209–14.

44. Oakley, *Omnipotence, Covenant, and Order,* chap. 3; William J. Courtenay, "The Critique on Natural Causality in the Mutakallimun and Nominalism." For a full discussion of divine omnipotence and its implications for natural philosophy, see Courtenay's *Capacity and Volition: A History of the Distinction of Absolute and Ordained Power;* Amos Funkenstein, *Theology and the Scientific Imagination from the Middle Ages to the Seventeenth Century,* pp. 117–201.

45. And it was generally held that those exceptions were built into the universe from the moment of creation; see above, chap. 9.

46. See the essays in Courtenay, *Covenant and Causality,* esp. chap. 4: "The Dialectic of Divine Omnipotence"; and chap. 5: "The Critique on Natural Causality in the Mutakallimun and Nominalism."

47. On the subtle connection between the doctrine of divine omnipotence and experimental method, see Funkenstein, *Theology and the Scientific Imagination,* pp. 152–79.

CHAPTER ELEVEN

1. I have decided not to employ theoretical classification schemes ("divisions of the sciences") developed during the Middle Ages, on the grounds that the scientific literature actually produced did not fit neatly into the categories thus defined. On

these schemes, see James A. Weisheipl, "Classification of the Sciences in Medieval Thought"; Weisheipl, "The Nature, Scope, and Classification of the Sciences."

2. Chaps. 7 and 9.

3. For a good example of twelfth-century cosmology, see *The Cosmographia of Bernardus Silvestris,* trans. with introduction and notes by Winthrop Wetherbee. See also, above, chap. 9. On medieval cosmology more generally, see C. S. Lewis, *The Discarded Image.*

4. A. C. Crombie, *Robert Grosseteste and the Origins of Experimental Science, 1100–1700.*

5. Grosseteste refers to this form as "first form" or "corporeal form." For more on corporeal form, see below, chap. 12.

6. On Grosseteste's cosmology, see the excellent study by James McEvoy, *The Philosophy of Robert Grosseteste,* pp. 149–88, 369–441. For a short version, see David C. Lindberg, "The Genesis of Kepler's Theory of Light: Light Metaphysics from Plotinus to Kepler," pp. 14–17.

7. Pierre Duhem, *Le système du monde,* 10 vols. Excerpts from these 10 volumes have been translated into English in Pierre Duhem, *Medieval Cosmology: Theories of Infinity, Place, Time, Void, and the Plurality of Worlds,* ed. and trans. Roger Ariew. I am especially indebted, in the account that follows, to the excellent summary of medieval cosmology in Edward Grant, "Cosmology"; also the articles collected in Grant's *Studies in Medieval Science and Natural Philosophy.* See also Olaf Pedersen, "The Corpus Astronomicum and the Traditions of Mediaeval Latin Astronomy." Thomas Aquinas's cosmology is nicely treated in the new Blackfriars edition of his *Summa Theologiae,* vol. 10: *Cosmogony,* ed. and trans. William A. Wallace. Edward Grant's forthcoming *The Medieval Cosmos 1200–1687* should contain the definitive treatment of medieval cosmology.

8. Edward Grant, "Medieval and Seventeenth-Century Conceptions of an Infinite Void Space beyond the Cosmos"; Grant, *Much Ado about Nothing,* esp. chaps. 5–6.

9. Edward Grant, "The Medieval Doctrine of Place: Some Fundamental Problems and Solutions," esp. pp. 72–79.

10. Grant, "Cosmology," pp. 275–79; Grant, "Celestial Orbs in the Latin Middle Ages," pp. 159–62; Grant, "Science and Theology in the Middle Ages," pp. 63–64.

11. For representative medieval texts, see those included in Lynn Thorndike, ed. and trans., *The Sphere of Sacrobosco and Its Commentators,* p. 206. For discussion, see Edward Grant, "Celestial Matter: A Medieval and Galilean Cosmological Problem"; Grant, "Celestial Orbs," pp. 167–72; Grant, "Cosmology," pp. 286–88.

12. James A. Weisheipl, "The Celestial Movers in Medieval Physics"; Grant, "Cosmology," pp. 284–86.

13. For this data, see Grant, "Cosmology," p. 292; Francis S. Benjamin and G. J. Toomer, eds. and trans., *Campanus of Novara and Medieval Planetary Theory: "Theorica planetarum",* pp. 356–63. Campanus defines a mile as the equivalent of 4,000 cubits and gives the circumference of the earth as 20,400 miles (Benjamin and Toomer, p. 147). For more on ideas of cosmic size, see Bernard R. Goldstein and Noel Swerdlow, "Planetary Distances and Sizes in an Anonymous Arabic Treatise Preserved in Bodleian MS Marsh 621"; Albert Van Helden, *Measuring the Universe: Cosmic Dimensions from Aristarchus to Halley.*

14. On the rainbow, see Edward Grant, ed., *A Source Book in Medieval Science,*

pp. 435–41; Carl B. Boyer, *The Rainbow: From Myth to Mathematics,* chaps. 3–5. For a convenient account of medieval meteorology, see John Kirtland Wright, *The Geographical Lore of the Time of the Crusades: A Study in the History of Medieval Science and Tradition in Western Europe,* pp. 166–81; Nicholas H. Steneck, *Science and Creation in the Middle Ages: Henry of Langenstein (d. 1397) on Genesis,* pp. 84–87.

15. The idea that Columbus was opposed by people who believed in the flatness of the earth is a modern legend; see Jeffrey B. Russell, *Inventing the Flat Earth: Columbus and Modern Historians.*

16. For a sketch of medieval geography, see Lewis, *Discarded Image,* pp. 139–46, from whom this point and the terminology for expressing it have been borrowed. For a full-length survey, see Wright, *Geographical Lore.* On cartography, see David Woodward, "Medieval *Mappaemundi.*"

17. William H. Stahl, *Roman Science,* pp. 115–19, 221–22.

18. On types of medieval maps and their functions, see the articles in *History of Cartography,* ed. J. B. Harley and David Woodward, vol. 1. On the two maps mentioned here, see Woodward "Medieval *Mappaemundi,*" pp. 290, 310.

19. On *mappaemundi,* see the thorough study by Woodward, "Medieval *Mappaemundi.*"

20. On portolan charts and Ptolemy's mapping techniques, see two articles in Harley and Woodward, *History of Cartography,* vol. 1: Tony Campbell, "Portolan Charts from the Late Thirteenth Century to 1500," pp. 317–463; O. A. W. Dilke, "The Culmination of Greek Cartography in Ptolemy," pp. 177–200.

21. The rotational speeds would, of course, have to be the same, whether it is the earth or a celestial sphere doing the rotating. However, owing to the smaller radius of the earth, a point on its surface would move more slowly than would a point on the surface of a celestial sphere.

22. Nicole Oresme, *Le Livre du ciel et du monde,* pp. 525, 531, with a variety of improvements and corrections. For analysis, see Marshall Clagett, *The Science of Mechanics in the Middle Ages,* pp. 583–88; Edward Grant, *Physical Science in the Middle Ages,* pp. 63–70; Grant McColley, "The Theory of the Diurnal Rotation of the Earth."

23. Oresme, *Livre du ciel,* p. 537 (with modified punctuation).

24. Ibid., pp. 537–39.

25. Pierre Duhem, *To Save the Phenomena: An Essay on the Idea of Physical Theory from Plato to Galileo* (1969); the work was first published in French in 1908. The same interpretation in a less developed form had been presented two years earlier by J. L. E. Dreyer, *History of the Planetary Systems from Thales to Kepler* (1906).

26. See G. E. R. Lloyd, "Saving the Appearances." I have also been influenced by chap. 1 of Bruce S. Eastwood's projected *Before Copernicus: Planetary Theory and the Circumsolar Idea from Late Antiquity to the Twelfth Century* (read in typescript).

27. Grant, "Cosmology," pp. 265–68.

28. On Islamic astronomy in general, see George Saliba, "The Development of Astronomy in Medieval Islamic Society"; Saliba, "Astrology/Astronomy, Islamic," *Dictionary of the Middle Ages,* 1:616–24; the collected articles of David A. King,

Islamic Mathematical Astronomy; A. I. Sabra, "The Scientific Enterprise"; Owen Gingerich, "Islamic Astronomy"; E. S. Kennedy, "The Arabic Heritage in the Exact Sciences"; and Noel M. Swerdlow and Otto Neugebauer, *Mathematical Astronomy in Copernicus's De Revolutionibus,* pp. 41–48. For an older attempt, see J. L. E. Dreyer, *History of Astronomy from Thales to Kepler,* 2d ed., chap. 11. On non-Ptolemaic systems, see A. I. Sabra, "The Andalusian Revolt against Ptolemaic Astronomy: Averroes and al-Biṭrūjī." For a large collection of useful articles on astronomical topics, see E. S. Kennedy (with colleagues and former students), *Studies in the Islamic Exact Sciences.*

29. On Greek and Arabic trigonometry, see E. S. Kennedy, "The History of Trigonometry: An Overview."

30. Aydin Sayili, *The Observatory in Islam and Its Place in the General History of the Observatory,* chaps. 6, 8; T. N. Kari-Niazov, "Ulugh Beg," *Dictionary of Scientific Biography,* 13:535–37. For a photograph of the still-impressive remains of Ulugh Beg's sextant, see Sabra, "Scientific Enterprise," p. 195.

31. For a lucid and reliable account of the astrolabe, see J. D. North, "The Astrolabe"; or North, *Chaucer's Universe,* pp. 38–86. For a more detailed analysis, see *The Planispheric Astrolabe.* On Islamic astronomical instruments in general, see David A. King, *Islamic Astronomical Instruments.*

32. On Ibn al-Haytham's astronomical work, see A. I. Sabra, "Ibn al-Haytham," *Dictionary of Scientific Biography,* 6:197–99; Sabra, "An Eleventh-Century Refutation of Ptolemy's Planetary Theory."

33. A. I. Sabra, "Andalusian Revolt against Ptolemaic Astronomy." Roger Arnaldez and Albert Z. Iskandar, "Ibn Rushd," *Dictionary of Scientific Biography,* 12:3–5. Al-Biṭrūjī, *On the Principles of Astronomy,* ed. and trans. Bernard R. Goldstein.

34. On early medieval astronomy, see the first few items in Bruce S. Eastwood, *Astronomy and Optics from Pliny to Descartes;* Eastwood, "Plinian Astronomical Diagrams in the Early Middle Ages"; Stephen C. McCluskey, "Gregory of Tours, Monastic Timekeeping, and Early Christian Attitudes to Astronomy"; and Claudia Kren, "Astronomy," in David L. Wagner, Ed., *The Seven Liberal Arts in the Middle Ages,* pp. 218–47.

35. My understanding of Western astronomy is heavily dependent on the work of Olaf Pedersen, especially his "Astronomy," in David C. Lindberg, ed., *Science in the Middle Ages,* pp. 303–36; "Corpus Astronomicum and the Traditions of Mediaeval Latin Astronomy"; and Olaf Pedersen and Mogens Pihl, *Early Physics and Astronomy: A Historical Introduction,* chap. 18.

36. On the *Toledan Tables,* see G. J. Toomer, "A Survey of the Toledan Tables"; Ernst Zinner, "Die Tafeln von Toledo."

37. Thorndike, *Sphere of Sacrobosco,* supplies the Latin text of this treatise, an English translation, and a very useful introduction. Sacrobosco also composed a treatise on arithmetic and another on the calendar, see ibid., pp. 3–4.

38. For the remaining planets, see Pedersen, "Astronomy," pp. 316–18; also Pedersen's translation of the *Theorica* in Edward Grant, ed., *A Source Book in Medieval Science,* pp. 451–65.

39. See, for example, Claudia Kren, "Homocentric Astronomy in the Latin West: The *De reprobatione ecentricorum et epiciclorum* of Henry of Hesse."

40. For Bacon's account, see Pierre Duhem, *Un fragment inédit de l'Opus*

tertium de Roger Bacon, précédé d'une étude sur ce fragment, pp. 128–37. On Bernard, see Claudia Kren, "Bernard of Verdun," *Dictionary of Scientific Biography,* 2:23–24. The reference to Guido de Marchia I owe to my colleague Michael Shank.

41. Extracts from the Alphonsine Tables have been translated and annotated by Victor E. Thoren, in Grant, *Source Book,* pp. 465–87. On the Toledan and Alphonsine Tables, see also North, *Chaucer's Universe,* pp. 147–53. Both Thoren and North provide a sample calculation.

42. No history of astronomy in the later Middle Ages exists. For useful glimpses, see the following works of J. D. North: *Richard of Wallingford, An Edition of His Writings with Introductions, English Translation and Commentary,* 3 vols.; "The Alphonsine Tables in England," in North's *Stars, Minds and Fate: Essays in Ancient and Medieval Cosmology,* pp. 327–59; and *Chaucer's Universe.* On medieval Jewish astronomy (which frequently interacted with Latin astronomy), see the papers collected in Bernard R. Goldstein, *Theory and Observation in Ancient and Medieval Astronomy.* On Regiomontanus and Copernicus, see Noel M. Swerdlow and Otto Neugebauer, *Mathematical Astronomy in Copernicus's De Revolutionibus.*

43. On early astrology, see Jim Tester, *A History of Western Astrology;* Olaf Pedersen, "Astrology," *Dictionary of the Middle Ages,* 1:604–10 (Pedersen includes a good bibliography); A. A. Long, "Astrology: Arguments Pro and Contra"; Theodore Otto Wedel, *The Mediaeval Attitude toward Astrology, Particularly in England;* Franz Cumont, *Astrology and Religion among the Greeks and Romans;* J. D. North, "Celestial Influence—the Major Premiss of Astrology"; North, "Astrology and the Fortunes of Churches"; Edward Grant, "Medieval and Renaissance Scholastic Conceptions of the Influence of the Celestial Region on the Terrestrial"; Lewis, *Discarded Image,* pp. 102–10; and the papers contained in Patrick Curry, ed., *Astrology, Science and Society: Historical Essays* (especially that of Richard Lemay, "The True Place of Astrology in Medieval Science and Philosophy").

44. On Mesopotamian astrology, see B. L. van der Waerden and Peter Huber, *Science Awakening, II: The Birth of Astronomy,* chap. 5; Richard Olson, *Science Deified and Science Defied: The Historical Significance of Science in Western Culture,* pp. 34–56.

45. Aristotle is quoted from *Meteorologica,* I.2, trans. E. W. Webster, in *The Complete Works of Aristotle,* ed. Jonathan Barnes, p. 555.

46. Ptolemy, *Tetrabiblos,* I.2, ed. and trans. F. E. Robbins, pp. 5–13 (with one change of wording). On Ptolemy's astrology, see also Tester, *History of Western Astrology,* chap. 4; Long, "Astrology: Arguments Pro and Contra," pp. 178–83.

47. Augustine, *City of God,* V.6, trans. William H. Green (London: Heinemann, 1963), vol. 2, p. 157. On Augustine's attitude toward astrology see also his *Confessions,* IV.3 and VII.6; Wedel, *Mediaeval Attitude toward Astrology,* pp. 20–24; Joshua D. Lipton, "The Rational Evaluation of Astrology in the Period of Arabo-Latin Translation, ca. 1126–1187 A.D.," pp. 133–35; Tester, *History of Western Astrology,* chap. 5.

48. Wedel, *Mediaeval Attitude toward Astrology,* chap. 2.

49. *The Didascalicon of Hugh of St. Victor: A Medieval Guide to the Arts,* trans. Jerome Taylor, p. 68. For the Latin text of the second quotation, see C. S. F. Burnett, "What is the *Experimentarius* of Bernardus Silvestris? A Preliminary Survey of the Material." For the third quotation (possibly from William of Conches), see Lipton,

"Rational Evaluation of Astrology," p. 145. Lipton's study contains a very useful analysis of twelfth-century astrology; see also Wedel, *Mediaeval Attitude toward Astrology*, pp. 60–63.

50. Lemay, *Abū Maʿshar*, pp. 41–132; David Pingree, "Abū Maʿshar al-Balkhī," *Dictionary of Scientific Biography*, 1:32–39.

51. See, for example, Nancy G. Siraisi, *Taddeo Alderotti and His Pupils: Two Generations of Italian Medical Learning*, pp. 140–45.

52. G. W. Coopland, *Nicole Oresme and the Astrologers*, pp. 53–57. On Oresme, see also Stefano Caroti, "Nicole Oresme's Polemic against Astrology in His 'Quodlibeta'," in Curry, *Astrology, Science and Society*, pp. 75–93.

CHAPTER TWELVE

1. On theories of "nature" and the "physical," see R. G. Collingwood, *The Idea of Nature;* Ivor Leclerc, *The Nature of Physical Existence.*

2. On the continuity between medieval and early modern science, see chap. 14, below.

3. On Aristotelian natural philosophy, see chap. 3, above, and the citations provided there. On subsequent developments within the Aristotelian tradition, see Harry Austryn Wolfson, *Crescas' Critique of Aristotle: Problems of Aristotle's "Physics" in Jewish and Arabic Philosophy;* Leclerc, *Nature of Physical Existence;* Norma E. Emerton, *The Scientific Reinterpretation of Form*, chaps. 2–3.

4. G. E. R. Lloyd, *Aristotle*, pp. 164–75; Anneliese Maier, "The Theory of the Elements and the Problem of Their Participation in Compounds," in Maier, *On the Threshold of Exact Science*, chap. 6.

5. Leclerc, *Nature of Physical Existence*, chaps. 8–9. For a very challenging discussion of Greek and medieval conceptions of matter, see the articles collected in Ernan McMullin, ed., *The Concept of Matter in Greek and Medieval Philosophy.*

6. See Wolfson, *Crescas' Critique*, pp. 580–90; Arthur Hyman, "Aristotle's 'First Matter' and Avicenna's and Averroes' 'Corporeal Form'," in *Harry Austryn Wolfson Jubilee Volume*, 1:385–406. On the significance of the idea of corporeal form within medieval Christian thought, see D. E. Sharp, *Franciscan Philosophy at Oxford in the Thirteenth Century*, pp. 186–89.

7. Leclerc, *Nature of Physical Existence*, pp. 125–29; Sharp, *Franciscan Philosophy at Oxford*, pp. 220–22, 292–95.

8. On the Aristotelian doctrine of *mixtio*, see Friedrich Solmsen, *Aristotle's System of the Physical World*, chap. 19; Waterlow, *Nature, Change, and Agency*, pp. 82–85; Emerton, *Scientific Reinterpretation of Form*, chap. 3.

9. On the medieval doctrine of *mixtio*, see E. J. Dijksterhuis, *The Mechanization of the World Picture*, pp. 200–204; Emerton, *Scientific Reinterpretation of Form*, pp. 77–85; Robert P. Multhauf, *The Origins of Chemistry*, pp. 149–52; and most usefully Anneliese Maier, *An der Grenze von Scholastik und Naturwissenschaft*, 2d ed., pp. 3–140, the introductory portion of which appears as "Theory of the Elements," in Maier, *Threshold*, trans. Sargent, chap. 6.

10. On *minima*, see Dijksterhuis, *Mechanization*, pp. 205–9; Emerton, *Scientific Reinterpretation of Form*, pp. 85–93.

11. For an excellent general introduction to problems and sources on medieval

alchemy, see Robert Halleux, *Les textes alchimiques;* also Claudia Kren, *Alchemy in Europe: A Guide to Research.* Older literature, still useful, includes: F. Sherwood Taylor, *The Alchemists;* E. J. Holmyard, *Alchemy;* and Multhauf, *Origins of Chemistry,* chaps. 5–9. For brief and more up-to-date sketches, see: Manfred Ullmann, "Al-Kīmiyā'," *The Encyclopaedia of Islam,* new ed., vol. 5, fasc. 79–80, pp. 110–15; and Robert Halleux, "Alchemy," *Dictionary of the Middle Ages,* 1:134–40. And for the latest word: William R. Newman, "The Genesis of the *Summa perfectionis*"; Newman, "Technology and Chemical Debate in the Late Middle Ages"; and Newman, *The "Summa perfectionis" of Pseudo-Geber: A Critical Edition, Translation, and Study* (portions of which Newman was kind enough to let me see in typescript).

12. The sulphur and mercury in question were not the common minerals by those names, but the pure essences thought to provide the various qualities needed to produce metals, sometimes referred to as "philosophical sulphur" and "philosophical mercury." Philosophical sulphur was frequently identified as the active, spiritual principle; philosophical mercury as the passive, material principle.

13. On alchemical apparatus and processes, see Holmyard, *Alchemy,* chap. 4.

14. On later alchemy, see Allen G. Debus, *Man and Nature in the Renaissance,* chap. 2; and Debus, *The Chemical Philosophy: Paracelsian Science and Medicine in the Sixteenth and Seventeenth Centuries,* 2 vols.

15. *Physics,* III.1, 200^b14–15.

16. I am indebted for this discussion of the nature of motion to John E. Murdoch and Edith D. Sylla, "The Science of Motion," pp. 213–22. See also the works of Anneliese Maier: *Zwischen Philosophie und Mechanik,* chaps. 1–3; *Die Vorläufer Galileis im 14. Jahrhundert,* 2d ed., chap. 1; and the English translation of the latter appearing in *Threshold of Exact Science,* trans. Sargent, chap. 1.

17. John E. Murdoch, "The Development of a Critical Temper: New Approaches and Modes of Analysis in Fourteenth-Century Philosophy, Science, and Theology," pp. 60–61; Murdoch and Sylla, "Science of Motion," pp. 216–17; Maier, *Threshold of Exact Science,* pp. 30–31.

18. Murdoch and Sylla, "Science of Motion," pp. 217–18; Maier, *Threshold of Exact Science,* pp. 33–38; Maier, *Zwischen Philosophie und Mechanik,* pp. 121–31.

19. Marshall Clagett, *The Science of Mechanics in the Middle Ages,* pp. 163–86.

20. On Gerard, see ibid., pp. 184–97; Clagett, "The *Liber de motu* of Gerard of Brussels and the Origins of Kinematics in the West"; Murdoch and Sylla, "Science of Motion," pp. 222–23; and Wilbur R. Knorr, "John of Tynemouth *alias* John of London: Emerging Portrait of a Singular Medieval Mathematician," pp. 312–22.

21. But velocity was treated as a scalar rather than a vector quantity. That is, it had magnitude, but was indifferent as to direction.

22. Clagett, *Science of Mechanics,* chap. 4.

23. We will not delve into the related medieval problem of explaining in physical terms how intensification and remission occur. The two principal theories were an *addition and subtraction* theory, according to which a form is intensified by addition of a new part of form, remitted by subtraction of a part of the original form; and a *replacement* theory, according to which the original form is annihilated and replaced by a new form of greater or lesser intensity. On this problem, see Edith D. Sylla, "Medieval Concepts of the Latitude of Forms: The Oxford Calculators," pp. 230–33; Murdoch and Sylla, "Science of Motion," pp. 231–33. On

the intensification and remission of qualities in general, see also Clagett, *Science of Mechanics,* pp. 205–6, 212–15; Murdoch and Sylla, "Science of Motion," pp. 233–37.

24. This notion goes back at least to Galen; see Marshall Clagett, *Giovanni Marliani and Late Medieval Physics,* pp. 34–36.

25. Clagett, *Science of Mechanics,* pp. 212–13.

26. If the equality of the triangle and the rectangle is not evident by inspection, draw a diagonal line from B to D, thus dividing rectangle BCDE into two equal triangles. Note then that both rectangle ACDF and triangle ACG have been subdivided into small, equal triangles—four in each case.

27. On the geometrical representation of qualities, see Marshall Clagett, *Nicole Oresme and the Medieval Geometry of Qualities and Motions,* pp. 50–121; Clagett, *Science of Mechanics,* chap. 6; Murdoch and Sylla, "Science of Motion," pp. 237–41. On the Merton Rule, see Clagett, *Science of Mechanics,* chap. 5.

28. On Galileo's relationship to the medieval mechanical tradition, see below, chap. 14, n. 36.

29. On this exceedingly technical question, see Richard Sorabji, *Matter, Space, and Motion: Theories in Antiquity and Their Sequel,* chap. 13; James A. Weisheipl, *Nature and Motion in the Middle Ages,* chaps. 4–5 (p. 92 for the quoted words). For the Aristotelian texts, see Aristotle's *Physics,* II.1, VII.1, and VIII.4.

30. On Philoponus, see Clagett, *Science of Mechanics,* pp. 508–10. For more recent studies, which do full justice to the radical Neoplatonic character of Philoponus's attack on Aristotelian dynamics, see Michael Wolff, "Philoponus and the Rise of Preclassical Dynamics"; and Sorabji, *Matter, Space, and Motion,* chap. 14.

31. For the latest word on this subject, see Fritz Zimmermann, "Philoponus' Impetus Theory in the Arabic Tradition"; and Sorabji, *Matter, Space, and Motion,* pp. 237–38. See also Clagett, *Science of Mechanics,* pp. 510–17.

32. On impetus theory, see Clagett, *Science of Mechanics,* pp. 521–25 (quotation from p. 524); Anneliese Maier, "Die naturphilosophische Bedeutung der scholastischen Impetustheorie," translated as "The Significance of the Theory of Impetus for Scholastic Natural Philosophy," in Maier, *On the Threshold of Exact Science,* pp. 76–102. Unknown to Buridan, Philoponus had anticipated his suggestion that impetus or impressed force could be used to explain celestial motion; see Sorabji, *Matter, Space, and Motion,* p. 237.

33. In an infinitely swift motion, no time would be required for the moving body to pass from one point to another. It follows that the body would be at both points simultaneously, and that is a physical impossibility.

34. Morris R. Cohen and I. E. Drabkin, *A Source Book in Greek Science,* p. 220, with several changes. See also Clagett, *Science of Mechanics,* pp. 433–35, 546–47.

35. The classic analyses of Bradwardine and his predecessors, still useful, are by Maier, *Die Vorläufer Galileis,* pp. 81–110 (partially translated in Maier's *On the Threshold of Exact Science,* pp. 61–75); and Ernest A. Moody, "Galileo and Avempace: The Dynamics of the Leaning Tower Experiment." For somewhat more recent scholarship, see Clagett, *Science of Mechanics,* chap. 7; and *Thomas of Bradwardine, His "Tractatus de Proportionibus": Its Significance for the Development of Mathematical Physics,* ed. and trans. H. Lamar Crosby, Jr.

36. A. G. Molland, "The Geometrical Background to the 'Merton School'," esp. pp. 116–21 (p. 120, for the quotation); Murdoch and Sylla, "Science of Motion,"

pp. 225–26; Edith D. Sylla, "Compounding Ratios: Bradwardine, Oresme, and the first edition of Newton's *Principia.*"

37. Murdoch and Sylla, "Science of Motion," pp. 227–30; Clagett, *Marliani,* chap. 6; Clagett, *Science of Mechanics,* p. 443. On Swineshead's work, see John E. Murdoch and Edith D. Sylla, "Swineshead, Richard," *Dictionary of Scientific Biography,* 13:184–213. On Oresme, see Nicole Oresme, *"De proportionibus proportionum" and "Ad pauca respicientes,"* ed. and trans. Edward Grant.

38. On medieval optics in general, see David C. Lindberg, *Theories of Vision from al-Kindi to Kepler;* Lindberg, "The Science of Optics"; Lindberg, "Optics, Western European," *Dictionary of the Middle Ages,* 9:247–53; the papers collected in Lindberg's *Studies in the History of Medieval Optics;* the optical papers contained in Bruce S. Eastwood, *Astronomy and Optics from Pliny to Descartes;* and A. Mark Smith, "Getting the Big Picture in Perspectivist Optics."

39. It could be convincingly argued that extramission of rays was a necessary feature of mathematical theories of vision, for it was the conical emanation of rays from the eye that defined the visual cone, which in turn made the mathematical analysis of vision possible.

40. See Aristotle's *Meteorology,* III.4–5; Lindberg, *Theories of Vision,* p. 217, n. 39.

41. On Alhazen's optical achievement, see the definitive translation and commentary by A. I. Sabra, ed. and trans., *The Optics of Ibn al-Haytham: Books I–III, On Direct Vision.* For shorter accounts, see Sabra, "Ibn al-Haytham," *Dictionary of Scientific Biography,* 6:189–210; Sabra, "Form in Ibn al-Haytham's Theory of Vision"; and Lindberg, *Theories of Vision,* chap. 4.

42. Above, chap. 5.

43. On the Western reception of Greek and Islamic optics, see (in addition to sources already cited) David C. Lindberg, "Roger Bacon and the Origins of *Perspectiva* in the West."

44. On Bacon's optics, see David C. Lindberg, ed. and trans., *Roger Bacon's Philosophy of Nature: A Critical Edition, with English Translation, Introduction, and Notes, of "De multiplicatione specierum" and "De speculis comburentibus"*; Lindberg, *Theories of Vision,* chap. 6; and Lindberg, "Bacon and the Origins of *Perspectiva*".

45. On Bacon's Neoplatonism, see David C. Lindberg, "The Genesis of Kepler's Theory of Light: Light Metaphysics from Plotinus to Kepler," pp. 12–23; Lindberg, *Bacon's Philosophy of Nature,* pp. liii–lxxi.

46. Lindberg, *Theories of Vision,* chap. 9; Katherine H. Tachau, *Vision and Certitude in the Age of Ockham: Optics, Epistemology and the Foundations of Semantics, 1250–1345.* Pecham's *Perspectiva communis* is available in David C. Lindberg, ed. and trans., *John Pecham and the Science of Optics.* A project to translate Witelo's massive *Perspectiva* is under way. Two volumes have been published: Sabetai Unguru, ed. and trans., *Witelonis Perspectivae liber primus;* and A. Mark Smith, ed. and trans., *Witelonis Perspectivae liber quintus.*

CHAPTER THIRTEEN

1. For the basic framework of this chapter, I am indebted to Nancy G. Siraisi, *Medieval and Early Renaissance Medicine: An Introduction to Knowledge and*

Practice; Michael McVaugh, "Medicine, History of," *Dictionary of the Middle Ages,* 8:247–54; and the general tutelage of my colleague Faye Getz. The reader may also wish to consult Charles H. Talbot, "Medicine," in David C. Lindberg, ed., *Science in the Middle Ages,* pp. 391–428; and Talbot, *Medicine in Medieval England.* For a useful review of recent literature on medieval medicine, see Getz, "Western Medieval Medicine." For an excellent set of translated medical texts (selected, annotated, and in some cases translated by Michael McVaugh), see Edward Grant, ed., *A Source Book in Medieval Science,* pp. 700–808. For medical illustrations, see Loren C. MacKinney, *Medical Illustrations in Medieval Manuscripts;* Peter M. Jones, *Medieval Medical Miniatures;* and Marie-José Imbault-Huart, *La médecine au moyen âge à travers les manuscrits de la Bibliothèque Nationale.*

2. On early medieval medicine, see especially John M. Riddle, "Theory and Practice in Medieval Medicine"; Henry E. Sigerist, "The Latin Medical Literature of the Early Middle Ages"; Linda E. Voigts, "Anglo-Saxon Plant Remedies and the Anglo-Saxons"; M. L. Cameron, "The Sources of Medical Knowledge in Anglo-Saxon England"; Siraisi, *Medieval and Early Renaissance Medicine,* pp. 5–13; and (older but still useful) Loren C. MacKinney, *Early Medieval Medicine, with Special Reference to France and Chartres.*

3. See *Isidore of Seville: The Medical Writings,* ed. and trans. William D. Sharpe; Celsus, *De medicina, with an English Translation.*

4. On Dioscorides, see John M. Riddle, *Dioscorides on Pharmacy and Medicine;* Riddle, "Dioscorides." See the latter, pp. 125–33, on *Ex herbis femininis* (a work not restricted to remedies for feminine ailments). On medical recipes, see also Voigts, "Anglo-Saxon Plant Remedies"; Sigerist, "Latin Medical Literature," pp. 136–41; MacKinney, *Early Medieval Medicine,* pp. 31–38.

5. MacKinney, *Early Medieval Medicine,* pp. 47–49, 61–73.

6. The relevant passage from Cassiodorus's *Institutiones* is quoted by MacKinney, *Early Medieval Medicine,* p. 51. On monastic medicine more generally, see ibid., pp. 50–58.

7. See especially Darrel W. Amundsen, "Medicine and Faith in Early Christianity"; Amundsen and Gary B. Ferngren, "The Early Christian Tradition," and Amundsen, "The Medieval Catholic Tradition," both in Ronald L. Numbers and Darrel W. Amundsen, eds., *Caring and Curing: Health and Medicine in the Western Religious Traditions;* and Siraisi, *Medieval and Early Renaissance Medicine,* pp. 7–9.

8. Amundsen, "Medieval Catholic Tradition," p. 79; Grant, *Source Book,* pp. 773–74.

9. Quoted by Siraisi, *Medieval and Early Renaissance Medicine,* p. 14, from Bernard of Clairvaux, *Letters,* no. 388, trans. Bruno Scott James (Chicago: Regnery, 1953), pp. 458–59.

10. Amundsen, "Medicine and Faith in Early Christianity," pp. 333–49 (p. 338 for the quotation from Basil). On Tertullian, see *De corona,* 8, and *Ad nationes,* II.5, in *The Ante-Nicene Fathers,* ed. Alexander Roberts and James Donaldson, rev. by A. Cleveland Coxe (Grand Rapids: Eerdmans, 1986), 3:97, 134. See also Siraisi, *Medieval and Early Renaissance Medicine,* p. 9.

11. IV.31, in *Baedae opera historica,* trans. J. E. King, 2 vols. (London: Heinemann, 1930), 2:191–93. On the cult of saints, see the brilliant study by Peter Brown, *The Cult of Saints: Its Rise and Function in Latin Christianity;* also

Amundsen, "Medieval Catholic Tradition," pp. 79–82. On miraculous cures, see Ronald C. Finucane, *Miracles and Pilgrims: Popular Beliefs in Medieval England,* esp. chaps. 4–5.

12. On Islamic medicine, see Michael W. Dols, *Medieval Islamic Medicine: Ibn Riḍwān's Treatise "On the Prevention of Bodily Ills in Egypt";* Manfred Ullmann, *Islamic Medicine;* Franz Rosenthal, "The Physician in Medieval Muslim Society"; articles by Max Meyerhof, collected in his *Studies in Medieval Arabic Medicine: Theory and Practice;* and Siraisi, *Medieval and Early Renaissance Medicine,* pp. 11–13. An older reference source, still useful, is Lucien Leclerc, *Histoire de la médecine arabe.*

13. The classic work on Salerno is Paul Oskar Kristeller, "The School of Salerno: Its Development and Its Contribution to the History of Learning." See also McVaugh, "Medicine," pp. 247–49; and Morris Harold Saffron, *Maurus of Salerno: Twelfth-century "Optimus Physicus" with his Commentary on the Prognostics of Hippocrates.*

14. Michael McVaugh, "Constantine the African," *Dictionary of Scientific Biography,* 3:393–95; McVaugh, "Medicine," pp. 248–49; above, chap. 9.

15. Siraisi, *Medieval and Early Renaissance Medicine,* pp. 17–21; Katharine Park, *Doctors and Medicine in Early Renaissance Florence,* pp. 58–76; Edward J. Kealey, *Medieval Medicus: A Social History of Anglo-Norman Medicine,* chap. 2. To be used with caution is Robert S. Gottfried, *Doctors and Medicine in Medieval England 1340–1530.*

16. Park, *Doctors and Medicine,* pp. 54–58.

17. On women healers, see Siraisi, *Medieval and Early Renaissance Medicine,* pp. 27, 34, 45–46; John Benton, "Trotula, Women's Problems, and the Professionalization of Medicine in the Middle Ages"; Edward J. Kealey, "England's Earliest Women Doctors"; Monica H. Green, "Women's Medical Practice and Medical Care in Medieval Europe." On Jewish practitioners, see Elliot N. Dorff, "The Jewish Tradition," in Numbers and Amundsen, *Caring and Curing;* Luis García Ballester, Lola Ferre, and Edward Feliu, "Jewish Appreciation of Fourteenth-Century Scholastic Medicine."

18. On medicine in the universities, see Siraisi, *Medieval and Early Renaissance Medicine,* chap. 3; McVaugh, "Medicine," pp. 249–52; Vern L. Bullough, *The Development of Medicine as a Profession: The Contribution of the Medieval University to Modern Medicine,* esp. chap. 3; and Faye M. Getz, "The Faculty of Medicine before 1500."

19. McVaugh, "Medicine," p. 247.

20. On the curriculum, see Siraisi, *Medieval and Early Renaissance Medicine,* pp. 65–77; Siraisi, *Taddeo Alderotti and His Pupils: Two Generations of Italian Medical Learning,* chaps. 4–5; Siraisi, *Avicenna in Renaissance Italy: The "Canon" and Medical Teaching in Italian Universities after 1500,* chap. 3; Getz, "Faculty of Medicine"; and McVaugh, "Medicine," pp. 248–52. A good idea of the content of the *Articella* collection can be obtained from the annotated translation of its most basic component, the *Isagoge* of Ḥunayn ibn Isḥāq (Johannitius), contained in Grant, *Source Book,* pp. 705–15.

21. The numerical data presented here have been drawn from Siraisi, *Medieval and Early Renaissance Medicine,* pp. 63–64. For quantitative data from Oxford, see Getz, "Faculty of Medicine."

22. Faye M. Getz, "Charity, Translation, and the Language of Medical Learning in Medieval England"; Getz, *Healing and Society in Medieval England.*

23. Siraisi, *Medieval and Early Renaissance Medicine,* pp. 101–6; Grant, *Source Book,* pp. 705–9; L. J. Rather, "The 'Six Things Non-Natural': A Note on the Origins and Fate of a Doctrine and a Phrase."

24. On the treatment of disease, see Siraisi, *Medieval and Early Renaissance Medicine,* chap. 5; Grant, *Source Book,* pp. 775–91.

25. On drug therapy, see Siraisi, *Medieval and Early Renaissance Medicine,* pp. 141–49 (p. 148 on the application of pig dung for nose-bleed); Jones, *Medieval Medical Miniatures,* chap. 4.

26. Translated by Michael McVaugh in Grant, *Source Book,* p. 788. This list of curative properties is followed by the recipe for theriac. On theriac, see also McVaugh, "Theriac at Montpellier."

27. See, for example, Michael McVaugh, "Arnald of Villanova and Bradwardine's Law"; McVaugh, "Quantified Medical Theory and Practice at Fourteenth-Century Montpellier." Also McVaugh's introduction to *Arnald de Villanova, Opera medica omnia,* vol. 2: *Aphorismi de gradibus.*

28. Translated by Michael McVaugh, in Grant, *Source Book,* p. 749. On urinalysis, see MacKinney, *Medical Illustrations,* pp. 9–14; Jones, *Medieval Medical Miniatures,* pp. 58–60.

29. Translated by Michael McVaugh in Grant, *Source Book,* p. 746. On pulse, see also MacKinney, *Medical Illustrations,* pp. 15–19.

30. On medical astrology, see Siraisi, *Alderotti,* pp. 140–45; Siraisi, *Medieval and Early Renaissance Medicine,* pp. 68, 111–12, 123, 128–29, 134–36, 149–52; Jones, *Medieval Medical Miniatures,* pp. 69–74.

31. On the black death, see McVaugh, "Medicine," p. 253; Siraisi, *Medieval and Early Renaissance Medicine,* pp. 128–29; Grant, *Source Book,* pp. 773–74. For a review of some of the recent literature on the black death, as of 1982, see Nancy G. Siraisi, introduction to Williman, Daniel, ed., *The Black Death: The Impact of the Fourteenth-Century Plague,* pp. 9–22.

32. On medieval surgery, see Siraisi, *Medieval and Early Renaissance Medicine,* chap. 6; MacKinney, *Medical Illustrations,* chap. 8. On Roger Frugard, see Siraisi, pp. 162–66; MacKinney, *Medical Illustrations,* passim (under the name "Rogerius"). On Guy de Chauliac, see Vern L. Bullough, "Chauliac, Guy de," *Dictionary of Scientific Biography,* 3:218–19.

33. On blood-letting, see Linda E. Voigts and Michael R. McVaugh, *A Latin Technical Phlebotomy and Its Middle English Translation;* MacKinney, *Medical Illustrations,* pp. 55–61.

34. Stupefactives capable of putting the patient to sleep were available, but it is not clear how widely they were used; see Linda E. Voigts and Robert P. Hudson, "'A drynke that men callen dwale to make a man to slepe whyle men kerven him': A Surgical Anesthetic from Late Medieval England." For the quoted passage see MacKinney, *Medical Illustrations,* pp. 80–81.

35. Vern L. Bullough, "Mondino de' Luzzi," *Dictionary of Scientific Biography,* 9:467–69; Bullough, *Development of Medicine as a Profession,* pp. 61–65; Siraisi, *Medieval and Early Renaissance Medicine,* pp. 86–97.

36. Quoted from Bullough, *Development of Medicine as a Profession,* p. 64, with minor changes.

37. On the origins of the hospital in the narrower sense, see especially Timothy S. Miller, *The Birth of the Hospital in the Byzantine Empire;* Miller, "The Knights of Saint John and the Hospitals of the Latin West"; Michael W. Dols, "The Origins of the Islamic Hospital: Myth and Reality"; and Kealey, *Medieval Medicus,* chaps. 4–5.

38. Miller, "Knights of Saint John," pp. 723–25.

39. It seems doubtful that the last word has been said on this complex subject. I have followed Dols, "Origins of the Islamic Hospital," pp. 382–84; and Miller, "Knights of Saint John," pp. 717–23, 726–33. On the Barmak family, see above, chap. 8.

40. On hospitals in the West, see Talbot, *Medicine in Medieval England,* chap. 14 (pp. 177–78 for the quotation).

41. For a good introduction to medieval botanical knowledge, herbals in particular, see Jerry Stannard, "Medieval Herbals and Their Development"; Stannard, "Natural History," pp. 443–49.

42. On Albert's botanical knowledge, see Karen Reeds, "Albert on the Natural Philosophy of Plant Life"; Jerry Stannard, "Albertus Magnus and Medieval Herbalism." On Albert's biological studies, see also above, chap. 10.

43. On medieval zoology, see Stannard, "Natural History," pp. 432–43. On Albert's contributions, see Joan Cadden, "Albertus Magnus' Universal Physiology: the Example of Nutrition"; Luke Demaitre and Anthony A. Travill, "Human Embryology and Development in the Works of Albertus Magnus"; and Robin S. Oggins, "Albertus Magnus on Falcons and Hawks." Selections from Albert's *De animalibus* appear in Albert the Great, *Man and the Beasts, De animalibus (books 22–26),* trans. James J. Scanlan.

44. Charles Homer Haskins, "Science at the Court of the Emperor Frederick II"; and Haskins, "The *De arte venandi cum avibus* of Frederick II."

45. On medieval bestiaries and the *Physiologus,* see the introduction to *Physiologus,* trans. Michael J. Curley, pp. ix–xxxviii; Stannard, "Natural History," pp. 430–43; C. S. Lewis, *The Discarded Image,* pp. 146–52; and Willene B. Clark and Meradith T. McMunn, eds., *Beasts and Birds of the Middle Ages.*

46. *The Bestiary: A Book of Beasts,* trans. T. H. White, p. 84. The examples in this paragraph are entirely from the twelfth-century bestiary translated by White.

47. See the wonderfully illuminating discussion of sixteenth-century zoological literature by William B. Ashworth, Jr., "Natural History and the Emblematic World View," pp. 304–6.

CHAPTER FOURTEEN

1. For recent discussions of the continuity debate, see David C. Lindberg, "Conceptions of the Scientific Revolution from Bacon to Butterfield"; Bruce S. Eastwood, "On the Continuity of Western Science from the Middle Ages." See the latter for additional bibliography.

The question of continuity between medieval and early modern science is surely a legitimate topic of investigation. But there are two dangers against which the historian must vigilantly guard. The first is the temptation to grade the ancient and medieval scientific tradition on the basis of its resemblance to modern science—the criteria of worth thus reducing to the anticipation or approximation of later developments. The second is that interdisciplinary squabbling among historians will be allowed to swallow up the serious scholarly issues presented by the continuity question—that historians of early science will overstate the ancient and medieval scientific achievement in order to defend their chosen specialty against attempts by practitioners of other specialties to denigrate it by understating its accomplishments. It is likely that the battle lines and argumentative strategies will then be determined as much by disciplinary loyalty as by the historical evidence.

2. Francis Bacon, *New Organon,* in *Works,* trans. James Spedding, Robert Ellis, and Douglas Heath, new ed., 15 vols. (New York: Hurd & Houghton, 1870–72), 4:77. François Marie Arouet de Voltaire, *Works,* trans. T. Smollett, T. Francklin, et al., 39 vols. (London: J. Newbery et al., 1761–74), 1:82. Marquis de Condorcet, *Sketch for a Historical Picture of the Progress of the Human Mind,* ed. Stuart Hampshire, trans. June Barraclough (London: Weidenfeld & Nicolson, 1955), p. 72.

3. Jacob Burckhardt, *The Civilization of the Renaissance in Italy,* pp. 371, 182. For an analysis of the concept of the Renaissance, see Wallace K. Ferguson, *The Renaissance in Historical Thought,* esp. chaps. 7–8; and Philip Lee Ralph, *The Renaissance in Perspective,* chap. 1. For the quotations from Burckhardt I am indebted to Edward Rosen, "Renaissance Science as Seen by Burckhardt and His Successors," in Tinsley Hilton, ed., *The Renaissance,* p. 78.

4. John Addington Symonds, *Renaissance in Italy,* part I: *The Age of the Despots,* pp. 13–15 (a passage to which I was led by Ralph, *Renaissance in Perspective,* p. 6).

5. Pierre Duhem, *Les origines de la statique,* vol. 1, p. iv. Duhem further articulated his view of the medieval contribution to science in his *Etudes sur Léonard de Vinci;* and his *Le système du monde.* On Duhem, see also R. N. D. Martin, "The Genesis of a Mediaeval Historian"; Stanley Jaki, *Uneasy Genius: The Life and Work of Pierre Duhem.*

6. Charles Homer Haskins, *Studies in the History of Mediaeval Science;* and *The Renaissance of the Twelfth Century.* Lynn Thorndike, *A History of Magic and Experimental Science;* and *Science and Thought in the Fifteenth Century.*

7. Anneliese Maier, "The Achievements of Late Scholastic Natural Philosophy," in Maier, *On the Threshold of Exact Science,* pp. 143–70; also Sargent's introduction, pp. 11–16; and John E. Murdoch and Edith D. Sylla, "Anneliese Maier and the History of Medieval Science." A bibliography of Anneliese Maier's publications on medieval science is contained in Maier's *Ausgehendes Mittelalters,* 3:617–26. A list of Marshall Clagett's publications is supplied in an appendix to Edward Grant and John E. Murdoch, eds., *Mathematics and its Applications to Science and Natural Philosophy in the Middle Ages,* pp. 325–28.

8. A. C. Crombie, *Augustine to Galileo: The History of Science A.D. 400–1650* (1952), p. 273. This book went through various revisions—also a change of name, when it appeared in 1959 as *Medieval and Early Modern Science.* For an assessment of Crombie's achievement, see Eastwood, "On the Continuity of Western Science."

9. A. C. Crombie, *Robert Grosseteste and the Origins of Experimental Science 1100–1700,* pp. 9–10.

10. Alexandre Koyré, "The Origins of Modern Science: A New Interpretation," pp. 13–14, 19.

11. Alexandre Koyré, "Galileo and Plato," in Koyré's *Metaphysics and Measure-*

ment: Essays in the Scientific Revolution, pp. 20–21. Although written in 1943, these words represented Koyré's continuing judgment.

12. A. Rupert Hall, "On the Historical Singularity of the Scientific Revolution of the Seventeenth Century," p. 213; cf. Hall, *The Revolution in Science 1500–1750,* p. 3. For earlier statements of Hall's position, see his *The Scientific Revolution 1500–1800,* introduction and first four chapters.

13. Ernan McMullin, "Medieval and Modern Science: Continuity or Discontinuity?" We are speaking here of the *theory,* rather than the *application,* of scientific method.

14. Thomas S. Kuhn, *The Structure of Scientific Revolutions.* Kuhn, "Mathematical versus Experimental Traditions in the Development of Physical Science."

15. The clearest statement of the "Yates thesis" is to be found in Frances A. Yates, *Giordano Bruno and the Hermetic Tradition;* and Yates, "The Hermetic Tradition in Renaissance Science." For analysis and criticism, see Brian P. Copenhaver, "Natural Magic, Hermetism, and Occultism in Early Modern Science"; Brian Vickers, ed., *Occult and Scientific Mentalities in the Renaissance;* and three volumes of collected essays by Charles B. Schmitt: *Reappraisals in Renaissance Thought; The Aristotelian Tradition and Renaissance Universities;* and *Studies in Renaissance Philosophy and Science.* The phrase "genuine science" is Yates's, quoted by Copenhaver, p. 261.

16. This point is nicely made by McMullin, "Medieval and Modern Science," pp. 103–4.

17. For the latest (but certainly not the last) word on the scientific revolution, see the essays contained in David C. Lindberg and Robert S. Westman, eds., *Reappraisals of the Scientific Revolution.*

18. On medieval methodology, see Crombie, *Grosseteste,* esp. chaps. 2–4; William A. Wallace, *Causality and Scientific Explanation,* vol. 1: *Medieval and Early Classical Science,* chaps. 1–4; McMullin, "Medieval and Modern Science"; and Eileen Serene, "Demonstrative Science."

19. Ernan McMullin, "Conceptions of Science in the Scientific Revolution," in *Reappraisals,* ed. Lindberg and Westman, pp. 27–86; McMullin, "Medieval and Modern Science," pp. 108–29. For a social-historical analysis of seventeenth-century experimental practice, see also Steven Shapin and Simon Schaffer, *Leviathan and the Air-Pump: Hobbes, Boyle, and the Experimental Life;* Peter Dear, "Jesuit Mathematical Science and the Reconstitution of Experience in the Early 17th Century."

20. See the forceful statement by Anneliese Maier, "The Theory of the Elements and the Problem of their Participation in Compounds," p. 125.

21. On essential natures, see above, chap. 3.

22. See Mordechai Feingold, *The Mathematicians' Apprenticeship: Science, Universities and Society in England, 1560–1640;* John Gascoigne, "A Reappraisal of the Role of the Universities in the Scientific Revolution."

23. How profoundly it altered the shape of the enterprise is a matter of hot dispute. There is a very large literature on this subject, ranging from the classic by Robert K. Merton, *Science, Technology and Society in Seventeenth-Century England,* to the recent book by Margaret C. Jacob, *The Cultural Meaning of the Scientific Revolution.* See also A. R. Hall, "Merton Revisited or Science and Society in the Seventeenth Century," and the collection of articles in George Basalla, ed., *The Rise of Modern Science: Internal or External Factors?*

24. See, for example, Albert Van Helden, *The Invention of the Telescope*.

25. On the process of scientific change, see Kuhn, *Structure of Scientific Revolutions*.

26. This statement requires clarification on two counts. First, reference to a synthesis of classical and Christian thought is not meant to suggest that all the problems had been solved or that bitter disagreements would not continue to surface. The synthesis in question was, like Aristotelian philosophy, a living tradition. Second, there is no intention here of suggesting that Christianity was a beneficial element (or a detrimental one) in the mix, or that the synthesis of classical and Christian thought was better (or worse) than some other conceivable synthesis. I am simply making the factual claim that medieval scholars did, in fact, produce a synthesis of classical and Christian thought and that within the resulting conceptual framework natural philosophy was productively pursued for several centuries.

27. Above, chaps. 10, 11.

28. Maier, "Theory of the Elements," esp. pp. 126–34; above, chap. 12.

29. See above, chap. 11.

30. See above, chap. 12.

31. Edward Grant, "Aristotelianism and the Longevity of the Medieval World View."

32. Alexandre Koyré, *From the Closed World to the Infinite Universe*, p. 4. Cf. Koyré, *Galileo Studies*, pp. 2–3; Koyré, *The Astronomical Revolution*, pp. 9–10.

33. Hall, "Historical Singularity," pp. 210–11.

34. See David C. Lindberg, "Continuity and Discontinuity in the History of Optics: Kepler and the Medieval Tradition," from which some of this phraseology is borrowed.

35. It may be instructive that the two examples of continuity that I am about to discuss are both taken from Kuhn's "classical sciences," which he considers the primary locus of revolutionary change.

36. The expression $v \propto t$ emerges directly from the medieval definition of uniformly accelerated motion, as that motion in which equal increments of velocity are acquired in equal units of time. And $s \propto t^2$ follows from a simple extension of the medieval theorem that states that the distances traversed in the first and second halves of a uniformly accelerated motion are in the ratio of $1:3$ (see above, chap. 11). On Galileo and the medieval mechanical tradition, see Marshall Clagett, *The Science of Mechanics in the Middle Ages*, pp. 251–53, 409–18, 576–82, 666–71; Clagett, *Nicole Oresme and the Medieval Geometry of Qualities and Motions*, pp. 71–73, 103–6; Edith D. Sylla, "Galileo and the Oxford *Calculatores*"; Christopher Lewis, *The Merton Tradition and Kinematics in Late Sixteenth and Early Seventeenth Century Italy*, pp. 279–83. The claim that Galileo incorporated medieval terms, concepts, and theories into his mechanics is in no way called into question by controversies over the precise channels by which medieval influence reached Galileo, nor by Galilean departures from the medieval tradition. For an attempt to separate Galileo from the medieval tradition, see Stillman Drake, "The Uniform Motion Equivalent of a Uniformly Accelerated Motion from Rest." On the quest for channels of influence, see Clagett, *Oresme and the Medieval Geometry of Qualities and Motions*, pp. 103–6; Sylla, "Galileo and the Oxford *Calculatores*"; Lewis, *Merton Tradition and Kinematics*; William A. Wallace, *Galileo and His Sources: The Heritage of the Collegio Romano in Galileo's Science*; and Wallace,

Prelude to Galileo: Essays on Medieval and Sixteenth-Century Sources of Galileo's Thought.

37. On these optical developments, see David C. Lindberg, *Theories of Vision from al-Kindi to Kepler,* esp. chap. 9; Lindberg, "Laying the Foundations of Geometrical Optics: Maurolico, Kepler, and the Medieval Tradition." Also three articles by Lindberg on radiation through apertures: "The Theory of Pinhole Images from Antiquity to the Thirteenth Century"; "A Reconsideration of Roger Bacon's Theory of Pinhole Images"; and "The Theory of Pinhole Images in the Fourteenth Century."

BIBLIOGRAPHY

Aaboe, Asger. "On Babylonian Planetary Theories." *Centaurus* 5 (1958): 209–77.

Ackrill, J. L. *Aristotle the Philosopher.* Oxford: Clarendon Press, 1981.

Adams, Marilyn McCord. *William Ockham,* 2 vols. Notre Dame, Ind.: University of Notre Dame Press, 1987.

Albert the Great. *Man and the Beasts, De animalibus (books 22–26),* trans. James J. Scanlan. Medieval & Renaissance Texts & Studies, no. 47. Binghamton: Center for Medieval and Early Renaissance Studies, 1987.

Amundsen, Darrel W. "Medicine and Faith in Early Christianity." *Bulletin of the History of Medicine* 56 (1982): 326–50.

———. "Medieval Canon Law on Medical and Surgical Practice by the Clergy." *Bulletin of the History of Medicine* 52 (1978): 22–44.

———. "The Medieval Catholic Tradition." In Numbers, Ronald L., and Amundsen, Darrel W., eds., *Caring and Curing: Health and Medicine in the Western Religious Traditions,* pp. 65–107. New York: Macmillan, 1986.

Amundsen, Darrel W., and Ferngren, Gary B. "The Early Christian Tradition." In Numbers, Ronald L., and Amundsen, Darrel W., eds., *Caring and Curing: Health and Medicine in the Western Religious Traditions,* pp. 40–64. New York: Macmillan, 1986.

Anawati, G. C. "Ḥunayn ibn Isḥāq." *Dictionary of Scientific Biography,* 15:230–34.

Anawati, G. C., and Iskandar, Albert Z. "Ibn Sīnā." *Dictionary of Scientific Biography,* 15:494–501.

Archimedes. *Archimedes in the Middle Ages,* ed. and trans. Marshall Clagett, 5 vols. Madison: University of Wisconsin Press, 1964; Philadephia: American Philosophical Society, 1976–1984.

———. *The Works of Archimedes: Edited in Modern Notation, with Introductory Chapters,* ed. Thomas L. Heath, 2d ed. Cambridge: Cambridge University Press, 1912.

Aristotle. *Complete Works,* ed. Jonathan Barnes, 2 vols. Princeton: Princeton University Press, 1984.

———. *Metaphysics,* trans. Hugh Tredennick, 2 vols. London: Heinemann, 1935.

Armstrong, A. H., ed. *The Cambridge History of Later Greek and Early Medieval Philosophy.* Cambridge: Cambridge University Press, 1970.

Armstrong, A. H., and Markus, R. A. *Christian Faith and Greek Philosophy.* London: Darton, Longman, & Todd, 1960.

Arnaldez, Roger, and Iskandar, Albert Z. "Ibn Rushd." *Dictionary of Scientific Biography,* 12:1–9.

Arts libéraux et philosophie au moyen âge: Actes du quatrième congrès international de philosophie médiévale, Université de Montréal, 27 août–2 septembre 1967. Montreal: Institut d'études médiévales, 1969.

Ashley, Benedict M. "St. Albert and the Nature of Natural Science." In Weisheipl, James A., ed., *Albertus Magnus and the Sciences: Commemorative Essays 1980,* pp. 73–102. Toronto: Pontifical Institute of Mediaeval Studies, 1980.

Ashworth, William B., Jr. "Natural History and the Emblematic World View." In Lindberg, David C., and Westman, Robert S., eds., *Reappraisals of the Scientific Revolution,* pp. 303–32. Cambridge: Cambridge University Press, 1990.

Asmis, Elizabeth. *Epicurus' Scientific Method.* Ithaca: Cornell University Press, 1984.

Bailey, Cyril. *The Greek Atomists and Epicurus.* Oxford: Clarendon Press, 1928.

Baldwin, John W. *Masters, Princes, and Merchants: The Social Views of Peter the Chanter and His Circle,* 2 vols. Princeton: Princeton University Press, 1970.

———. *The Scholastic Culture of the Middle Ages.* Lexington, Mass.: D. C. Heath, 1971.

Balme, D. M. "The Place of Biology in Aristotle's Philosophy." In Gotthelf, Allan, and Lennox, James G., eds., *Philosophical Issues in Aristotle's Biology,* pp. 9–20. Cambridge: Cambridge University Press, 1987.

Barnes, Jonathan. *Aristotle.* Oxford: Oxford University Press, 1982.

———. "Aristotle's Theory of Demonstration." In Barnes, Schofield, and Sorabji, *Articles on Aristotle,* I: *Science,* pp. 65–87. London: Duckworth, 1975.

———. *The Presocratic Philosophers,* 2 vols. London: Routledge & Kegan Paul, 1979.

Barnes, Jonathan; Brunschwig, Jacques; Burnyeat, Myles; and Schofield, Malcolm, eds. *Science and Speculation: Studies in Hellenistic Theory and Practice.* Cambridge: Cambridge University Press, 1982.

Barnes, Jonathan; Schofield, Malcolm; and Sorabji, Richard, eds., *Articles on Aristotle,* I: *Science.* London: Duckworth, 1975.

Barrow, Robin. *Greek and Roman Education.* London: Macmillan, 1967.

Basalla, George, ed. *The Rise of Modern Science: Internal or External Factors?* Lexington, Mass.: D. C. Heath, 1968.

Beaujouan, Guy. "Motives and Opportunities for Science in the Medieval Universities." In Crombie, A.C., ed., *Scientific Change,* pp. 219–36. London: Heinemann, 1963.

———. "The Transformation of the Quadrivium." In Benson, Robert L., and Constable, Giles, eds., *Renaissance and Renewal in the Twelfth Century,* pp. 463–87. Cambridge, Mass.: Harvard University Press, 1982.

Benjamin, Francis S., and Toomer, G. J., eds. and trans. *Campanus of Novara and Medieval Planetary Theory: "Theorica planetarum."* Madison: University of Wisconsin Press, 1971.

Benson, Robert L., and Constable, Giles, eds. *Renaissance and Renewal in the Twelfth Century.* Cambridge, Mass.: Harvard University Press, 1982.

Benton, John. "Trotula, Women's Problems, and the Professionalization of Medicine in the Middle Ages." *Bulletin of the History of Medicine* 59 (1985): 30–53.

Berggren, J. L. "History of Greek Mathematics: A Survey of Recent Research." *Historia Mathematica* 11 (1984): 394–410.

Biggs, Robert. "Medicine in Ancient Mesopotamia." *History of Science* 8 (1969): 94–105.

Birkenmajer, Aleksander. "Le rôle joué par les médecins et les naturalistes dans la réception d'Aristote au XIIᵉ et XIIIᵉ siècles." In Birkenmajer, Aleksander, *Etudes d'histoire des sciences et de philosophie du moyen âge,* pp. 73–87. Studia Copernicana, no. 1. Wrocław: Ossolineum, 1970.

Al-Biṭrūjī. *On the Principles of Astronomy,* ed. and trans. Bernard R. Goldstein, 2 vols. New Haven: Yale University Press, 1971.

Blair, Peter Hunter. *The World of Bede.* Cambridge: Cambridge University Press, 1970.

Boethius of Dacia. *On the Supreme Good, On the Eternity of the World, On Dreams,* trans. John F. Wippel. Mediaeval Sources in Translation, no. 30. Toronto: Pontifical Institute of Mediaeval Studies, 1987.

Bonner, Stanley F. *Education in Ancient Rome: From the Elder Cato to the Younger Pliny.* Berkeley and Los Angeles: University of California Press, 1977.

Boyer, Carl B. *A History of Mathematics.* New York: John Wiley, 1968.

———. *The Rainbow: From Myth to Mathematics.* New York: Yoseloff, 1959.

Brain, Peter. *Galen on Bloodletting: A Study of the Origins, Development and Validity of His Opinions, with a Translation of the Three Works.* Cambridge: Cambridge University Press, 1986.

Brandon, S. G. F. *Creation Legends of the Ancient Near East.* London: Hodder and Stoughton, 1963.

Breasted, James Henry. *Development of Religion and Thought in Ancient Egypt.* New York: Scribner's, 1912.

———. *The Edwin Smith Surgical Papyrus,* 2 vols. University of Chicago, Oriental Institute Publications, 3–4. Chicago: University of Chicago Press, 1930.

Brehaut, Ernest. *An Encyclopedist of the Dark Ages: Isidore of Seville.* New York: Columbia University, 1912.

Brown, Peter. *Augustine of Hippo: A Biography.* Berkeley and Los Angeles: University of California Press, 1969.

———. *The Cult of Saints: Its Rise and Function in Latin Christianity.* Chicago: University of Chicago Press, 1981.

Bullough, Vern L. "Chauliac, Guy de." *Dictionary of Scientific Biography,* 3:218–19.

———. *The Development of Medicine as a Profession: The Contribution of the Medieval University to Modern Medicine.* Basel: Karger, 1966.

———. "Mondino de' Luzzi." *Dictionary of Scientific Biography,* 9:467–69.

Burckhardt, Jacob. *The Civilization of the Renaissance in Italy,* trans. S. G. C. Middlemore. New York: Modern Library, 1954.

Burnett, Charles S. F., ed. *Adelard of Bath: An English Scientist and Arabist of the Early Twelfth Century.* Warburg Institute Surveys and Texts, no. 14. London: The Warburg Institute, 1987.

———. "Scientific Speculations." In Dronke, Peter, ed., *A History of Twelfth-*

Century Western Philosophy, pp. 155–66. Cambridge: Cambridge University Press, 1988.

———. "Translation and Translators, Western European." *Dictionary of the Middle Ages,* 12:136–42.

———. "What is the *Experimentarius* of Bernardus Silvestris? A Preliminary Survey of the Material." *Archives d'histoire doctrinale et littéraire du moyen âge* 44 (1977): 79–125.

Bynum, Caroline Walker. "Did the Twelfth Century Discover the Individual?" *Journal of Ecclesiastical History* 31 (1980): 1–17.

Cadden, Joan. "Albertus Magnus' Universal Physiology: the Example of Nutrition." In Weisheipl, James A., ed., *Albertus Magnus and the Sciences: Commemorative Essays 1980,* pp. 321–29. Toronto: Pontifical Institute of Mediaeval Studies, 1980.

Callus, D. A. "Introduction of Aristotelian Learning to Oxford." *Proceedings of the British Academy* 29 (1943): 229–81.

———, ed. *Robert Grosseteste, Scholar and Bishop: Essays in Commemoration of the Seventh Centenary of His Death.* Oxford: Clarendon Press, 1955.

Cameron, M. L. "The Sources of Medical Knowledge in Anglo-Saxon England." *Anglo-Saxon England* 11 (1983): 135–52.

Caroti, Stefano. "Nicole Oresme's Polemic against Astrology in His 'Quodlibeta'." In Curry, Patrick, ed., *Astrology, Science and Society: Historical Essays,* pp. 75–93. Woodbridge, Suffolk: Boydell, 1987.

Carré, Meyrick H. *Realists and Nominalists.* Oxford: Clarendon Press, 1946.

Catto, J. I., ed. *The Early Oxford Schools.* Vol. 1 of *The History of the University of Oxford,* general ed. T. H. Aston. Oxford: Clarendon Press, 1984.

Celsus, Aulus Cornelius. *De medicina, with an English Translation,* trans. W. G. Spencer, 3 vols. London: Heinemann, 1935–38.

Chadwick, Henry. *Early Christian Thought and the Classical Tradition: Studies in Justin, Clement, and Origen.* New York: Oxford University Press, 1966.

———. *The Early Church.* Harmondsworth: Penguin, 1967.

Chenu, M.-D. *Nature, Man, and Society in the Twelfth Century: Essays on New Theological Perspectives in the Latin West,* trans. Jerome Taylor and Lester K. Little. Chicago: University of Chicago Press, 1968. Originally published as *La théologie au douzième siècle.* Paris: J. Vrin, 1957.

———. *Toward Understanding St. Thomas,* ed. and trans. A.-M. Landry and D. Hughes. Chicago: Henry Regnery, 1964.

Cherniss, Harold. *The Riddle of the Early Academy.* Berkeley and Los Angeles: University of California Press, 1945.

Clagett, Marshall. *Ancient Egyptian Science: A Source Book,* vol. 1. Philadelphia: American Philosophical Society, 1989.

———, ed. *Critical Problems in the History of Science.* Madison: University of Wisconsin Press, 1962.

———. *Giovanni Marliani and Late Medieval Physics.* New York: Columbia University Press, 1941.

———. *Greek Science in Antiquity.* London: Abelard-Schuman, 1957.

———. "The *Liber de motu* of Gerard of Brussels and the Origins of Kinematics in the West." *Osiris,* 1st ser., 12 (1956): 73–175.

———, ed. and trans. *Nicole Oresme and the Medieval Geometry of Qualities and Motions.* Madison: University of Wisconsin Press, 1968.

———. *The Science of Mechanics in the Middle Ages.* Madison: University of Wisconsin Press, 1959.

———. "Some Novel Trends in the Science of the Fourteenth Century." In Singleton, Charles S., ed., *Art, Science, and History in the Renaissance,* pp. 275–303. Baltimore: Johns Hopkins University Press, 1968.

———. *Studies in Medieval Physics and Mathematics.* London: Variorum, 1979.

Clark, Willene B., and McMunn, Meradith T., eds. *Beasts and Birds of the Middle Ages.* Philadelphia: University of Pennsylvania Press, 1989.

Clarke, M. L. *Higher Education in the Ancient World.* London: Routledge & Kegan Paul, 1971.

Cobban, Alan B. *The Medieval English Universities: Oxford and Cambridge to c. 1500.* Aldershot: Scolar Press, 1988.

———. *The Medieval Universities: Their Development and Organization.* London: Methuen, 1975.

Cochrane, Charles N. *Christianity and Classical Culture: A Study of Thought and Action from Augustus to Augustine.* Oxford: Clarendon Press, 1940.

Cohen, Morris R., and Drabkin, I. E., eds. *A Source Book in Greek Science.* Cambridge, Mass.: Harvard University Press, 1958.

Colish, Marcia L. *The Stoic Tradition from Antiquity to the Early Middle Ages,* 2 vols. Leiden: Brill, 1985.

Collingwood, R. G. *The Idea of Nature.* Oxford: Clarendon Press, 1945.

Contreni, John J. *The Cathedral School of Laon from 850 to 930: Its Manuscripts and Masters.* Münchener Beiträge zur Mediävistik und Renaissance-Forschung, vol. 29. Munich: Arbeo-Gesellschaft, 1978.

———. "Schools, Cathedral." *Dictionary of the Middle Ages,* 11:59–63.

Coopland, G. W. *Nicole Oresme and the Astrologers: A Study of His Livre de divinacions.* Cambridge, Mass.: Harvard University Press, 1952.

Copenhaver, Brian P. "Natural Magic, Hermetism, and Occultism in Early Modern Science." In Lindberg, David C., and Westman, Robert S., eds., *Reappraisals of the Scientific Revolution,* pp. 261–301. Cambridge: Cambridge University Press, 1990.

Courtenay, William J. *Capacity and Volition: A History of the Distinction of Absolute and Ordained Power.* Quodlibet: Ricerche e strumenti di filosofia medievale, no. 8. Bergamo: Pierluigi Lubrina, 1990.

———. *Covenant and Causality in Medieval Thought.* London: Variorum, 1984.

———. "The Critique on Natural Causality in the Mutakallimun and Nominalism." *Harvard Theological Review* 66 (1973): 77–94.

———. "The Dialectic of Divine Omnipotence." In Courtenay, William J., *Covenant and Causality in Medieval Thought,* chap. 4.

———. "Nature and the Natural in Twelfth-Century Thought." In Courtenay, William J., *Covenant and Causality in Medieval Thought,* chap. 3.

———. "Ockham, William of." *Dictionary of the Middle Ages,* 9:209–14.

———. *Schools and Scholars in Fourteenth-Century England.* Princeton: Princeton University Press, 1987.

———. *Teaching Careers at the University of Paris in the Thirteenth and Four-*

teenth Centuries. Texts and Studies in the History of Mediaeval Education, no. 18. Notre Dame, Ind.: United States Subcommission for the History of Universities, University of Notre Dame, 1988.

Courtenay, William J., and Tachau, Katherine H. "Ockham, Ockhamists, and the English-German Nation at Paris, 1339–1341." *History of Universities* 2 (1982): 53–96.

Crombie, A. C. *Augustine to Galileo: The History of Science A.D. 400–1650.* London: Falcon, 1952. Reissued as *Medieval and Early Modern Science,* 2 vols. Garden City: Doubleday Anchor, 1959.

————. *Robert Grosseteste and the Origins of Experimental Science, 1100–1700.* Oxford: Clarendon Press, 1953.

————. *Science, Optics and Music in Medieval and Early Modern Thought.* London: Hambledon, 1990.

Crosby, H. Lamar, Jr., ed. and trans. *Thomas of Bradwardine, His "Tractatus de Proportionibus": Its Significance for the Development of Mathematical Physics.* Madison: University of Wisconsin Press, 1961.

Crowe, Michael J. *Theories of the World from Antiquity to the Copernican Revolution.* New York: Dover, 1990.

Crowley, Theodore. *Roger Bacon: The Problem of the Soul in His Philosophical Commentaries.* Dublin: James Duffy/Louvain: Editions de l'Institut Supérieur de Philosophie, 1950.

Cumont, Franz. *Astrology and Religion among the Greeks and Romans.* New York: Putnam's Sons, 1912.

Curley, Michael J., trans. *Physiologus.* Austin, Tex.: University of Texas Press, 1979.

Curry, Patrick, ed. *Astrology, Science and Society: Historical Essays.* Woodbridge, Suffolk: Boydell, 1987.

Dales, Richard C. *The Intellectual Life of Western Europe in the Middle Ages.* Washington, D.C.: University Press of America, 1980.

————. "Marius 'On the Elements' and the Twelfth-Century Science of Matter." *Viator* 3 (1972): 191–218.

————. *Medieval Discussions of the Eternity of the World.* Leiden: Brill, 1990.

————. "Time and Eternity in the Thirteenth Century." *Journal of the History of Ideas* 49 (1988): 27–45.

d'Alverny, Marie-Thérèse. "Translations and Translators." In Benson, Robert L., and Constable, Giles, eds., *Renaissance and Renewal in the Twelfth Century,* pp. 421–62. Cambridge, Mass.: Harvard University Press, 1982.

Dear, Peter. "Jesuit Mathematical Science and the Reconstitution of Experience in the Early 17th Century." *Studies in History and Philosophy of Science* 18 (1987): 133–75.

Debus, Allen G. *The Chemical Philosophy: Paracelsian Science and Medicine in the Sixteenth and Seventeenth Centuries,* 2 vols. New York: Science History Publications, 1977.

————. *Man and Nature in the Renaissance.* Cambridge: Cambridge University Press, 1978.

De Lacy, Phillip. "Galen's Platonism." *American Journal of Philology* 93 (1972): 27–39.

Demaitre, Luke E. *Doctor Bernard de Gordon: Professor and Practitioner.* Toronto: Pontifical Institute of Mediaeval Studies, 1980.

Demaitre, Luke E., and Travill, Anthony A. "Human Embryology and Development in the Works of Albertus Magnus." In Weisheipl, James A., ed., *Albertus Magnus and the Sciences: Commemorative Essays 1980,* pp. 405–40. Toronto: Pontifical Institute of Mediaeval Studies, 1980.

de Santillana, Giorgio. *The Origins of Scientific Thought: From Anaximander to Proclus, 600 B.C. to A.D. 500.* Chicago: University of Chicago Press, 1961.

de Vaux, Carra. "Astronomy and Mathematics." In Arnold, Thomas, and Guillaume, Alfred, eds., *The Legacy of Islam,* pp. 376–97. London: Oxford University Press, 1931.

Dicks, D. R. *Early Greek Astronomy to Aristotle.* Ithaca: Cornell University Press, 1970.

————. "Eratosthenes." *Dictionary of Scientific Biography,* 4:388–93.

Dictionary of Scientific Biography, 16 vols. New York: Scribner's, 1970-80.

Dijksterhuis, E. J. *Archimedes,* trans. C. Dikshoorn. Copenhagen: Munksgaard, 1956.

————. *The Mechanization of the World Picture,* trans. C. Dikshoorn. Oxford: Clarendon Press, 1961.

Diogenes Laertius. *Lives of Eminent Philosophers,* trans. R. D. Hicks, 2 vols. London: Heinemann, 1925.

Dodge, Bayard. *Muslim Education in Medieval Times.* Washington, D.C.: The Middle East Institute, 1962.

Dols, Michael W., trans. *Medieval Islamic Medicine: Ibn Riḍwān's Treatise "On the Prevention of Bodily Ills in Egypt",* with an Arabic text edited by Adil S. Gamal. Berkeley and Los Angeles: University of California Press, 1984.

————. "The Origins of the Islamic Hospital: Myth and Reality." *Bulletin of the History of Medicine* 61 (1987): 367–90.

Dorff, Elliot N. "The Jewish Tradition." In Numbers, Ronald L., and Amundsen, Darrel W., eds., *Caring and Curing: Health and Medicine in the Western Religious Traditions,* pp. 5–39. New York: Macmillan, 1986.

Drake, Stillman. "The Uniform Motion Equivalent of a Uniformly Accelerated Motion from Rest." *Isis* 63 (1972): 28–38.

Dreyer, J. L. E. *History of the Planetary Systems from Thales to Kepler.* Cambridge: Cambridge University Press, 1906. Reissued as *A History of Astronomy from Thales to Kepler,* ed. W. H. Stahl. New York: Dover, 1953.

Dronke, Peter, ed. *A History of Twelfth-Century Western Philosophy.* Cambidge: Cambridge University Press, 1988.

————. "Thierry of Chartres." In Dronke, Peter, ed., *Twelfth-Century Western Philosophy,* pp. 358–85.

Duhem, Pierre. *Etudes sur Léonard de Vinci,* 3 vols. Paris: Hermann, 1906–13.

————, ed. *Un fragment inédit de l'Opus tertium de Roger Bacon, précédé d'une étude sur ce fragment.* Quaracchi: Collegium S. Bonaventurae, 1909.

————. *Medieval Cosmology: Theories of Infinity, Place, Time, Void, and the Plurality of Worlds,* ed. and trans. Roger Ariew. Chicago: University of Chicago Press, 1985.

————. *Les origines de la statique,* 2 vols. Paris: Hermann, 1905–6.

————. *Le système du monde,* 10 vols. Paris: Hermann, 1913–59.

————. *To Save the Phenomena: An Essay on the Idea of Physical Theory from Plato to Galileo,* trans. Edmund Doland and Chaninah Maschler. Chicago: University of Chicago Press, 1969. Originally published in French in 1908.

Düring, Ingemar. "The Impact of Aristotle's Scientific Ideas in the Middle Ages." *Archiv für Geschichte der Philosophie* 50 (1968): 115–33.

Easton, Stewart C. *Roger Bacon and His Search for a Universal Science.* Oxford: Basil Blackwell, 1952.

Eastwood, Bruce S. *Astronomy and Optics from Pliny to Descartes.* London: Variorum, 1989.

————. "Kepler as Historian of Science: Precursors of Copernican Heliocentrism according to *De revolutionibus,* I, 10." *Proceedings of the American Philosophical Society* 126 (1982): 367–94.

————. "On the Continuity of Western Science from the Middle Ages." *Isis,* forthcoming.

————. "Plinian Astronomical Diagrams in the Early Middle Ages." In Grant, Edward, and Murdoch, John E., eds., *Mathematics and Its Applications to Science and Natural Philosophy in the Middle Ages: Essays in Honor of Marshall Clagett,* pp. 141–72. Cambridge: Cambridge University Press, 1987.

————. "Plinian Astronomy in the Middle Ages and Renaissance." In French, Roger, and Greenaway, Frank, eds., *Science in the Early Roman Empire: Pliny the Elder, His Sources and Influence,* chap. 11. Totawa, N.J.: Barnes & Noble, 1986.

Ebbell, B. *The Papyrus Ebers, the Greatest Egyptian Medical Document.* Copenhagen: Munksgaard, 1939.

Edel, Abraham. *Aristotle and His Philosophy.* Chapel Hill: University of North Carolina Press, 1982.

Edelstein, Emma J., and Edelstein, Ludwig. *Asclepius: A Collection and Interpretation of the Testimonies,* 2 vols. Baltimore, Johns Hopkins University Press, 1945.

Edelstein, Ludwig. *Ancient Medicine: Selected Papers of Ludwig Edelstein,* ed. Owsei Temkin and C. Lilian Temkin. Baltimore: Johns Hopkins University Press, 1967.

————. "The Distinctive Hellenism of Greek Medicine." *Bulletin of the History of Medicine* 40 (1966): 197–255.

————. "Empiricism and Skepticism in the Teaching of the Greek Empiricist School." In Edelstein, *Ancient Medicine,* pp. 195–203.

————. "Greek Medicine and Its Relation to Religion and Magic." *Bulletin of the Institute of the History of Medicine* 5 (1937): 201–46.

————. "The Methodists." In Edelstein, *Ancient Medicine,* pp. 173–91.

————. "The Relation of Ancient Philosophy to Medicine." *Bulletin of the History of Medicine* 26 (1952): 299–316.

Elford, Dorothy. "William of Conches." In Dronke, Peter, ed., *A History of Twelfth-Century Western Philosophy,* pp. 308–27. Cambridge: Cambridge University Press, 1988.

Emerton, Norma E. *The Scientific Reinterpretation of Form.* Ithaca: Cornell University Press, 1984.

Epp, Ronald H., ed. *Recovering the Stoics.* Supplement to *The Southern Journal of*

Philosophy, vol. 23. Memphis: Department of Philosophy, Memphis State University, 1985.

Euclid. *The Elements,* trans. Thomas Heath, 3 vols. Cambridge: Cambridge University Press, 1908.

Evans, Gillian R. *Anselm and a New Generation.* Oxford: Clarendon Press, 1980.

———. "The Influence of Quadrivium Studies in the Eleventh- and Twelfth-Century Schools." *Journal of Medieval History* 1 (1975): 151–64.

———. *Old Arts and New Theology: The Beginnings of Theology as an Academic Discipline.* Oxford: Clarendon Press, 1980.

———. *The Thought of Gregory the Great.* Cambridge: Cambridge University Press, 1986.

Fakhry, Majid. *A History of Islamic Philosophy.* New York: Columbia University Press, 1970.

Farrington, Benjamin. *Greek Science,* rev. ed. Harmondsworth: Penguin, 1961.

Feingold, Mordechai. *The Mathematicians' Apprenticeship: Science, Universities and Society in England, 1560–1640.* Cambridge: Cambridge University Press, 1984.

Ferguson, Wallace K. *The Renaissance in Historical Thought.* Boston: Houghton Mifflin, 1948.

Ferruolo, Stephen C. *The Origins of the University: The Schools of Paris and Their Critics, 1100–1215.* Stanford: Stanford University Press, 1985.

Fichtenau, Heinrich. *The Carolingian Empire,* trans. Peter Munz. Oxford: Basil Blackwell, 1957.

La filosofia della natura nel medioevo: Atti del Terzo Congresso Internazionale di Filosofia Medioevale, 31 August–5 September 1964. Milan: Società Editrice Vita e Pensiero, 1966.

Finley, M. I. *The World of Odysseus,* rev. ed. New York: Viking Press, 1965.

Finucane, Ronald C. *Miracles and Pilgrims: Popular Beliefs in Medieval England.* Totowa, N.J.: Rowman and Littlefield, 1977.

Flint, Valerie I. J. *The Rise of Magic in Early Medieval Europe.* Princeton: Princeton University Press, 1991.

Fontaine, Jacques. *Isidore de Séville et la culture classique dans l'Espagne wisigothique,* 2d ed., 3 vols. Paris: Etudes Augustiniennes, 1983.

Frankfort, H.; Frankfort, H. A.; Wilson, John A.; and Jacobsen, Thorkild. *Before Philosophy: The Intellectual Adventure of Ancient Man.* Baltimore: Penguin, 1951.

Fraser, P. M. *Ptolemaic Alexandria,* 3 vols. Oxford: Clarendon Press, 1972.

Frede, Michael. "The Method of the So-Called Methodical School of Medicine." In Barnes, Jonathan; Brunschwig, Jacques; Burnyeat, Myles; and Schofield, Malcolm, eds., *Science and Speculation: Studies in Hellenistic Theory and Practice,* pp. 1–23. Cambridge: Cambridge University Press, 1982.

French, Roger, and Greenaway, Frank, eds. *Science in the Early Roman Empire: Pliny the Elder, His Sources and Influence.* Totawa, N.J.: Barnes & Noble, 1986.

Frend, W. H. C. *The Rise of the Monophysite Movement: Chapters in the History of the Church in the Fifth and Sixth Centuries.* Cambridge: Cambridge University Press, 1972.

Funkenstein, Amos. *Theology and the Scientific Imagination from the Middle Ages to the Seventeenth Century.* Princeton: Princeton University Press, 1986.

Furley, David. *Cosmic Problems: Essays on Greek and Roman Philosophy of Nature.* Cambridge: Cambridge University Press, 1989.

———. *The Greek Cosmologists,* vol. 1: *The Formation of the Atomic Theory and Its Earliest Critics.* Cambridge: Cambridge University Press, 1987.

———. *Two Studies in the Greek Atomists.* Princeton: Princeton University Press, 1967.

Gabriel, Astrik L. "Universities." *Dictionary of the Middle Ages,* 12:282–300.

Galen. *On Respiration and the Arteries,* ed. and trans. David J. Furley and J. S. Wilkie. Princeton: Princeton University Press, 1984.

———. *On the Natural Faculties,* trans. A. J. Brock. London: Heinemann, 1963.

———. *On the Usefulness of the Parts of the Body,* ed. and trans. Margaret T. May, 2 vols. Ithaca: Cornell University Press, 1968.

———. *Three Treatises on the Nature of Science,* ed. and trans. Richard Walzer and Michael Frede. Indianapolis: Hackett, 1985.

García Ballester, Luis. "Galen as a Medical Practitioner: Problems in Diagnosis." In Nutton, Vivian, ed., *Galen: Problems and Prospects,* pp. 13–46. London: Wellcome Institute for the History of Medicine, 1981.

García Ballester, Luis; Fere, Lola; and Feliu, Edward. "Jewish Appreciation of Fourteenth-Century Scholastic Medicine." *Osiris,* n.s. 6 (1990): 85–117.

Gascoigne, John. "A Reappraisal of the Role of the Universities in the Scientific Revolution." In Lindberg, David C., and Westman, Robert S., eds., *Reappraisals of the Scientific Revolution,* pp. 207–60. Cambridge: Cambridge University Press, 1990.

Gersh, Stephen. *Middle Platonism and Neoplatonism: The Latin Tradition,* 2 vols. Notre Dame, Ind.: University of Notre Dame Press, 1986.

Getz, Faye M. "Charity, Translation, and the Language of Medical Learning in Medieval England." *Bulletin of the History of Medicine* 64 (1990): 1–17.

———. "The Faculty of Medicine before 1500." In Catto, J. I., ed., vol. 2 of *The History of the University of Oxford,* ed. T. H. Aston. Oxford: Clarendon Press, forthcoming.

———. *Healing and Society in Medieval England: A Middle English Translation of the Pharmaceutical Writings of Gilbert Anglicus.* Madison: University of Wisconsin Press, 1991.

———. "Western Medieval Medicine." *Trends in History* 4, nos. 2–3 (1988): 37–54.

Ghalioungui, Paul. *The House of Life, Per Ankh: Magic and Medical Science in Ancient Egypt,* 2d ed. Amsterdam: B. M. Israel, 1973.

———. *The Physicians of Pharaonic Egypt.* Cairo: Al-Ahram Center for Scientific Translations, 1983.

Gillings, R. J. "The Mathematics of Ancient Egypt." *Dictionary of Scientific Biography,* 15:681–705.

Gilson, Etienne. *The Christian Philosophy of St. Thomas Aquinas,* trans. L. K. Shook. New York: Random House, 1956.

Gimpel, Jean. *The Medieval Machine: The Industrial Revolution of the Middle Ages.* New York: Holt, Rinehart and Winston, 1976.

Gingerich, Owen. "Islamic Astronomy." *Scientific American* 254, no. 4 (April 1986): 74–83.

Goldstein, Bernard R. *The Arabic Version of Ptolemy's "Planetary Hypotheses"*

Transactions of the American Philosophical Society, n.s., vol. 57, pt. 4. Philadelphia: American Philosophical Society, 1967.

———. *Theory and Observation in Ancient and Medieval Astronomy*. London: Variorum, 1985.

Goldstein, Bernard R., and Bowen, Alan C. "A New View of Early Greek Astronomy." *Isis* 74 (1983): 330–40.

Goldstein, Bernard R., and Swerdlow, Noel. "Planetary Distances and Sizes in an Anonymous Arabic Treatise Preserved in Bodleian MS Marsh 621." *Centaurus* 15 (1970): 135–70.

Goody, Jack. *The Domestication of the Savage Mind*. Cambridge: Cambridge University Press, 1977.

Goody, Jack, and Watt, Ian. "The Consequences of Literacy." *Comparative Studies in Society and History* 5 (1962–63): 304–45.

Gottfried, Robert S. *Doctors and Medicine in Medieval England 1340–1530*. Princeton: Princeton University Press, 1986.

Gotthelf, Allan, and Lennox, James G., eds. *Philosophical Issues in Aristotle's Biology*. Cambridge: Cambridge University Press, 1987.

Gottschalk, H. B. "Strato of Lampsacus." *Dictionary of Scientific Biography*, 13:91–95.

Grant, Edward. "Aristotelianism and the Longevity of the Medieval World View." *History of Science* 16 (1978): 93–106.

———. "Celestial Matter: A Medieval and Galilean Cosmological Problem." *Journal of Medieval and Renaissance Studies* 13 (1983): 157–86.

———. "Celestial Orbs in the Latin Middle Ages." *Isis* 78 (1987): 153–73.

———. "The Condemnation of 1277, God's Absolute Power, and Physical Thought in the Late Middle Ages." *Viator* 10 (1979): 211–44.

———. "Cosmology." In Lindberg, David C., ed., *Science in the Middle Ages*, pp. 265–302. Chicago: University of Chicago Press, 1978.

———. "Late Medieval Thought, Copernicus, and the Scientific Revolution." *Journal of the History of Ideas* 23 (1962): 197–220.

———. "Medieval and Renaissance Scholastic Conceptions of the Influence of the Celestial Region on the Terrestrial." *The Journal of Medieval and Renaissance Studies* 17 (1987): 1–23.

———. "Medieval and Seventeenth-Century Conceptions of an Infinite Void Space beyond the Cosmos." *Isis* 60 (1969): 39–60.

———. *The Medieval Cosmos 1200–1687*. Forthcoming.

———. "The Medieval Doctrine of Place: Some Fundamental Problems and Solutions." In Maierù, A., and Paravicini Bagliani, A., eds., *Studi sul XIV secolo in memoria di Anneliese Maier,* pp. 57–79. Storia e Letteratura, Raccolta di studi e testi, no. 151. Rome: Edizioni di Storia e Letteratura, 1981.

———. *Much Ado about Nothing: Theories of Space and Vacuum from the Middle Ages to the Scientific Revolution*. Cambridge: Cambridge University Press, 1981.

———, ed. and trans. *Nicole Oresme and the Kinematics of Circular Motion: Tractatus de commensurabilitate vel incommensurabilitate motuum celi*. Madison: University of Wisconsin Press, 1971.

———. *Physical Science in the Middle Ages*. New York: Wiley, 1971.

———. "Science and the Medieval University." In Kittelson, James M., and Transue,

Pamela J., eds., *Rebirth, Reform and Resilience: Universities in Transition 1300–1700,* pp. 68–102. Columbus: Ohio State University Press, 1984.

———. "Science and Theology in the Middle Ages." In Lindberg, David C., and Numbers, Ronald L., eds., *God and Nature: Historical Essays on the Encounter between Christianity and Science,* pp. 49–75. Berkeley and Los Angeles: University of California Press, 1986.

———, ed. *A Source Book in Medieval Science.* Cambridge, Mass.: Harvard University Press, 1974.

———. *Studies in Medieval Science and Natural Philosophy.* London: Variorum, 1981.

Grant, Edward, and Murdoch, John E., eds. *Mathematics and Its Applications to Science and Natural Philosophy in the Middle Ages: Essays in Honor of Marshall Clagett.* Cambridge: Cambridge University Press, 1987.

Grant, Robert M. *Miracle and Natural Law in Graeco-Roman and Early Christian Thought.* Amsterdam: North-Holland, 1952.

Grayeff, Felix. *Aristotle and His School.* London: Duckworth, 1974.

Green, Monica H. "Women's Medical Practice and Medical Care in Medieval Europe." *Signs* 14 (1989): 434–73.

Gregory, Tullio. *Anima mundi: La filosofia di Guglielmo di Conches e la scuola di Chartres.* Florence: G. C. Sansoni, 1955.

———. "La nouvelle idée de nature et de savoir scientifique au XIIe siècle." In Murdoch, John E., and Sylla, Edith D., eds., *The Cultural Context of Medieval Learning,* pp. 193–212. Boston Studies in the Philosophy of Science, 27. Dordrecht: Reidel, 1975.

———. "The Platonic Inheritance." In Dronke, Peter, ed., *A History of Twelfth-Century Western Philosophy,* pp. 54–80. Cambridge: Cambridge University Press, 1988.

Grendler, Paul F. *Schooling in Renaissance Italy: Literacy and Learning, 1300–1600.* Baltimore: Johns Hopkins University Press, 1989.

Grene, Marjorie. *A Portrait of Aristotle.* Chicago: University of Chicago Press, 1963.

Griffin, Jasper. *Homer.* Oxford: Oxford University Press, 1980.

Hackett, M. B. "The University as a Corporate Body." In Catto, J. I., ed., *The Early Oxford Schools,* vol. 1 of *The History of the University of Oxford,* general ed. T. H. Aston, pp. 37–95. Oxford: Clarendon Press, 1984.

Hadot, Ilsetraut, ed. *Simplicius: sa vie, son oeuvre, sa survie.* Actes du Colloque international de Paris, 28 Sept.–1 Oct. 1985. Berlin: Walter de Gruyter, 1987.

Hahm, David E. *The Origins of Stoic Cosmology.* Columbus: Ohio State University Press, 1977.

Hall, A. Rupert. "Merton Revisited or Science and Society in the Seventeenth Century." *History of Science* 2 (1963): 1–16.

———. "On the Historical Singularity of the Scientific Revolution of the Seventeenth Century." In Elliott, J. H., and Koenigsberger, H. G., eds., *The Diversity of History: Essays in Honour of Sir Herbert Butterfield,* pp. 199–221. London: Routledge & Kegan Paul, 1970.

———. *The Revolution in Science 1500–1750. London: Longman, 1983.*

———. *The Scientific Revolution 1500–1800.* London: Longmans, Green, 1954.

Halleux, Robert. "Alchemy." *Dictionary of the Middle Ages,* 1:134–40.

————. *Les textes alchimiques.* Typologie des sources du moyen âge occidental, no. 32. Turnhout: Brepols, 1979.

Hamilton, Edith. *Mythology.* Boston: Little, Brown, 1942.

Hansen, Bert. *Nicole Oresme and the Marvels of Nature: A Study of his "De causis mirabilium" with Critical Edition, Translation, and Commentary.* Toronto: Pontifical Institute of Mediaeval Studies, 1985.

Hare, R. M. *Plato.* Oxford: Oxford University Press, 1982.

Hargreave, David. "Reconstructing the Planetary Motions of the Eudoxean System." *Scripta Mathematica* 28 (1970): 335–45.

Häring, Nikolaus. "Chartres and Paris Revisited." In O'Donnell, J. Reginald, ed., *Essays in Honour of Anton Charles Pegis,* pp. 268–329. Toronto: Pontifical Institute of Mediaeval Studies, 1974.

————. "The Creation and Creator of the World according to Thierry of Chartres and Clarenbaldus of Arras." *Archives d'histoire doctrinale et littéraire du moyen âge* 22 (1955): 137–216.

Harley, J. B., and Woodward, David, eds. *The History of Cartography,* vol. 1: *Cartography in Prehistoric, Ancient, and Medieval Europe and the Mediterranean.* Chicago: University of Chicago Press, 1987.

Harris, John R. "Medicine." In Harris, John R., ed., *The Legacy of Egypt,* 2d ed., pp. 112–37. Oxford: Clarendon Press, 1971.

Hartner, Willy. "Al-Battānī." *Dictionary of Scientific Biography,* 1:507–16.

Haskins, Charles Homer. "The *De arte venandi cum avibus* of Frederick II." *English Historical Review* 36 (1921): 334–55. Reprinted in Haskins, *Studies in the History of Mediaeval Science,* pp. 299–326.

————. *The Renaissance of the Twelfth Century.* Cambridge, Mass.: Harvard University Press, 1927.

————. *The Rise of Universities.* Providence: Brown University Press, 1923.

————. "Science at the Court of the Emperor Frederick II." *American Historical Review* 27 (1922): 669–94. Reprinted in Haskins, *Studies in the History of Mediaeval Science,* pp. 242–71.

————. *Studies in the History of Mediaeval Science.* Cambridge, Mass.: Harvard University Press, 1924.

Heath, Thomas L. *Aristarchus of Samos, The Ancient Copernicus: A History of Greek Astronomy to Aristarchus.* Oxford: Clarendon Press, 1913.

————. *A History of Greek Mathematics,* 2 vols. Oxford: Clarendon Press, 1921.

Helton, Tinsley, ed. *The Renaissance: A Reconsideration of the Theories and Interpretations of the Age.* Madison: University of Wisconsin Press, 1961.

Hesiod. *The Poems of Hesiod,* trans. R. M. Frazer. Norman: University of Oklahoma Press, 1983.

————. *Theogony and Works and Days,* trans., with introduction and notes, by M. L. West. Oxford: Oxford University Press, 1988.

Hildebrandt, M. M. *The External School in Carolingian Society.* Leiden: Brill, 1991.

Hillgarth, J. N. "Isidore of Seville, St." *Dictionary of the Middle Ages,* 6:563–66.

Hippocrates, with an English Translation, trans. W. H. S. Jones, E. T. Withington, and Paul Potter, 6 vols. London: Heinemann, 1923–88.

Hissette, Roland. *Enquête sur les 219 articles condamnés à Paris le 7 mars 1277.* Philosophes médiévaux, no. 22. Louvain: Publications universitaires, 1977.

Hitti, Philip K. *History of the Arabs from the Earliest Times to the Present,* 7th ed. London: Macmillan, 1961.

Holmyad, E. J. *Alchemy.* Harmondsworth: Penguin, 1957.

Hopkins, Jasper. *A Companion to the Study of St. Anselm.* Minneapolis: University of Minnesota Press, 1972.

Hoskin, Michael, and Molland, A. G. "Swineshead on Falling Bodies: An Example of Fourteenth-Century Physics." *British Journal for the History of Science* 3 (1966): 150–82.

Hugh of St. Victor. *The Didascalicon of Hugh of St. Victor: A Medieval Guide to the Arts,* ed. and trans. Jerome Taylor. New York: Columbia University Press, 1961.

Hyman, Arthur. "Aristotle's 'First Matter' and Avicenna's and Averroes' 'Corporeal Form'." In *Harry Austryn Wolfson Jubilee Volume,* 1:385–406. Jerusalem: American Academy for Jewish Research, 1965.

Imbault-Huart, Marie-José. *La médecine au moyen âge à travers les manuscrits de la Bibliothèque Nationale.* Paris: Editions de la Porte Verte, 1983.

Iskandar, Albert Z. "Ḥunayn the Translator; Ḥunayn the Physician." *Dictionary of Scientific Biography,* 15:234–39.

Jackson, Ralph. *Doctors and Diseases in the Roman Empire.* Norman, Oklahoma: University of Oklahoma Press, 1988.

Jacob, Margaret C. *The Cultural Meaning of the Scientific Revolution.* New York: Knopf, 1988.

Jaki, Stanley. *Uneasy Genius: The Life and Work of Pierre Duhem.* The Hague: Nijhoff, 1984.

Jolivet, Jean. "The Arabic Inheritance." In Dronke, Peter, ed., *A History of Twelfth Century Western Philosophy,* pp. 113–48. Cambridge: Cambridge University Press, 1988.

Jones, Charles W. "Bede." *Dictionary of the Middle Ages,* 2:153–56.

Jones, Peter M. *Medieval Medical Miniatures.* London: The British Library in association with the Wellcome Institute for the History of Medicine, 1984.

Kahn, Charles H. *Anaximander and the Origins of Greek Cosmology.* New York: Columbia University Press, 1960.

Kaiser, Christopher. *Creation and the History of Science.* Grand Rapids: Eerdmans, 1991.

Kari-Niazov, T. N. "Ulugh Beg." *Dictionary of Scientific Biography,* 13:535–37.

Kealey, Edward J. "England's Earliest Women Doctors." *Journal of the History of Medicine* 40 (1985): 473–77.

———. *Medieval Medicus: A Social History of Anglo-Norman Medicine.* Baltimore: Johns Hopkins University Press, 1981.

Kennedy, E. S. "The Arabic Heritage in the Exact Sciences." *Al-Abhath: A Quarterly Journal for Arab Studies* 23 (1970): 327–44.

———. "The Exact Sciences." In *The Cambridge History of Iran,* vol. 4: *The Period from the Arab Invasion to the Saljuqs,* ed. R. N. Frye, pp. 378–95. Cambridge: Cambridge University Press, 1975.

———. "The History of Trigonometry: An Overview." In *Historical Topics for the Mathematics Classroom.* Washington, D.C.: National Council of Teachers of Mathematics, 1969. Reprinted in Kennedy, E. S., et al., *Studies in the Islamic Exact Sciences,* pp. 3–29.

Kennedy, E. S., with colleagues and former students. *Studies in the Islamic Exact Sciences.* Beirut: American University of Beirut, 1983.

Kibre, Pearl. "'Astronomia' or 'Astrologia Ypocratis'." In Hilfstein, Erna; Czartoryski, Paweł; and Grande, Frank D., eds., *Science and History: Studies in Honor of Edward Rosen,* pp. 133–56. Studia Copernicana, no. 16. Wrocław: Ossolineum, 1978.

———. "The *Quadrivium* in the Thirteenth Century Universities (with Special Reference to Paris)." In *Arts libéraux et philosophie au moyen âge: Actes du quatrième congrès international de philosophie médiévale, Université de Montréal, 27 août–2 septembre 1967,* pp. 175–91. Montreal: Institut d'études médiévales, 1969.

———. *Scholarly Privileges in the Middle Ages.* Cambridge, Mass.: Mediaeval Academy of America, 1962.

———. *Studies in Medieval Science: Alchemy, Astrology, Mathematics and Medicine.* London: Hambledon Press, 1984.

Kibre, Pearl, and Siraisi, Nancy G. "The Institutional Setting: The Universities." In Lindberg, David C., ed., *Science in the Middle Ages,* pp. 120–44. Chicago: University of Chicago Press, 1978.

Kieckhefer, Richard. *Magic in the Middle Ages.* Cambridge: Cambridge University Press, 1990.

King, David A. *Islamic Astronomical Instruments.* London: Variorum, 1987.

———. *Islamic Mathematical Astronomy.* London: Variorum, 1986.

Kirk, G. S., and Raven, J. E. *The Presocratic Philosophers: A Critical History with a Selection of Texts.* Cambridge: Cambridge University Press, 1960.

Knorr, Wilbur. "Archimedes and the Pseudo-Euclidean *Catoptrics:* Early Stages in the Ancient Geometric Theory of Mirrors." *Archives internationales d'histoire des sciences* 35 (1985): 28–105.

———. *The Evolution of the Euclidean Elements: A Study of the Theory of Incommensurable Magnitudes and Its Significance for Early Greek Geometry.* Dordrecht: D. Reidel, 1975.

———. "John of Tynemouth *alias* John of London: Emerging Portrait of a Singular Medieval Mathematician." *British Journal for the History of Science* 23 (1990): 293–330.

Knowles, David. *The Evolution of Medieval Thought.* New York: Vintage, 1964.

Kogan, Barry S. *Averroes and the Metaphysics of Causation.* Albany: State University of New York Press, 1985.

Kovach, Francis J., and Shahan, Robert W., eds. *Albert the Great: Commemorative Essays.* Norman: University of Oklahoma Press, 1980.

Koyré, Alexandre. *The Astronomical Revolution: Copernicus, Kepler, Borelli,* trans. R. E. W. Maddison. Paris: Hermann, 1973.

———. *From the Closed World to the Infinite Universe.* Baltimore: Johns Hopkins University Press, 1957.

———. *Galileo Studies,* trans. John Mepham. Atlantic Highlands, N.J.: Humanities Press, 1978.

———. *Metaphysics and Measurement: Essays in the Scientific Revolution.* London: Chapman & Hall, 1968.

———. "The Origins of Modern Science: A New Interpretation." *Diogenes* 16 (Winter 1956): 1–22.

Kren, Claudia. *Alchemy in Europe: A Guide to Research.* New York: Garland, 1990.

———. "Astronomy." In Wagner, David L., ed., *The Seven Liberal Arts in the Middle Ages,* pp. 218–47. Bloomington: Indiana University Press, 1983.

———. "Bernard of Verdun." *Dictionary of Scientific Biography,* 2:23–24.

———. "Homocentric Astronomy in the Latin West: The *De reprobatione ecentricorum et epiciclorum* of Henry of Hesse." *Isis* 59 (1968): 269–81.

———. *Medieval Science and Technology: A Selected, Annotated Bibliography.* New York: Garland, 1985.

Kretzmann, Norman, ed. *Infinity and Continuity in Ancient and Medieval Thought.* Ithaca: Cornell University Press, 1982.

Kretzmann, Norman; Kenny, Anthony; and Pinborg, Jan, eds. *The Cambridge History of Later Medieval Philosophy.* Cambridge: Cambridge University Press, 1982.

Kristeller, Paul Oskar. "The School of Salerno: Its Development and Its Contribution to the History of Learning." *Bulletin of the History of Medicine* 17 (1945): 138–94.

Kudlien, Fridolf. "Early Greek Primitive Medicine." *Clio medica* 3 (1968): 305–36.

Kudlien, Fridolf, and Durling, Richard J., eds. *Galen's Method of Healing.* Leiden: Brill, 1991.

Kuhn, Thomas S. *The Copernican Revolution: Planetary Astronomy in the Development of Western Thought.* Cambridge: Harvard University Press, 1957.

———. "Mathematical versus Experimental Traditions in the Development of Physical Science." *Journal of Interdisciplinary History* 7 (1976): 1–31. Reprinted in Kuhn, *The Essential Tension: Selected Studies in Scientific Tradition and Change,* pp. 31–65. Chicago: University of Chicago Press, 1977.

———. *The Structure of Scientific Revolutions.* Chicago: University of Chicago Press, 1962.

Laistner, M. L. W. *Christianity and Pagan Culture in the Later Roman Empire.* Ithaca: Cornell University Press, 1951.

———. *Thought and Letters in Western Europe,* A.D. *500–900,* new ed. London: Methuen, 1957.

Lattin, Harriet Pratt, ed. and trans. *The Letters of Gerbert with His Papal Privileges as Sylvester II.* New York: Columbia University Press, 1961.

Lear, Jonathan. *Aristotle: The Desire to Understand.* Cambridge: Cambridge University Press, 1988.

Leclerc, Ivor. *The Nature of Physical Existence.* London: George Allen & Unwin, 1972.

Leclerc, Lucien. *Histoire de la médecine arabe,* 2 vols. Paris: Ernest Leroux, 1876.

Leclercq, Jean, O.S.B. *The Love of Learning and the Desire for God: A Study of Monastic Culture,* trans. Catherine Misrahi. New York: Fordham University Press, 1961.

———. "The Renewal of Theology." In Benson, Robert L., and Constable, Giles, eds., *Renaissance and Renewal in the Twelfth Century,* pp. 68–87. Cambridge, Mass.: Harvard University Press, 1982.

Lejeune, Albert. *Euclide et Ptolémée: Deux stades de l'optique géométrique grecque.* Louvain: Bibliothèque de l'Université, 1948.

———. *Recherches sur la catoprique grecque.* Brussels: Palais des Académies, 1957.

Lemay, Richard. *Abu Ma'shar and Latin Aristotelianism in the Twelfth Century: The*

Recovery of Aristotle's Natural Philosophy through Arabic Astrology. Beirut: American University of Beirut, 1962.

————. "Gerard of Cremona." *Dictionary of Scientific Biography,* 15:173–92.

————. "The True Place of Astrology in Medieval Science and Philosophy." In Curry, Patrick, ed., *Astrology, Science and Society: Historical Essays,* pp. 57–73. Woodbridge, Suffolk: Boydell, 1987.

Lévy-Bruhl, Lucien. *How Natives Think,* trans. Lilian A. Clare. London: George Allen & Unwin, 1926.

Lewis, Bernard, ed. *Islam and the Arab World: Faith, People, Culture.* New York: Knopf, 1976.

Lewis, C. S. *The Discarded Image: An Introduction to Medieval and Renaissance Literature.* Cambridge: Cambridge University Press, 1964.

Lewis, Christopher. *The Merton Tradition and Kinematics in Late Sixteenth and Eary Seventeenth Century Italy.* Padua: Antenore, 1980.

Liebeschütz, H. "Boethius and the Legacy of Antiquity." In Armstrong, A. H., ed., *The Cambridge History of Later Greek and Early Medieval Philosophy,* pp. 538–64. Cambridge: Cambridge University Press, 1970.

Lindberg, David C. "Alhazen's Theory of Vision and Its Reception in the West." *Isis* 58 (1967): 321–41.

————. "Conceptions of the Scientific Revolution from Bacon to Butterfield: A Preliminary Sketch." In Lindberg and Westman, *Reappraisals of the Scientific Revolution,* pp. 1–26.

————. "Continuity and Discontinuity in the History of Optics: Kepler and the Medieval Tradition." *History and Technology* 4 (1987): 423–40.

————. "The Genesis of Kepler's Theory of Light: Light Metaphysics from Plotinus to Kepler." *Osiris* n.s., 2 (1986): 5–42.

————, ed. and trans. *John Pecham and the Science of Optics: "Perspectiva communis," edited with an Introduction, English Translation, and Critical Notes.* Madison: University of Wisconsin Press, 1970.

————. "Laying the Foundations of Geometrical Optics: Maurolico, Kepler, and the Medieval Tradition." In Lindberg, David C., and Cantor, Geoffrey, *The Discourse of Light from the Middle Ages to the Enlightenment,* pp. 1–65. Los Angeles: William Andrews Clark Memorial Library, 1985.

————. "On the Applicability of Mathematics to Nature: Roger Bacon and His Predecessors." *British Journal for the History of Science* 15 (1982): 3–25.

————. "Optics, Western European." *Dictionary of the Middle Ages,* 9:247–53.

————. "A Reconsideration of Roger Bacon's Theory of Pinhole Images." *Archive for History of Exact Sciences* 6 (1970): 214–23.

————. "Roger Bacon and the Origins of *Perspectiva* in the West." In Grant, Edward, and Murdoch, John E., eds., *Mathematics and Its Applications to Science and Natural Philosophy in the Middle Ages: Essays in Honor of Marshall Clagett,* pp. 249–68. Cambridge: Cambridge University Press, 1987.

————, ed. and trans. *Roger Bacon's Philosophy of Nature: A Critical Edition, with English Translation, Introduction, and Notes, of "De multiplicatione specierum" and "De speculis comburentibus."* Oxford: Clarendon Press, 1983.

————. "Science and the Early Church." In Lindberg and Numbers, *God and Na-*

ture: Historical Essays on the Encounter between Christianity and Science, pp. 19–48.

———. "Science as Handmaiden: Roger Bacon and the Patristic Tradition." *Isis* 78 (1987): 518–36.

———, ed. *Science in the Middle Ages.* Chicago: University of Chicago Press, 1978.

———. "The Science of Optics." In Lindberg, *Science in the Middle Ages,* pp. 338–68.

———. *Studies in the History of Medieval Optics.* London: Variorum, 1983.

———. *Theories of Vision from al-Kindi to Kepler.* Chicago: University of Chicago Press, 1976.

———. "The Theory of Pinhole Images from Antiquity to the Thirteenth Century." *Archive for History of Exact Sciences* 5 (1968): 154–76.

———. "The Theory of Pinhole Images in the Fourteenth Century." *Archive for History of Exact Sciences* 6 (1970): 299–325.

———. "The Transmission of Greek and Arabic Learning to the West." In Lindberg, *Science in the Middle Ages,* pp. 52–90.

Lindberg, David C., and Numbers, Ronald L., eds. *God and Nature: Historical Essays on the Encounter between Christianity and Science.* Berkeley and Los Angeles: University of California Press, 1986.

Lindberg, David C., and Westman, Robert S., eds. *Reappraisals of the Scientific Revolution.* Cambridge: Cambridge University Press, 1990.

Lindgren, Uta. *Gerbert von Aurillac und das Quadrivium: Untersuchungen zur Bildung im Zeitalter der Ottonen.* Sudhoffs Archiv: Zeitschrift für Wissenschaftsgeschichte, Beiheft 18. Wiesbaden: Franz Steiner, 1976.

Lipton, Joshua D. "The Rational Evaluation of Astrology in the Period of Arabo-Latin Translation, ca. 1126–1187 A.D." Ph.D. dissertation, University of California at Los Angeles, 1978.

Little, A. G., ed. *Roger Bacon Essays.* Oxford: Clarendon Press, 1914.

Livesey, Steven J. *Theology and Science in the Fourteenth Century: Three Questions on the Unity and Subalternation of the Sciences from John of Reading's Commentary on the Sentences.* Studien und Texte zur Geistesgeschichte des Mittelalters, vol. 25. Leiden: Brill, 1989.

Lloyd, G. E. R. *Aristotle: The Growth and Structure of His Thought.* Cambridge: Cambridge University Press, 1968.

———. *Demystifying Mentalities.* Cambridge: Cambridge University Press, 1990.

———. *Early Greek Science: Thales to Aristotle.* London: Chatto & Windus, 1970.

———. *Greek Science after Aristotle.* London: Chatto & Windus, 1973.

———, ed. *Hippocratic Writings.* Harmondsworth: Penguin, 1978.

———. *Magic, Reason and Experience: Studies in the Origins and Development of Greek Science.* Cambridge: Cambridge University Press, 1979.

———. *The Revolutions of Wisdom: Studies in the Claims and Practice of Ancient Greek Science.* Berkeley and Los Angeles: University of California Press, 1987.

———. "Saving the Appearances." *Classical Quarterly* 28 (1978): 202–22.

Locher, A. "The Structure of Pliny the Elder's Natural History." In French, Roger, and Greenaway, Frank, eds., *Science in the Early Roman Empire,* pp. 20–29. Totawa, N.J.: Barnes & Noble, 1986.

Long, A. A. "Astrology: Arguments Pro and Contra." In Barnes, Jonathan; Brunschwig, Jacques; Burnyeat, Myles; and Schofield, Malcolm, eds., *Science and Specu-lation: Studies in Hellenistic Theory and Practice,* pp. 165–92. Cambridge: Cambridge University Press, 1982.

———. *Hellenistic Philosophy: Stoics, Epicureans, Sceptics,* 2d ed. London: Duck-worth, 1974.

———. "The Stoics on World-Conflagration and Everlasting Recurrence." In Epp, Ronald H., ed., *Recovering the Stoics,* pp. 13–37. Supplement to *The Southern Journal of Philosophy,* vol. 23. Memphis: Department of Philosophy, Memphis State University, 1985.

Long, A. A., and Sedley, D. N. *The Hellenistic Philosophers,* 2 vols. Cambridge: Cam-bridge University Press, 1987.

Longrigg, James. "Anatomy in Alexandria in the Third Century B.C." *British Journal for the History of Science* 21 (1988): 455–88.

———. "Erasistratus." *Dictionary of Scientific Biography,* 4:382–86.

———. "Presocratic Philosophy and Hippocratic Medicine." *History of Science* 27 (1989): 1–39.

———. "Superlative Achievement and Comparative Neglect: Alexandrian Medical Science and Modern Historical Research." *History of Science* 19 (1981): 155–200.

Lones, Thomas E. *Aristotle's Researches in Natural Science.* London: West, Newman, 1912.

Lucretius. *De rerum natura,* trans. W. H. D. Rouse and M. F. Smith, rev. 2d ed. Lon-don: Heinemann, 1982.

Luscombe, David E. *Peter Abelard.* London: Historical Association, 1979.

———. "Peter Abelard." In Dronke, Peter, ed., *A History of Twelfth-Century West-ern Philosophy,* pp. 279–307. Cambridge: Cambridge University Press, 1988.

Lutz, Cora E. *Schoolmasters of the Tenth Century.* Hamden, Conn.: Archon, 1977.

Lynch, John Patrick. *Aristotle's School: A Study of a Greek Educational Institution.* Berkeley and Los Angeles: University of California Press, 1972.

Lytle, Guy Fitch. "The Careers of Oxford Students in the Later Middle Ages." In Kit-telson, James M., and Transue, Pamela J., eds., *Rebirth, Reform and Resilience: Universities in Transition 1300–1700,* pp. 213–53. Columbus: Ohio State Uni-versity Press, 1984.

———. "Patronage Patterns and Oxford Colleges, c. 1300–c. 1530." In Stone, Law-rence, ed., *The University in Society,* 1:111–49. Princeton: Princeton University Press, 1974.

MacKinney, Loren C. *Early Medieval Medicine, with Special Reference to France and Chartres.* Baltimore: Johns Hopkins University Press, 1937.

———. *Medical Illustrations in Medieval Manuscripts.* London: Wellcome Histori-cal Medical Library, 1965.

Macrobius. *Commentary on the Dream of Scipio,* trans. with introduction and notes by William H. Stahl. New York: Columbia University Press, 1952.

Mahoney, Michael S. "Another Look at Greek Geometrical Analysis." *Archive for History of Exact Sciences* 5 (1968): 318–48.

———. "Mathematics." In Lindberg, David C., ed., *Science in the Middle Ages,* pp. 145–78. Chicago: University of Chicago Press, 1978.

Maier, Anneliese. "The Achievements of Late Scholastic Natural Philosophy." In Maier, *On the Threshold of Exact Science,* pp. 143–70.

————. *An der Grenze von Scholastik und Naturwissenschaft,* 2d ed. Rome: Edizioni di Storia e Letteratura, 1952.

————. *Ausgehendes Mittelalter: Gesammelte Aufsätze zur Geistesgeschichte des 14. Jahrhunderts,* 3 vols. Rome: Edizioni di Storia e Letteratura, 1964–77.

————. "Die naturphilosophische Bedeutung der scholastischen Impetustheorie." *Scholastik* 30 (1955): 321–43. Translated as "The Significance of the Theory of Impetus for Scholastic Natural Philosophy," in Maier, *On the Threshold of Exact Science,* pp. 76–102.

————. *Die Vorläufer Galileis im 14. Jahrhundert,* 2d ed. Rome: Edizioni di Storia e Letteratura, 1966.

————. *Metaphysische Hintergründe der spätscholastischen Naturphilosophie.* Rome: Edizioni di Storia e Letteratura, 1955.

————. *On the Threshold of Exact Science: Selected Writings of Anneliese Maier on Late Medieval Natural Philosophy,* trans. Steven D. Sargent. Philadelphia: Univ. of Pennsylvania Press, 1982.

————. "The Theory of the Elements and the Problem of their Participation in Compounds." In Maier, *On the Threshold of Exact Science,* pp. 124–42.

————. *Zwischen Philosophie und Mechanik.* Rome: Edizioni di Storia e Letteratura, 1958.

Maierù, A., and Paravicini Bagliani, A., eds. *Studi sul XIV secolo in memoria di Anneliese Maier.* Storia e Letteratura, Raccolta di studi e testi, no. 151. Rome: Edizioni di Storia e Letteratura, 1981.

Majno, Guido. *The Healing Hand: Man and Wound in the Ancient World.* Cambridge, Mass.: Harvard University Press, 1975.

Makdisi, George. *The Rise of Colleges: Institutions of Learning in Islam and the West.* Edinburgh: Edinburgh University Press, 1981.

Malinowski, Bronislaw. *Myth in Primitive Psychology.* New York: W. W. Norton, 1926.

Marenbon, John. *Early Medieval Philosophy (480–1150): An Introduction.* London: Routledge & Kegan Paul, 1983.

————. *From the Circle of Alcuin to the School of Auxerre: Logic, Theology and Philosophy in the Early Middle Ages.* Cambridge: Cambridge University Press, 1981.

Marrou, H. I. *A History of Education in Antiquity,* trans. George Lamb. New York: Sheed and Ward, 1956.

Martin, R. N. D. "The Genesis of a Mediaeval Historian: Pierre Duhem and the Origins of Statics." *Annals of Science* 33 (1976): 119–29.

McCluskey, Stephen C. "Gregory of Tours, Monastic Timekeeping, and Early Christian Attitudes to Astronomy." *Isis* 81 (1990): 9–22.

McColley, Grant. "The Theory of the Diurnal Rotation of the Earth." *Isis* 26 (1937): 392–402.

McDiarmid, J. B. "Theophrastus." *Dictionary of Scientific Biography,* 13:328–34.

McEvoy, James. *The Philosophy of Robert Grosseteste.* Oxford: Clarendon Press, 1982.

McInerny, Ralph. *St. Thomas Aquinas.* Notre Dame, Ind.: University of Notre Dame Press, 1982.

McKitterick, Rosamond. *The Carolingians and the Written Word.* Cambridge: Cambridge University Press, 1989.

McMullin, Ernan, ed. *The Concept of Matter in Greek and Medieval Philosophy.* Notre Dame, Ind.: University of Notre Dame Press, 1963.

———. "Conceptions of Science in the Scientific Revolution." In Lindberg, David C., and Westman, Robert S., eds., *Reappraisals of the Scientific Revolution,* pp. 27–86. Cambridge: Cambridge University Press, 1990.

———. "Medieval and Modern Science: Continuity or Discontinuity?" *International Philosophical Quarterly* 5 (1965): 103–29.

McVaugh, Michael. "Arnald of Villanova and Bradwardine's Law." *Isis* 58 (1967): 56–64.

———, ed. *Arnald de Villanova, Opera medica omnia,* vol. 2: *Aphorismi de gradibus.* Granada: Seminarium historiae medicae Granatensis, 1975.

———. "Constantine the African." *Dictionary of Scientific Biography,* 3:393–95.

———. "The *Experimenta* of Arnald of Villanova." *Journal of Medieval and Renaissance Studies* 1 (1971): 107–18.

———. "The Nature and Limits of Medical Certitude." *Osiris,* n.s. 6 (1990): 62–84.

———. "Theriac at Montpellier." *Sudhoffs Archiv: Zeitschrift für Wissenschaftsgeschichte* 56 (1972): 113–44.

———. "Medicine, History of." *Dictionary of the Middle Ages,* 8:247–54.

———. "Quantified Medical Theory and Practice at Fourteenth-Century Montpellier." *Bulletin of the History of Medicine* 43 (1969): 397–413.

Melling, David J. *Understanding Plato.* Oxford: Oxford University Press, 1987.

Merton, Robert K. *Science, Technology and Society in Seventeenth Century England.* Originally published in *Osiris* 4 (1938): 360–632. Reissued, with a new introduction, New York: Harper and Row, 1970.

Meyerhof, Max. "Science and Medicine." In Arnold, Thomas, and Guillaume, Alfred, eds., *The Legacy of Islam,* pp. 311–55. London: Oxford University Press, 1931.

———. *Studies in Medieval Arabic Medicine: Theory and Practice,* ed. Penelope Johnstone. London: Variorum, 1984.

Millas-Vallicrosa, J. M. "Translations of Oriental Scientific Works." In Métraux, Guy S., and Crouzet, François, eds., *The Evolution of Science,* pp. 128–67. New York: Mentor, 1963.

Miller, Timothy S. *The Birth of the Hospital in the Byzantine Empire.* Baltimore: Johns Hopkins University Press, 1985.

———. "The Knights of Saint John and the Hospitals of the Latin West." *Speculum* 53 (1978): 709–33.

Minio-Paluello, Lorenzo. "Boethius, Anicius Manlius Severinus." *Dictionary of Scientific Biography,* 2:228–36.

———. "Michael Scot." *Dictionary of Scientific Biography,* 9:361–65.

———. "Moerbeke, William of." *Dictionary of Scientific Biography,* 9:434–40.

Mohr, Richard D. *The Platonic Cosmology.* Leiden: Brill, 1985.

Moline, Jon. *Plato's Theory of Understanding.* Madison: University of Wisconsin Press, 1981.

Molland, A. G. "Aristotelian Holism and Medieval Mathematical Physics." In Caroti,

Stefano, ed., *Studies in Medieval Natural Philosophy,* pp. 227–35. Florence: Olschki, 1989.

———. "Continuity and Measure in Medieval Natural Philosophy." *Miscellanea Mediaevalia* 16 (1983): 132–44.

———. "An Examination of Bradwardine's Geometry." *Archive for History of Exact Sciences* 19 (1978): 113–75.

———. "The Geometrical Background to the 'Merton School'." *British Journal for the History of Science* 4 (1968–69): 108–25.

———. "Nicole Oresme and Scientific Progress." *Miscellanea Mediaevalia* 9 (1974): 206–20.

Moody, Ernest A. "Galileo and Avempace: The Dynamics of the Leaning Tower Experiment." *Journal of the History of Ideas* 12 (1951): 163–93, 375–422.

———. *Studies in Medieval Philosophy, Science, and Logic: Collected Papers 1933–1969.* Berkeley and Los Angeles: University of California Press, 1975.

Moody, Ernest A., and Clagett, Marshall, eds. and trans. *The Medieval Science of Weights.* Madison: University of Wisconsin Press, 1960.

Morris, Colin. *The Discovery of the Individual, 1050–1200.* New York: Harper and Row, 1972.

Multhauf, Robert P. *The Origins of Chemistry.* New York: Franklin Watts, 1967.

Murdoch, John E. *Album of Science: Antiquity and the Middle Ages.* New York: Scribner's, 1984.

———. "The Development of a Critical Temper: New Approaches and Modes of Analysis in Fourteenth-Century Philosophy, Science, and Theology." *Medieval and Renaissance Studies* 7 (1978): 51–79.

———. "From Social into Intellectual Factors: An Aspect of the Unitary Character of Late Medieval Learning." In Murdoch and Sylla, eds., *The Cultural Context of Medieval Learning,* pp. 271–348.

———. "*Mathesis in philosophiam scholasticam introducta:* The Rise and Development of the Application of Mathematics in Fourteenth Century Philosophy and Theology." In *Arts libéraux et philosophie au moyen âge: Actes du quatrième congrès international de philosophie médiévale, Université de Montréal, 27 août–2 septembre 1967,* pp. 215–54. Montreal: Institut d'études médiévales, 1969.

———. "Philosophy and the Enterprise of Science in the Later Middle Ages." In Elkana, Yehuda, ed., *The Interaction between Science and Philosophy,* pp. 51–74. Atlantic Highlands, N.J.: Humanities Press, 1974.

Murdoch, John E., and Sylla, Edith D. "Anneliese Maier and the History of Medieval Science." In Maierù, A., and Paravicini Bagliani, A., eds., *Studi sul XIV secolo in memoria di Anneliese Maier,* pp. 7–13. Storia e letteratura: Raccolta di studi e testi, no. 151. Rome: Edizioni Storia e Letteratura, 1981.

———, eds. *The Cultural Context of Medieval Learning.* Boston Studies in the Philosophy of Science, no. 26. Dordrecht: D. Reidel, 1975.

———. "The Science of Motion." In Lindberg, David C., ed., *Science in the Middle Ages,* pp. 206–64. Chicago: University of Chicago Press, 1978.

———. "Swineshead, Richard." *Dictionary of Scientific Biography,* 13:184–213.

Murray, Alexander. *Reason and Society in the Middle Ages.* Oxford: Clarendon Press, 1978.

Nakosteen, Mehdi. *History of Islamic Origins of Western Education,* A.D. *800– 1359, with an Introduction to Medieval Muslim Education.* Boulder: University of Colorado Press, 1964.

Nasr, Seyyed Hossein. *An Introduction to Islamic Cosmological Doctrines.* Cambridge, Mass.: Belknap Press of Harvard University Press, 1964.

———. *Science and Civilization in Islam.* Cambridge, Mass.: Harvard University Press, 1968.

Neugebauer, Otto. "Apollonius' Planetary Theory." *Communications on Pure and Applied Mathematics* 8 (1955): 641–48.

———. *Astronomy and History: Selected Essays.* New York: Springer, 1983.

———. *The Exact Sciences in Antiquity.* Princeton: Princeton University Press, 1952.

———. *A History of Ancient Mathematical Astronomy,* 3 pts. New York: Springer, 1975.

———. "On the Allegedly Heliocentric Theory of Venus by Heraclides Ponticus." *American Journal of Philology* 93 (1972): 600–601.

———. "On the 'Hippopede' of Eudoxus." *Scripta Mathematica* 19 (1953): 225–29.

Neugebauer, Otto, and Sachs, A., eds. *Mathematical Cuneiform Texts.* American Oriental Series, vol. 29. New Haven: American Oriental Society, 1945.

Newman, William R. "The Genesis of the *Summa perfectonis.*" *Archives internationales d'histoire des sciences* 35 (1985): 240–302.

———. *The "Summa perfectionis" of Pseudo-Geber: A Critical Edition, Translation and Study.* Leiden: Brill, 1991.

———. "Technology and Chemical Debate in the Late Middle Ages." *Isis* 80 (1989): 423–45.

North, J. D. "The Alphonsine Tables in England." In North, J. D., *Stars, Minds and Fate,* pp. 327–59.

———. "The Astrolabe." *Scientific American* 230, no. 1 (January 1974): 96–106.

———. "Astrology and the Fortunes of Churches." *Centaurus* 24 (1980): 181–211.

———. "Celestial Influence—the Major Premiss of Astrology." In Zambelli, P., ed., *Astrologi hallucinati,* pp. 45–100. Berlin: Walter de Gruyter, 1986.

———. *Chaucer's Universe.* Oxford: Clarendon Press, 1988.

———, ed. and trans. *Richard of Wallingford, An Edition of His Writings with Introductions, English Translation and Commentary,* 3 vols. Oxford: Clarendon Press, 1976.

———. *Stars, Minds and Fate: Essays in Ancient and Medieval Cosmology.* London: Hambledon, 1989.

———. *The Universal Frame: Historical Essays in Astronomy, Natural Philosophy and Scientific Method.* London: Hambledon, 1989.

Numbers, Ronald L., and Amundsen, Darrel W., eds. *Caring and Curing: Health and Medicine in the Western Religious Traditions.* New York: Macmillan, 1986.

Nussbaum, Martha Craven. *Aristotle's "De motu animalium": Text with Translation, Commentary, and Interpretive Essays.* Princeton: Princeton University Press, 1978.

Nutton, Vivian. "The Chronology of Galen's Early Career." *Classical Quarterly* 23 (1973): 158–71.

———. *From Democedes to Harvey: Studies in the History of Medicine.* London: Variorum, 1988.

————. "Galen in the Eyes of His Contemporaries." *Bulletin of the History of Medicine* 58 (1984): 315–24.

Oakley, Francis. *Omnipotence, Covenant, and Order: An Excursion in the History of Ideas from Abelard to Leibniz.* Ithaca: Cornell University Press, 1984.

O'Donnell, James J. *Cassiodorus.* Berkeley and Los Angeles: University of California Press, 1979.

Oggins, Robin S. "Albertus Magnus on Falcons and Hawks." In Weisheipl, James A., ed., *Albertus Magnus and the Sciences: Commemorative Essays 1980,* pp. 441–62. Toronto: Pontifical Institute of Mediaeval Studies, 1980.

O'Leary, De Lacy. *How Greek Science Passed to the Arabs.* London: Routledge & Kegan Paul, 1949.

Olson, Richard. *Science Deified and Science Defied: The Historical Significance of Science in Western Culture from the Bronze Age to the Beginnings of the Modern Era ca. 3500 B.C. to ca. A.D. 1640.* Berkeley and Los Angeles: University of California Press, 1982.

O'Meara, Dominic J. *Pythagoras Revived: Mathematics and Philosophy in Late Antiquity.* Oxford: Clarendon Press, 1989.

O'Meara, John J. *Eriugena.* Oxford: Clarendon Press, 1988.

Oresme, Nicole. *"De proportionibus porportionum" and "Ad pauca respicientes,"* ed. and trans. Edward Grant. Madison: University of Wisconsin Press, 1966.

————. *Le livre du ciel et du monde,* ed. and trans. A. D. Menut and A. J. Denomy. Madison: University of Wisconsin Press, 1968.

Orme, Nicholas. *English Schools of the Middle Ages.* London: Methuen, 1973.

Overfield, James H. "University Studies and the Clergy in Pre-Reformation Germany." In Kittelson, James M., and Transue, Pamela J., eds., *Rebirth, Reform and Resilience: Universities in Transition 1300–1700,* pp. 254–92. Columbus: Ohio State University Press, 1984.

Owen, G. E. L. *Logic, Science and Dialectic: Collected Papers in Greek Philosophy,* ed. Martha Nussbaum. Ithaca: Cornell University Press, 1986.

Parent, J. M. *La doctrine de la création dans l'école de Chartres.* Paris: J. Vrin, 1938.

Park, Katharine. "Albert's Influence on Medieval Psychology." In Weisheipl, James A., ed., *Albertus Magnus and the Sciences: Commemorative Essays 1980,* pp. 501–35. Toronto: Pontifical Institute of Mediaeval Studies, 1980.

————. *Doctors and Medicine in Early Renaissance Florence.* Princeton: Princeton University Press, 1985.

Parker, Richard. "Egyptian Astronomy, Astrology, and Calendrical Reckoning." *Dictionary of Scientific Biography,* 15:706–27.

Pedersen, Olaf. "Astrology." *Dictionary of the Middle Ages,* 1:604–10.

————. "Astronomy." In Lindberg, David C., ed., *Science in the Middle Ages,* pp. 303–36. Chicago: University of Chicago Press, 1978.

————. "The Corpus Astronomicum and the Traditions of Mediaeval Latin Astronomy: A Tentative Interpretation." In *Astronomy of Copernicus and Its Background,* pp. 57–96. Colloquia Copernicana, no. 3; Studia Copernicana, no. 13. Wrocław: Ossolineum, 1975.

————. "The Development of Natural Philosophy 1250–1350." *Classica et Mediaevalia* 14 (1953): 86–155.

————. "Some Astronomical Topics in Pliny." In French, Roger, and Greenaway,

Frank, eds., *Science in the Early Roman Empire: Pliny the Elder, His Sources and Influence,* chap. 10. Totawa, N.J.: Barnes & Noble, 1986.

————. *A Survey of the Almagest.* Acta Historica Scientiarum Naturalium et Medicinalium, vol. 30. Odense: Odense University Press, 1974.

Pedersen, Olaf, and Pihl, Mogens. *Early Physics and Astronomy: A Historical Introduction.* New York: Science History Publications, 1974.

Pegis, Anton C. *St. Thomas and the Problem of the Soul in the Thirteenth Century.* Toronto: Pontifical Institute of Mediaeval Studies, 1934.

Pellegrin, Pierre. *Aristotle's Classification of Animals: Biology and the Conceptual Unity of the Aristotelian Corpus,* trans. Anthony Preus. Berkeley and Los Angeles: University of California Press, 1986.

Peters, F. E. *Allah's Commonwealth: A History of Islam in the Near East, 600–1100 A.D.* New York: Simon and Schuster, 1973.

————. *Aristotle and the Arabs: The Aristotelian Tradition in Islam.* New York: New York University Press, 1968.

————. *The Harvest of Hellenism: A History of the Near East from Alexander the Great to the Triumph of Christianity.* New York: Simon and Schuster, 1970.

Phillips, E. D. *Greek Medicine.* London: Thames and Hudson, 1973.

Philoponus, John. *Against Aristotle on the Eternity of the World,* trans. Christian Wildberg. Ithaca: Cornell University Press, 1987.

Pingree, David. "Abū Ma'shar al-Balkhī." *Dictionary of Scientific Biography,* 1:32–39.

————. "Hellenophilia versus the History of Science." Manuscript of a paper delivered at Harvard University, 14 November 1990.

The Planispheric Astrolabe. Greenwich: National Maritime Museum, 1976.

Plato. *Plato, with an English Translation,* 10 vols. London: Loeb, 1914–29.

————. *Plato's Cosmology: The "Timaeus" of Plato,* trans. with commentary by Francis M. Cornford. London: Routledge & Kegan Paul, 1957.

————. *Plato's Theory of Knowledge: The "Theaetetus" and the "Sophist" of Plato,* trans. with commentary by Francis M. Cornford. London: Routledge & Kegan Paul, 1935.

————. *The "Republic" of Plato,* trans. Francis M. Cornford. Oxford: Oxford University Press, 1941.

Pliny the Elder. *Natural History,* trans. H. Rackham, 10 vols. London: Heinemann, 1938–62.

Pliny the Younger. *Letters,* with an English translation by William Melmoth, revised by W. M. L. Hutchinson, 2 vols. London: Heinemann, 1961.

Powell, Barry. *Homer and the Origin of the Greek Alphabet.* Cambridge: Cambridge University Press, 1991.

Preus, Anthony. *Science and Philosophy in Aristotle's Biological Works.* Hildesheim: Georg Olms, 1975.

Ptolemy, Claudius. *L'Optique de Claude Ptolémée,* ed. and trans. Albert Lejeune. Leiden: Brill, 1989.

————. *Ptolemy's Almagest,* ed. and trans. G. J. Toomer. New York: Springer, 1984.

————. *Tetrabiblos,* ed. and trans. F. E. Robbins. London: Heinemann, 1948.

Quinn, John Francis. *The Historical Constitution of St. Bonaventure's Philosophy.* Toronto: Pontifical Institute of Mediaeval Studies, 1973.

Rahman, Fazlur. *Islam,* 2d ed. Chicago: University of Chicago Press, 1979.

Randall, John Herman, Jr. *The School of Padua and the Emergence of Modern Science.* Padua: Antenore, 1961.

Ralph, Philip Lee. *The Renaissance in Perspective.* New York: St. Martin's Press, 1973.

Rashdall, Hastings. *The Universities of Europe in the Middle Ages,* ed. F. M. Powicke and A. B. Emden, 3 vols. Oxford: Clarendon Press, 1936.

Rather, L. J. "The 'Six Things Non-Natural': A Note on the Origins and Fate of a Doctrine and a Phrase." *Clio Medica* 3 (1968): 337–47.

Rawson, Elizabeth. *Intellectual Life in the Late Roman Republic.* Baltimore: Johns Hopkins University Press, 1985.

Reeds, Karen. "Albert on the Natural Philosophy of Plant Life." In Weisheipl, James A., ed., *Albertus Magnus and the Sciences: Commemorative Essays 1980,* pp. 341–54. Toronto: Pontifical Institute of Mediaeval Studies, 1980.

Reymond, Arnold. *History of the Sciences in Greco-Roman Antiquity,* trans. Ruth Gheury de Bray. London: Methuen, 1927.

Reynolds, Terry S. *Stronger than a Hundred Men: A History of the Vertical Water Wheel.* Baltimore: Johns Hopkins University Press, 1983.

Riché, Pierre. *Education and Culture in the Barbarian West, Sixth through Eighth Centuries,* trans. John J. Contreni. Columbia, S.C.: University of South Carolina Press, 1976.

Riddle, John M. "Dioscorides." In Cranz, F. Edward, and Kristeller, Paul O., eds., *Catalogus translationum et commentariorum: Mediaeval and Renaissance Latin Translations and Commentaries. Annotated Lists and Guides,* vol. 6, pp. 1–143. Washington, D.C.: Catholic University of America Press, 1980.

———. *Dioscorides on Pharmacy and Medicine.* Austin: University of Texas Press, 1985.

———. "Theory and Practice in Medieval Medicine." *Viator* 5 (1974): 157–70.

Rosen, Edward. "Renaissance Science as Seen by Burckhardt and His Successors." In Helton, Tinsley, ed., *The Renaissance: A Reconsideration of the Theories and Interpretations of the Age,* pp. 77–103. Madison: University of Wisconsin Press, 1961.

Rosenfeld, B. A., and Grigorian, A. T. "Thābit ibn Qurra." *Dictionary of Scientific Biography,* 13:288–95.

Rosenthal, Franz. "The Physician in Medieval Muslim Society." *Bulletin of the History of Medicine* 52 (1978): 475–91.

Ross, W. D. *Aristotle: A Complete Exposition of His Works and Thought,* 5th ed. Cleveland: Meridian, 1959.

Rothschuh, Karl E. *History of Physiology,* trans. Guenter B. Risse. Huntington, N.Y.: Krieger, 1973.

Russell, Bertrand. *A History of Western Philosophy,* 2d ed. London: George Allen & Unwin, 1961.

Russell, Jeffrey B. *Inventing the Flat Earth: Columbus and Modern Historians.* Westport, Conn.: Praeger, 1991.

Sabra, A. I. "The Andalusian Revolt against Ptolemaic Astronomy: Averroes and al-Biṭrūjī." In Mendelsohn, Everett, ed., *Transformation and Tradition in the Sciences: Essays in Honor of I. Bernard Cohen,* pp. 133–53. Cambridge: Cambridge University Press, 1984.

————. "The Appropriation and Subsequent Naturalization of Greek Science in Medieval Islam: A Preliminary Statement." *History of Science* 25 (1987): 223–43.

————. "An Eleventh-Century Refutation of Ptolemy's Planetary Theory." In Hilfstein, Erna; Czartoryski, Paweł; and Grande, Frank D., eds., *Science and History: Studies in Honor of Edward Rosen,* pp. 117–31. Studia Copernicana, no. 16. Wrocław: Ossolineum, 1978.

————. "Al-Farghānī." *Dictionary of Scientific Biography,* 4:541–45.

————. "Form in Ibn al-Haytham's Theory of Vision." *Zeitschrift für Geschichte der arabisch-islamischen Wissenschaften* 5 (1989): 115–40.

————. "Ibn al-Haytham." *Dictionary of Scientific Biography,* 6:189–210.

————, ed. and trans. *The Optics of Ibn al-Haytham: Books I–III, On Direct Vision,* 2 vols. London: Warburg Institute, 1989.

————. "Science, Islamic." *Dictionary of the Middle Ages,* 11:81–88.

————. "The Scientific Enterprise." In Lewis, Bernard, ed., *Islam and the Arab World,* pp. 181–92. New York: Knopf, 1976.

Sacrobosco, John of. *The Sphere of Sacrobosco and Its Commentators,* ed. and trans. Lynn Thorndike. Chicago: University of Chicago Press, 1949.

Sa'di, Lufti M. "A Bio-Bibliographical Study of Hunayn ibn Is-haq al-Ibadi (Johannitius)." *Bulletin of the Institute of the History of Medicine* 2 (1934): 409–46.

Saffron, Morris Harold. *Maurus of Salerno: Twelfth-century "Optimus Physicus" with his Commentary on the Prognostics of Hippocrates.* Transactions of the American Philosophical Society, vol. 62, pt. 1. Philadelphia: American Philosophical Society, 1972.

Saliba, George. "Astrology/Astronomy, Islamic." *Dictionary of the Middle Ages,* 1:616–24.

————. "The Development of Astronomy in Medieval Islamic Society." *Arab Studies Quarterly* 4 (1982): 211–25.

Sambursky, S. *The Physical World of Late Antiquity.* London: Routledge & Kegan Paul, 1962.

————. *The Physical World of the Greeks,* trans. Merton Dagut. London: Routledge & Kegan Paul, 1956.

————. *Physics of the Stoics.* London: Routledge & Kegan Paul, 1959.

Sandbach, F. H. *The Stoics.* London: Chatto & Windus, 1975.

Sarton, George. *Galen of Pergamon.* Lawrence, Kansas: University of Kansas Press, 1954.

————. *Introduction to the History of Science,* 3 vols. Washington, D.C.: Williams and Wilkins, 1927–48.

Sayili, Aydin. *The Observatory in Islam and Its Place in the General History of the Observatory.* Publications of the Turkish Historical Society, series 7, no. 38. Ankara: Türk Tarih Kurumu Basimevi, 1960.

Scarborough, John. "Classical Antiquity: Medicine and Allied Sciences, An Update." *Trends in History* 4, nos. 2–3 (1988): 5–36.

————, ed. *Folklore and Folk Medicines.* Madison, Wis.: American Institute of the History of Pharmacy, 1987.

————. "Galen Redivivus: An Essay Review." *Journal of the History of Medicine* 43 (1988): 313–21.

————. "The Galenic Question." *Sudhoffs Archiv* 65 (1981): 1–31.

————. *Roman Medicine.* Ithaca: Cornell University Press, 1969.

Schmitt, Charles B. *The Aristotelian Tradition and Renaissance Universities.* London: Variorum, 1984.

————. *Aristotle and the Renaissance.* Cambridge, Mass.: Harvard University Press, 1983.

————. *Reappraisals in Renaissance Thought.* London: Variorum, 1989.

————. *Studies in Renaissance Philosophy and Science.* London: Variorum, 1981.

Seneca, Lucius Annaeus. *Physical Science in the Time of Nero: Being a Translation of the "Quaestiones naturales" of Seneca,* trans. John Clarke, notes by Archibald Giekie. London: Macmillan, 1910.

Serene, Eileen. "Demonstrative Science." In Kretzmann, Norman; Kenny, Anthony; and Pinborg, Jan, eds., *The Cambridge History of Later Medieval Philosophy,* pp. 496–517. Cambridge: Cambridge University Press, 1982.

Shank, Michael H. *"Unless You Believe, You Shall Not Understand": Logic, University, and Society in Late Medieval Vienna.* Princeton: Princeton University Press, 1988.

Shapin, Steven, and Schaffer, Simon. *Leviathan and the Air-Pump: Hobbes, Boyle, and the Experimental Life.* Princeton: Princeton University Press, 1985.

Sharp, D. E. *Franciscan Philosophy at Oxford in the Thirteenth Century.* Oxford: Clarendon Press, 1930.

Sharpe, William D., ed. and trans. *Isidore of Seville: The Medical Writings.* Transactions of the American Philosophical Society, vol. 54, pt. 2. Philadelphia: American Philosophical Society, 1964.

Sigerist, Henry E. *A History of Medicine,* vol. 1: *Primitive and Archaic Medicine;* vol. 2: *Early Greek, Hindu, and Persian Medicine.* Oxford: Oxford University Press, 1951–61.

————. "The Latin Medical Literature of the Early Middle Ages." *Journal of the History of Medicine* 13 (1958): 127–46.

Singer, Charles. *A Short History of Anatomy and Physiology from the Greeks to Harvey.* New York: Dover, 1957.

Singleton, Charles S., ed. *Art, Science, and History in the Renaissance.* Baltimore: Johns Hopkins University Press, 1968.

Siraisi, Nancy G. *Arts and Sciences at Padua: The* Studium *of Padua before 1350.* Toronto: Pontifical Institute of Mediaeval Studies, 1973.

————. *Avicenna in Renaissance Italy: The "Canon" and Medical Teaching in Italian Universities after 1500.* Princeton: Princeton University Press, 1987.

————. "Introduction." In Williman, Daniel, ed., *The Black Death: The Impact of the Fourteenth-Century Plague,* pp. 9–22. Binghamton: Center for Medieval & Early Renaissance Studies, 1982.

————. *Medieval and Early Renaissance Medicine: An Introduction to Knowledge and Practice.* Chicago: University of Chicago Press, 1990.

————. *Taddeo Alderotti and His Pupils: Two Generations of Italian Medical Learning.* Princeton: Princeton University Press, 1981.

Smalley, Beryl. *The Study of the Bible in the Middle Ages.* Oxford: Basil Blackwell, 1952.

Smith, A. Mark. "Getting the Big Picture in Perspectivist Optics." *Isis* 72 (1981): 568–89.

————. "Ptolemy's Search for a Law of Refraction: A Case-Study in the Classical Methodology of 'Saving the Appearances' and Its Limitations." *Archive for History of Exact Sciences* 26 (1982): 221–40.

————. "Saving the Appearances of the Appearances: The Foundations of Classical Geometrical Optics." *Archive for History of Exact Sciences* 24 (1981): 73–100.

Smith, Wesley D. *The Hippocratic Tradition.* Ithaca: Cornell University Press, 1979.

Solmsen, Friedrich. *Aristotle's System of the Physical World: A Comparison with His Predecessors.* Ithaca: Cornell University Press, 1960.

————. *Hesiod and Aeschylus.* Ithaca: Cornell University Press, 1949.

————. *Plato's Theology.* Ithaca: Cornell University Press, 1942.

Sorabji, Richard, ed. *Aristotle Transformed: The Ancient Commentators and Their Influence.* Ithaca: Cornell University Press, 1990.

————. *Matter, Space, and Motion: Theories in Antiquity and Their Sequel.* Ithaca: Cornell University Press, 1988.

————, ed. *Philoponus and the Rejection of Aristotelian Science.* London: Duckworth, 1987.

————. *Necessity, Cause, and Blame: Perspectives on Aristotle's Theory.* Ithaca: Cornell University Press, 1980.

Southern, Richard W. "From Schools to University." In Catto, J. I., ed., *The Early Oxford Schools,* vol. 1 of *The History of the University of Oxford,* general ed. T. H. Aston, pp. 1–36. Oxford: Clarendon Press, 1984.

————. "Humanism and the School of Chartres." In Southern, Richard W., *Medieval Humanism and Other Studies,* pp. 61–85. New York: Harper Torchbooks, 1970.

————. *Robert Grosseteste: The Growth of an English Mind in Medieval Europe.* Oxford: Clarendon Press, 1986.

————. *Saint Anselm: A Portrait in a Landscape.* Cambridge: Cambridge University Press, 1990.

————. "The Schools of Paris and the School of Chartres." In Benson, Robert L., and Constable, Giles, eds., *Renaissance and Renewal in the Twelfth Century,* pp. 113–37. Cambridge, Mass.: Harvard University Press, 1982.

Stahl, William H. "Aristarchus of Samos." *Dictionary of Scientific Biography,* 1:246.

————. *Roman Science: Origins, Development and Influence to the Later Middle Ages.* Madison: University of Wisconsin Press, 1962.

Stahl, William H.; Johnson, Richard; and Burge, E. L. *Martianus Capella and the Seven Liberal Arts,* 2 vols. New York: Columbia University Press, 1971–77.

Stannard, Jerry. "Albertus Magnus and Medieval Herbalism." In Weisheipl, James A., ed., *Albertus Magnus and the Sciences: Commemorative Essays, 1980,* pp. 355–77. Toronto: Pontifical Institute of Mediaeval Studies, 1980.

————. "Medieval Herbals and Their Development." *Clio Medica* 9 (1974): 23–33.

————. "Natural History." In Lindberg, David C., ed., *Science in the Middle Ages,* pp. 429–60. Chicago: University of Chicago Press, 1978.

Steneck, Nicholas H. *Science and Creation in the Middle Ages: Henry of Langenstein (d. 1397) on Genesis.* Notre Dame, Ind.: University of Notre Dame Press, 1976.

Stevens, Wesley M. *Bede's Scientific Achievement.* Jarrow upon Tyne: Parish of Jarrow, 1986.

Stock, Brian. *The Implications of Literacy: Written Language and Models of Interpretation in the Eleventh and Twelfth Centuries.* Princeton: Princeton University Press, 1983.

―――. *Myth and Science in the Twelfth Century: A Study of Bernard Silvester.* Princeton: Princeton University Press, 1972.

―――. "Science, Technology, and Economic Progress." In Lindberg, David C., ed., *Science in the Middle Ages,* pp. 1–51. Chicago: University of Chicago Press, 1978.

Swerdlow, Noel M., and Neugebauer, Otto. *Mathematical Astronomy in Copernicus's De Revolutionibus,* 2 pts. New York: Springer, 1984.

Sylla, Edith Dudley. "Compounding Ratios: Bradwardine, Oresme, and the first Edition of Newton's *Principia.*" In Mendelsohn, Everett, ed., *Transformation and Tradition in the Sciences: Essays in Honor of I. Bernard Cohen,* pp. 11–43. Cambridge: Cambridge University Press, 1984.

―――. "Galileo and the Oxford *Calculatores:* Analytical Languages and the Mean-Speed Theorem for Accelerated Motion." In Wallace, William A., ed., *Reinterpreting Galileo,* pp. 53–108. Washington, D.C.: Catholic University of America Press, 1986.

―――. "Medieval Concepts of the Latitude of Forms: The Oxford Calculators." *Archives d'histoire doctrinale et littéraire du moyen âge* 40 (1973): 225–83.

―――. "Medieval Quantifications of Qualities: The 'Merton School'." *Archive for History of Exact Sciences* 8 (1971): 9–39.

―――. "Science for Undergraduates in Medieval Universities." In Long, Pamela O., ed., *Science and Technology in Medieval Society,* pp. 171–86. Annals of the New York Academy of Sciences, vol. 441. New York: New York Academy of Sciences, 1985.

Symonds, John Addington. *Renaissance in Italy,* pt. I: *The Age of the Despots;* pt. II: *The Revival of Learning.* New York: Henry Holt, 1888.

Tachau, Katherine H. *Vision and Certitude in the Age of Ockham: Optics, Epistemology and the Foundations of Semantics, 1250–1345.* Leiden: Brill, 1988.

Talbot, Charles H. "Medicine." In Lindberg, David C., ed., *Science in the Middle Ages,* pp. 391–428. Chicago: University of Chicago Press, 1978.

―――. *Medicine in Medieval England.* London: Oldbourne, 1967.

Taylor, F. Sherwood. *The Alchemists.* New York: Henry Schuman, 1949.

Temkin, Owsei. *The Double Face of Janus and Other Essays in the History of Medicine.* Baltimore: Johns Hopkins University Press, 1977.

―――. *Galenism: Rise and Decline of a Medical Philosophy.* Ithaca: Cornell University Press, 1973.

―――. "Greek Medicine as Science and Craft." *Isis* 44 (1953): 213–25.

―――. "On Galen's Pneumatology." *Gesnerus* 8 (1951): 180–89.

Tester, Jim. *A History of Western Astrology.* Woodbridge, Suffolk: Boydell, 1987.

Thomas Aquinas. *Faith, Reason and Theology: Questions I–IV of His Commentary on the De Trinitate of Boethius,* trans. Armand Maurer. Toronto: Pontifical Institute of Mediaeval Studies, 1987.

―――. *Summa Theologiae* (Blackfriars Edition), vol. 10: *Cosmogony,* ed. and trans. William A. Wallace. New York: McGraw-Hill, 1967.

Thomas Aquinas, Siger of Brabant, and Bonaventure. *On the Eternity of The World,* trans. Cyril Vollert, Lottie H. Kendzierski, and Paul M. Byrne. Mediaeval Philosophical Texts in Translation, no. 16. Milwaukee: Marquette University Press, 1964.

Thorndike, Lynn. *A History of Magic and Experimental Science,* 8 vols. New York: Columbia University Press, 1923–58.

———. *Michael Scot.* London: Nelson, 1965.

———. *Science and Thought in the Fifteenth Century.* New York: Columbia University Press, 1929.

———. *University Records and Life in the Middle Ages.* New York: Columbia University Press, 1944.

Toomer, G. J. "Heraclides Ponticus." *Dictionary of Scientific Biography,* 15:202–5.

———. "Hipparchus." *Dictionary of Scientific Biography,* 15:205–24.

———. "Mathematics and Astronomy." In Harris, J. R., ed., *The Legacy of Egypt,* 2d ed., pp. 27–54. Oxford: Clarendon Press, 1971.

———. "Ptolemy." *Dictionary of Scientific Biography,* 11:186–206.

———. "A Survey of the Toledan Tables." *Osiris* 15 (1968): 1–174.

Toulmin, Stephen, and Goodfield, June. *The Fabric of the Heavens: The Development of Astronomy and Dynamics.* New York: Harper, 1961.

Ullmann, Manfred. "Al-Kīmiyāʾ." *The Encyclopaedia of Islam,* new ed., vol. 5, fasc. 79–80, pp. 110–15.

———. *Islamic Medicine,* trans. Jean Watt. Edinburgh: Edinburgh University Press, 1978.

Unguru, Sabetai. "History of Ancient Mathematics: Some Reflections on the State of the Art." *Isis* 70 (1979): 555–65.

———. "On the Need to Rewrite the History of Greek Mathematics." *Archive for History of Exact Sciences* 15 (1975): 67–114.

van der Waerden, B. L. "Mathematics and Astronomy in Mesopotamia." *Dictionary of Scientific Biography,* 15:667–80.

———. *Science Awakening: Egyptian, Babylonian and Greek Mathematics,* trans. Arnold Dresden. New York: John Wiley, 1963.

van der Waerden, B. L., with Huber, Peter. *Science Awakening II: The Birth of Astronomy.* Leyden: Noordhoff, 1974.

Van Helden, Albert. *The Invention of the Telescope.* Transactions of the American Philosophical Society, vol. 67, pt. 4. Philadelphia: American Philosophical Society, 1977.

———. *Measuring the Universe: Cosmic Dimensions from Aristarchus to Halley.* Chicago: University of Chicago Press, 1985.

Vansina, Jan. *The Children of Woot: A History of the Kuba Peoples.* Madison: University of Wisconsin Press, 1978.

———. *Oral Tradition as History.* Madison: University of Wisconsin Press, 1985.

Van Steenberghen, Fernand. *Aristotle in the West,* trans. Leonard Johnston. Louvain: Nauwelaerts, 1955.

———. *Les oeuvres et la doctrine de Siger de Brabant.* Paris: Palais des Académies, 1938.

———. *The Philosophical Movement in the Thirteenth Century.* London: Nelson, 1955.

———. *Thomas Aquinas and Radical Aristotelianism.* Washington, D.C.: Catholic University of America Press, 1980.

Verbeke, G. "Simplicius." *Dictionary of Scientific Biography,* 12:440–43.

———. "Themistius." *Dictionary of Scientific Biography,* 13:307–9.

Veyne, Paul. *Did the Greeks Believe in Their Myths?* Trans. Paula Wissing. Chicago: University of Chicago Press, 1988.

Vickers, Brian, ed. *Occult and Scientific Mentalities in the Renaissance.* Cambridge: Cambridge University Press, 1984.

Vlastos, Gregory. *Plato's Universe.* Seattle: University of Washington Press, 1975.

Voigts, Linda E. "Anglo-Saxon Plant Remedies and the Anglo-Saxons." *Isis* 70 (1979): 250–68.

Voigts, Linda E., and Hudson, Robert P. "'A drynke that men callen dwale to make a man to slepe whyle men kerven hem': A Surgical Anesthetic from Late Medieval England." In Campbell, Sheila, ed., *Health, Disease and Healing in Medieval Culture.* New York: St. Martin's Press, forthcoming.

Voigts, Linda E., and McVaugh, Michael R. *A Latin Technical Phlebotomy and Its Middle English Translation.* Transactions of the American Philosophical Society, vol. 74, pt. 2. Philadelphia: American Philosophical Society, 1984.

von Grunebaum, G. E. *Classical Islam: A History 600–1258,* trans. Katherine Watson. Chicago: Aldine, 1970.

———. *Islam: Essays in the Nature and Growth of a Cultural Tradition,* 2d ed. London: Routledge & Kegan Paul, 1961.

von Staden, Heinrich. "Hairesis and Heresy: The Case of the *haireseis iatrikai.*" In Meyer, Ben F., and Sanders, E. P., eds., *Jewish and Christian Self-Definition,* vol. 3: *Self-Definition in the Graeco-Roman World,* pp. 76–100, 199–206. London: SCM Press, 1982.

———. *Herophilus: The Art of Medicine in Early Alexandria.* Cambridge: Cambridge University Press, 1989.

Vööbus, Arthur. *History of the School of Nisibis.* Corpus scriptorum Christianorum orientalium, vol. 266. Louvain: Secrétariat du CorpusSCO, 1965.

Wagner, David L., ed. *The Seven Liberal Arts in the Middle Ages.* Bloomington: Indiana University Press, 1983.

Wallace, Willam A. "Aristotle in the Middle Ages." *Dictionary of the Middle Ages,* 1:456–69.

———. *Causality and Scientific Explanation,* 2 vols. Ann Arbor: University of Michigan Press, 1972–1974.

———. *Galileo and His Sources: The Heritage of the Collegio Romano in Galileo's Science.* Princeton: Princeton University Press, 1984.

———. "The Philosophical Setting of Medieval Science." In Lindberg, David C., ed., *Science in the Middle Ages,* pp. 91–119. Chicago: University of Chicago Press, 1978.

———. *Prelude to Galileo: Essays on Medieval and Sixteenth-Century Sources of Galileo's Thought.* Boston Studies in the Philosophy of Science, vol. 62. Dordrecht: Reidel, 1981.

———, ed. *Reinterpreting Galileo.* Studies in Philosophy and the History of Science, no. 15. Washington, D.C.: Catholic University of America Press, 1986.

———. "Thomism and Its Opponents." *Dictionary of the Middle Ages,* 12:38–45.

Walzer, Richard. "Arabic Transmission of Greek Thought to Medieval Europe." *Bulletin of the John Rylands Library* 29 (1945–46): 160–83.

Waterlow, Sarah. *Nature, Change, and Agency in Aristotle's "Physics."* Oxford: Clarendon Press, 1982.

Wedel, Theodore Otto. *The Mediaeval Attitude toward Astrology, Particularly in England.* New Haven: Yale University Press, 1920.

Weinberg, Julius. *A Short History of Medieval Philosophy.* Princeton: Princeton University Press, 1964.

Weisheipl, James A., ed. *Albertus Magnus and the Sciences: Commemorative Essays, 1980.* Toronto: Pontifical Institute of Mediaeval Studies, 1980.

———. "The Celestial Movers in Medieval Physics." *The Thomist* 24 (1961): 286–326.

———. "Classification of the Sciences in Medieval Thought." *Mediaeval Studies* 27 (1965): 54–90.

———. "The Concept of Nature." *The New Scholasticism* 28 (1954): 377–408.

———. "Curriculum of the Faculty of Arts at Oxford in the Fourteenth Century." *Mediaeval Studies* 26 (1964): 143–85.

———. *The Development of Physical Theory in the Middle Ages.* New York: Sheed and Ward, 1959.

———. "Developments in the Arts Curriculum at Oxford in the Early Fourteenth Century." *Mediaeval Studies* 28 (1966): 151–75.

———. *Friar Thomas d'Aquino: His Life, Thought and Works.* Garden City: Doubleday, 1974.

———. "The Life and Works of St. Albert the Great." In Weisheipl, James A., ed., *Albertus Magnus and the Sciences: Commemorative Essays, 1980,* pp. 13–51. Toronto: Pontifical Institute of Mediaeval Studies, 1980.

———. *Nature and Motion in the Middle Ages,* ed. William E. Carroll. Washington: Catholic University of America Press, 1985.

———. "The Nature, Scope, and Classification of the Sciences." In Lindberg, David C., ed., *Science in the Middle Ages,* pp. 461–82. Chicago: University of Chicago Press, 1978.

———. "The Principle *Omne quod movetur ab alio movetur* in Medieval Physics." *Isis* 56 (1965): 26–45.

———. "Science in the Thirteenth Century." In Catto, J. I., ed., *The Early Oxford Schools,* vol. 1 of *The History of the University of Oxford,* general ed. T. H. Aston, pp. 435–69. Oxford: Clarendon Press, 1984.

Wetherbee, Winthrop, trans. *The Cosmographia of Bernardus Silvestris.* New York: Columbia University Press, 1973.

———. "Philosophy, Cosmology, and the Twelfth-Century Renaissance." In Dronke, Peter, ed., *A History of Twelfth-Century Western Philosophy,* pp. 21–53. Cambridge: Cambridge University Press, 1988.

White, Lynn, Jr. *Medieval Technology and Social Change.* Oxford: Oxford University Press, 1962.

White, T. H., trans. *The Bestiary: A Book of Beasts.* New York: G. P. Putnam's Sons, 1954.

Whitney, Elspeth. *Paradise Restored: The Mechanical Arts from Antiquity through the Thirteenth Century.* Transactions of the American Philosophical Society, vol. 80, pt. 1. Philadelphia: American Philosophical Society, 1990.

Williman, Daniel, ed. *The Black Death: The Impact of the Fourteenth-Century*

Plague. Binghamton: Center for Medieval & Early Renaissance Studies, 1982.

Wilson, Curtis. *William Heytesbury: Medieval Logic and the Rise of Mathematical Physics.* Madison: University of Wisconsin Press, 1960.

Wilson, N. G. *Scholars of Byzantium.* Baltimore: Johns Hopkins University Press, 1983.

Wippel, John F. "The Condemnations of 1270 and 1277 at Paris." *Journal of Medieval and Renaissance Studies* 7 (1977): 169–201.

Witelo. *Witelonis Perspectivae liber primus: Book I of Witelo's "Perspectiva": An English Translation with Introduction and Commentary and Latin Edition of the Mathematical Book of Witelo's "Perspectiva,"* ed. and trans. Sabetai Unguru. Studia Copernicana, no. 15. Wrocław: Ossolineum, 1977.

———. *Witelonis Perspectivae liber quintus: Book V of Witelo's Perspectiva: An English Translation with Introduction and Commentary and Latin Edition of the First Catoptrical Book of Witelo's Perspectiva,* ed. and trans. A. Mark Smith. Studia Copernicana, no. 23. Wrocław: Ossolineum, 1983.

Wolff, Michael. "Philoponus and the Rise of Preclassical Dynamics." In Sorabji, Richard, ed., *Philoponus and the Rejection of Aristotelian Science,* pp. 84–120. London: Duckworth, 1987.

Wolfson, Harry Austryn. *Crescas' Critique of Aristotle: Problems of Aristotle's "Physics" in Jewish and Arabic Philosophy.* Cambridge, Mass.: Harvard University Press, 1929.

Woodward, David. "Medieval *Mappaemundi.*" In Harley, J. B., and Woodward, David, eds., *The History of Cartography,* vol. 1: *Cartography in Prehistoric, Ancient, and Medieval Europe and the Mediterranean,* pp. 286–370. Chicago: University of Chicago Press, 1987.

Wright, John Kirtland. *The Geographical Lore of the Time of the Crusades: A Study in the History of Medieval Science and Tradition in Western Europe.* New York: American Geographical Society, 1925.

Yates, Frances A. *Giordano Bruno and the Hermetic Tradition.* London: Routledge & Kegan Paul, 1964.

———. "The Hermetic Tradition in Renaissance Science." In Singleton, Charles S., ed., *Art, Science, and History in the Renaissance,* pp. 255–74. Baltimore: Johns Hopkins University Press, 1968.

Zimmermann, Fritz. "Philoponus' Impetus Theory in the Arabic Tradition." In Sorabji, Richard, ed., *Philoponus and the Rejection of Aristotelian Science,* pp. 121–29. London: Duckworth, 1987.

Zinner, Ernst. "Die Tafeln von Toledo." *Osiris* 1 (1936): 747–74.

Index

Abacus, 190

ʿAbbās, al-, 168

ʿAbbasids, 168, 176, 180–81

ʿAbd al-Raḥmān, 203

Abelard, Peter, 195–97, 206; *Sic et non,*
195–96

Abū Maʿshar (Albumasar): *Introduction to
the Science of Astrology,* 278–79

Academy (Plato's), 35, 47, 71–73, 80, 212

Acceleration, 367–68; of falling bodies, 303

Accommodation: principle of, in biblical
exegesis, 260–61

Achilles, 292

Active principles: in Stoic thought, 81

Actuality and potentiality, 51–52, 223

Adelard of Bath, 200, 268

Aether, *See* Quintessence

Agriculture: beginnings of, 5; relationship
of, to astronomy, 16

Air: as an element, 31; as the underlying
reality, 29

Albategnius. *See* Battānī, al-

Albert the Great, 227–31, 234; and the
Aristotelian tradition, 229–31; biological
knowledge of, 230; *On Animals,* 350; on
the nature of motion, 292; *On
Vegetables,* 350

Albumasar. *See* Abū Maʿshar

Alchemy, 56, 287–90, 368

Alcuin of York, 185

Alexander of Aphrodisias, 77

Alexander of Tralles, 318

Alexander the Great, 47, 74; military
campaigns of, 69, 133, 163, 165–66

Alexandria, 69, 120, 125, 133, 163, 169;
library of, 74, 98–99; Museum of, 74, 99;
schools of, 74

Alfonsine Tables, 272–73

Alfonso X of Castile, 272

Algebra, 370n.24

Alhazen. *See* Ibn al-Haytham

ʿAlī ibn ʿAbbās al-Majūsī (Haly Abbas):
medical writings of, 331, 342; *Pantegni,*
325, 327

Allah, 166

Alpetragius. *See* Biṭrūjī, al-

Alteration, 291

Amyntas II, 47

Anatomy, human, 119–21, 126, 342;
relevance of, for medical practice,
119-20, 124, 126, 342

Anaxagoras, 26, 35

Anaximander, 26–27, 29

Anaximenes, 27, 29

Andrew of St. Victor, 200

Andronicus of Rhodes, 76–77

Anesthetics, 342

Angels, 250–51

Animals: Aristotle's classification of, 63

Animism, 42

Anselm of Bec (Canterbury), 193–97

Anticrates of Cnidos, 113

Antiochus of Ascalon, 139

Antipodes, 254

Antoninus Pius, 74

Apeiron: as the ultimate reality, 29

Apollonius of Perga, 89; on planetary
motion, 98

Appropriation thesis, 174

Aquinas, Thomas: 227–28, 231–34, 236–37,
240–41; and the Aristotelian tradition,
231–34; on chemical combination, 286;
on faith and reason, 231–34; on motion,
302–3; *On the Eternity of the World,* 233;

Aquinas, Thomas (*continued*)
  On the Unicity of the Intellect, against the Averroists, 233–34
Arabic numerals, 190
Aramaic, 165
Aratus of Soli: *The Phaenomena,* 137, 140, 147
Archimedes, 170; as a mathematician, 88–89; mathematical writings of, 205; *On Floating Bodies,* 110; on the application of mathematics to nature, 108–10, 294; *On the Equilibrium of Planes,* 109–10; on the law of the lever, 108–10
Aristarchus of Samos: on cosmic dimensions, 97–98; on planetary motion, 97
Aristotelian tradition, 74–77, 162, 212, 216–18, 222–23, 229–34, 237–39, 241, 243, 247, 279, 282, 364–66
Aristotle, 35, 47–68, 73–77, 150, 212, 281, 297; banning of, 216–17; in the medieval curriculum, 216–18, 226, 241; library of, 76; life and education of, 47; medieval appraisal of, 364–66; on animal physiology, 64; on biology, 62–67, 224, 290; on causation, 53–54; on celestial influence, 275, 279; on change and stability, 51–52; on chemical combination, 285–86; on cosmology, 54–58, 177; on epistemology, 50–51; on form and matter, 49–51, 64–65, 282–84; on his predecessors, 27–29, 31; on human anatomy, 63; on light and vision, 67–68, 308, 312–13; on marine life, 63; on motion, 56, 58–60, 292, 294, 301–2, 304, 306; on natural place, 57–58; on organic generation, 65–66; on planetary motion, 61–62, 95–96, 104–5, 267; on potentiality and actuality, 51–52; on the classification of animals, 63; on the distinction between mathematics and physics, 85–86; on the elements, 55, 283; on the function of the lungs, 66; on the heart, 66; on the incubation of a bird's egg, 64; on the middle (or mixed) sciences, 86; on the quantification of motion, 59–60, 294, 304; on the rainbow, 309; on the shape of the earth, 58; on the soul, 65, 68, 162, 169, 226; on the underlying reality, 49–50; on zoology, 62–66; response of, to Plato's theory of forms, 48; scientific achievements of, appraised, 67–68. Works: *Ethics,* 229; *History of Animals,* 63; logical writings, 149, 165, 188, 193, 215, 223; *Metaphysics,* 169, 224, 226, 262; *Meteorology,* 67, 138, 162, 205, 224; *On Animals,* 226; *On Generation and Corruption,* 162, 169, 205, 226; *On Sense and the Sensible,* 226; *On the Generation of Animals,* 64; *On the Heavens,* 55, 59, 67, 162, 205, 262; *Physics,* 59, 67, 162, 169, 205, 224, 226, 229, 290, 293–94, 302–3; *Posterior Analytics,* 224; *The Parts of Animals,* 130; zoological writings, 350. *See also* Pseudo-Aristotle
Arithmetic, 146
Arteries, 120–22, 127–29, 337
Articella, 331, 335
Asclepiades of Bithynia, 124
Asclepius, 112–13, 118, 131
Assisi, 222
Astrolabe, 190, 204, 264–66; transmission of, to Christendom, 268
Astrology, 140, 158, 205, 211, 220, 226–27, 230, 237, 239, 269, 274–75, 276–80, 368; Babylonian, 12, 17; horoscopic, 17, 274, 277; judicial, 17; medieval, 202; origins of, 12; relationship of, to astronomy, 17, 263; relationship of, to medicine, 320, 330, 338–39; relationship of, to plague, 339; relationship of, to religion, 17
Astronomical instruments, 263–66
Astronomical observatories, 263
Astronomical tables, 268
Astronomy: Aristotle on, 61; as a quantitative art, 95, 99, 268, 272; as a university subject, 211; as the art of constructing mathematical models, 94–95, 104–5, 261–63, 267; as physical explanation, 95, 104–5, 261–62, 264, 267; Babylonian, 12, 16–18, 98–99; early medieval, 190; Greek, 89–105; in the university curriculum, 270–72; Indian, 177, 263; instrumentalist interpretation of, 261; Islamic, 177–78, 263–67; medieval European, 267–73; of Plato, 42; origins of, 12; Persian, 263; quantitative goals of, 95; relationship of, to agriculture, 16–17; relationship of, to astrology, 17, 263; relationship of, to the calendar, 89, 263, 268; relationship of, to cosmology, 251, 262; relationship of, to religion, 16–17, 55; Roman, 144, 146–47
Athena, 23
Athens, 80, 138
Atomism, 78–79, 287, 361–62; geometrical, 40–41, 85; in medicine, 124

Atomists, 30; on change and stability, 34; on light and vision, 105, 308
Atoms, 30, 34, 78–80, 287
Atto, 188, 190
Augmentation and diminution, 291
Augustine, 197–98; and the scientific tradition, 150–51, 155, 226–27; on astrology, 202, 276–77
Aurillac, 188
Autolycus of Pitane, 294
Averroes. *See* Ibn Rushd
Avicebron (Ibn Gabirol), 284
Avicenna. *See* Ibn Sīnā

Babylonian science. *See* Mesopotamian science
Bacon, Francis: *New Organon,* 355–56
Bacon, Roger, 224–27, 229–30; lectures of, on Aristotle, 218, 226; on light and vision, 313; on motion, 303; on planetary spheres, 272; *Opus maius,* 226
Bactria, 69, 163, 168
Baghdad, 168–70, 176, 180–81, 323
Bakhtīshuᶜ family, 168–69
Balance, 294
Barbary Ape, 126
Barmak family, 168, 346
Basil of Caeserea, 321
Battānī al- (Albategnius): on astronomy, 177, 252, 263
Beatus map, 256
Bede, 158, 184–85; *Ecclesiastical History of the English People,* 159, 322; *On the Nature of Things,* 159
Benedict (St.) of Nursia, 153, 231
Benedictine Rule, 153
Bernard of Chartres, 193
Bernard of Clairvaux, 196, 320–21
Bernard of Verdun: on planetary spheres, 272
Bernard Sylvester, 200
Bestiaries, 351–53
Biology. *See* Botany; Natural history; Medicine; Anatomy, human; Zoology
Bīrūnī, al-, 176
Biṭrūjī, al- (Alpetragius): as critic of Ptolemaic planetary models, 267
Black Death, 339
Bladder stone, 341
Blood, 121, 127–28
Blood-letting, 116, 122, 333, 339, 341
Bobbio, 188

Boethius, 184–85, 188–89, 193, 197, 203, 223; *Arithmetic,* 187, as a translator, 148, 205; logical works of, 188
Boethius of Dacia: *On the Eternity of the World,* 235
Boethius of Sidon, 77
Bologna, 297: as an educational center, 193, 206; university of, 208–9, 330, 332, 341–42
Bonaventure, 227, 233, 236
Botany, 75, 230, 348–50
Boyle, Robert, 361
Bradwardine, Thomas, 248, 295, 367; on motion, 305–7
Brain, 127, 129; anatomy of, 120–21
Burckhardt, Jacob: *The Civilization of the Renaissance in Italy,* 356
Buridan, John, 251, 258; on motion, 293, 303; on the possible rotation of the earth, 365; *Questions on Aristotle's Physics,* 241
Byzantium, 161, 169

Cairo, 177, 180
Calcidius, 148, 193, 197, 245
Calendar, 156, 159, 184, 211, 226; Babylonian, 14; Egyptian, 14; Greek, 89
Callippus of Cyzicus: on planetary motion, 95
Cambridge: university of, 297
Campanus of Novara: on planetary spheres, 272; on the dimensions of the cosmos, 252
Canterbury, archbishop of, 240
Carolingian Empire, 184–86
Carthage, 150
Cartography, 256–58
Cassiodorus, 156, 158, 318
Cataracts, 340–41
Causation: Aristotle's theory of, 52–54, 223; divine, 23–27, 29, 31, 39, 40, 42–44, 111, 113, 115, 119, 231; early Greek conceptions of, 24–26; Epicurean theory of, 79; natural vs. supernatural, 198, 200; prehistoric conceptions of, 7–8; theories of, 223, 237, 242, 247
Cautery, 341
Cave, Plato's allegory of the, 37
Celestial equator, 91
Celestial intelligences, 62, 286
Celestial movers, 43, 61–62, 237, 241, 251
Celestial region, 54–55, 61, 82, 162, 247–48, 250–51, 275; divinity of, 274

Celestial spheres, 42–43, 61, 90–91, 246–51, 264, 266–68, 271; Aristotelian, 95–96; Eudoxan, 92–94; of Callippus, 95

Celsus, 120, 140; on medicine, 317

Cerebral palsy, 115

Change: Aristotle's theory of, 51–52, 290; kinds of, 290–91; the problem of, 32–34, 37, 41, 44, 51–52

Charlemagne, 185

Charles the Bald, 186, 188

Chartres, 194; cathedral school of, 192, 198; medical studies at, 330

Christianity: and education, 149–53; and literacy, 150; and the Aristotelian tradition, 219–21, 224, 226–27, 229; and the classical tradition, 150–51, 153, 155–58, 185, 216–17, 364, 405n.26; relationship of, to science, 149–51, 157, 161, 163–65, 198–201, 217, 356–57, 363–64; relationship of, to the Aristotelian tradition, 230–34, 237–38, 240, 247–50

Chryssipus of Soli, 80

Cicero, 135–36, 138–40, 147, 189, 193; *Dream of Scipio,* 145; logical works of, 188; *On Divination,* 83; *On the Nature of the Gods,* 197

Circumference of the earth, 58, 98, 138, 144, 245, 253

Clagett, Marshall: on the medieval scientific achievement, 357

Clarembald of Arras, 200

Cleanthes of Assos, 80

Climatic zones, 254, 256

Clovis, 318

Cnidus, 90

Cohesion: Stoic explanation of, 81–82

Cologne: Dominican school at, 228

Columbus, Christopher, 392n.15

Combination, chemical, 285–86

Comets, 252; Aristotle on, 67

Commodus, 125

Compass, 274

Complexion, 277, 332–33

Computus, 159

Condemnations of 1270 and 1277, 236–41, 248, 280, 293, 365

Condorcet, Marquis de, 356

Conic sections, 89

Constantine the African, 204, 230, 326–27

Constantinople, 161, 346

Continents, 253–55

Contingency of nature, 243

Continuity debate, 281–82, 355, 357–63, 366–68, 402n.1

Contraries, 56, 224

Copernicus, Nicolaus, 97, 102–3, 147, 177, 252, 272, 368

Cordoba, 180–81, 203

Corinth, 125, 133

Cosmogony, 7–9; Homeric, 23; medieval, 198, 200, 245–61; of Robert Grosseteste, 224

Cosmology, 26–27, 36, 97, 252; Aristotelian, 54–58, 224, 247–49, 251, 258, 275; early medieval, 158; Epicurean, 78–80; medieval, 198, 200; Platonic, 39–42, 197, 246–47, 275; Pythagorean, 97; relationship of, to astronomy, 251, 262; Robert Grosseteste's, 246; Stoic, 82–83, 275; twelfth-century, 246

Creation, 198, 219–20, 224, 235, 237, 246–47, 249

Creation myths, 9

Critical days, 339

Crombie, Alistair: and the continuity debate, 358; on the medieval scientific achievement, 361

Crusades, 190, 204, 346

Curriculum: of Greek schools, 71; of medieval schools, 191–93, 197; of monastic schools, 155–56; of medieval universities, 211–12, 216–18, 222, 241, 364; of Roman schools, 151

Cuthbert, 322

Damascus, 168

Dark ages: conception of, 183, 355–57

Deduction, 50

Deferent. *See* Epicycle-on-deferent model of planetary motion

Demetrius Phaleron, 74

Demiurge, 36, 40, 42–43, 130, 150, 202, 275

Democritus, 26, 30, 34, 79

Demonstration: Aristotelian, 50, 361; Euclidean, 87–88

Dentistry, 329

Descartes, René, 361

Determinism, 237–39, 242, 365; astrological, 202, 227, 276–77, 280; in ancient atomism, 31, 79–80, 83; in Aristotle's natural philosophy, 220, 239, 365; in Stoic philosophy, 83

Diagnosis (medical), 18, 117, 125–26, 333, 335–38

Diet: as a medical therapy, 116
Digest, 193
Digestion, 121
Diopter, 98
Dioscorides: *De materia medica,* 318–19, 323, 334, 348
Disease: as imbalance, 116, 118, 126, 332–33; causes of, 10, 20, 332–33, 339; Erasistratus' theory of, 122; Galen's theory of, 126; Hippocratic theories of, 115–19; Plato's theory of, 42; supernatural theories of, 10, 18, 20, 111, 320
Dissection: of a pig, 342; human, 120, 126, 342–45. *See also* Anatomy, human
Dogmatists (school of physicians), 124
Dominican order, 223, 227, 231
Doxographic tradition, 75, 139, 148
Duhem, Pierre, 175, 238, 248, 358; on realism and instrumentalism in astronomy, 261–62; on the medieval scientific achievement, 357
Dumbleton, John of, 295
Dura mater, 120
Dye-making, 290
Dynamics, 58–61, 301–5, 365, 367; "law" of, 59–60, 306–7; defined, 295

Earth: as an element, 31; circumference of, 58, 98, 138, 144, 245, 253; possible diurnal rotation of, 97, 258, 260–61, 365; shape of, 27, 42, 58, 145, 158, 245, 253
Earthquake, 26, 67
Ebers papyrus, 18
Ebstorf map, 254
Eccentric model of planetary motion, 100, 103
Eclipses, 26, 58, 98, 144, 158, 270; alleged prediction of, by Thales, 27
Ecliptic, 270; defined, 42, 91
Edessa, 164
Education: Carolingian, 184–85; Greek, 21, 25, 35, 47, 69–77, 80, 138; Islamic, 174; medieval, 145, 153, 155–58, 184, 191–94, 197, 206–13, 216–18, 222–23; monastic, 153, 155–58, 185, 191–93; Roman, 136, 151–53. *See also* Curriculum; Universities
Edwin Smith papyrus, 18
Egyptian science, 9, 13–14, 18–19; influence of, on Greek science, 20
Elea, 32, 134
Elements, 31, 34, 40, 55, 85, 224, 252, 283, 287; Aristotle's theory of, 55–56, 283; Em-

pedocles' theory of, 31; in medical theory, 332; Plato's theory of, 40, 41; Pythagorean theory of, 31; Stoic theory of, 82; Theophrastus' theory of, 75
Elixir, 288, 290
Emanation, 224
Empedocles, 26, 31, 34–35, 40–41; on the elements, 283
Empiricism, 33–35, 38, 50, 53, 64, 66–67, 79, 118, 243, 334. *See also* Experience; Experiment; Observation; Sense perception
Empiricists (school of physicians), 124
Empyreum, 250
Encyclopedias, 138, 140–42, 144–47, 149, 158–59
Ennius, 137
Ephesus, 27, 29–30
Epicurean tradition, 83, 140
Epicurus, 73, 79, 150; on cosmology, 80; on the aim of natural philosophy, 78; possible influence of, on Strato, 77
Epicycle-on-deferent model of planetary motion, 100–3, 251, 264, 266–67, 270–71
Epidaurus: as center of a healing cult, 113–14
Epilepsy, 115
Epistemology, 44, 220, 362; Aristotle's, 50–51; Parmenides', 33; Plato's, 37–39; Presocratic, 34–35. *See also* Empiricism; Experience; Experiment; Rationalism; Sense perception
Equant model of planetary motion, 102–3, 271
Equinoxes, 43, 91
Erasistratus of Ceos, 120–22, 124–25, 129; influence of, on Galen, 127
Eratosthenes, 98, 144–46
Eriugena, John Scotus, 186, 197, 205; *On Nature,* 188
Eternity of the universe, 54, 82, 162, 219, 227, 230, 233–35, 237, 365
Euclid, 89, 109, 230, 297; *Catoptrics,* 205; *Elements,* 87–88, 149, 169, 205, 216; on light and vision, 107, 308, 312–13; *Optics,* 105–6, 205, 308
Euclidean demonstration, 87–88
Eudoxus, 88, astronomical writings of, 137; on the demand for uniform circular motion in the heavens, 94; on planetary motion, 61, 90–96, 99
Evaporation, 252
Ex herbis feminis, 318

Exhaustion, method of: in Archimedes, 89; in Euclid, 88

Experience, 118, 307. *See also* Empiricism; Epistemology; Experiment; Observation; Sense perception

Experiment, 2, 107–8, 118, 307, 358–59, 361–62; Aristotle's attitude toward, 53. *See also* Empiricism; Experience; Observation; Sense perception

Experimental science, 226, 243–44

Eye: anatomy and physiology of, 120, 127, 308–9, 311

Faith and reason, 193–97, 221, 227, 231–38, 241–42. *See also* Rationalism

Falconry, 350

Falling bodies, 52, 162, 304–5, 367; acceleration of, 303; Aristotle on, 59, 301–2, 304–5; cause of, 301–2, 305; Strato on the acceleration of, 76

Fallopian tubes, 121

Farghānī, al-, 177, 252, 263; *Rudiments of Astronomy,* 269

Fatimids, 180

Ferrara: university of, 330

Fever, 122, 126, 333

Final cause, 62, 79; defined, 53–54; in Aristotle's biology, 64, 66. *See also* Teleology

Fire: as an active element, 82; as an element, 31, 40, 55; as the underlying reality, 29

Firmament, 249–50

Fluxus formae, 292–93

Force: and motion, 60, 304, 306–7; impetus theory of, 303–4; impressed, 302–3; incorporeal motive, 302

Form: accidental, 283–84, 286, 288; as the mover in natural motion, 302; corporeal, 284, 391n.5; elemental, 284; intensification and remission of, 296, 335; in the heavens, 250; medieval discussions of, 282–84; Plato's theory of, 36–37; substantial, 282–83, 285–88. *See also* Form and matter

Form and matter, 223, 227, 250, 285, 362; applied to the soul, 233; Aristotle's theory of, 49–51, 64, 65, 282–84, 365; Averroes' theory of, 284; Avicenna's theory of, 284; medieval discussions of, 282–84

Forma fluens, 292–93

Four causes: in Aristotle, 53, 64–65

Fractures, 118; of the skull, 342

Francis (St.) of Assisi, 222

Franciscan order, 223–24, 226–27, 240, 284, 297

Franciscus de Marchia: on motion, 303

Frederick II: *On the Art of Hunting with Birds,* 350

Free will, human, 80

Gaia, 23–24

Galen of Pergamum, 125–31, 171, 230, 332, 364; anatomical writings of, 342; influence of, in Islam, 323–24; influence of, on medieval medicine, 318; life and education of, 133; medical writings of, 169, 204–5, 331, 334–35, 344; *On Anatomical Procedures,* 127; on human anatomy, 345; on the anatomy and physiology of sight, 308–10; *On the Art of Healing,* 126; on the importance of anatomical knowledge, 126, 342; *On the Usefulness of the Parts of the Body,* 130

Galileo Galilei, 39, 260, 303, 357–58, 361, 365, 367–68; and the medieval tradition, 405n.36

Garden of Epicurus, 72–73, 212

Gassendi, Pierre, 361

Geber. *See* Jābir ibn Ḥayyān

Generation and corruption, 279, 291

Generation, organic: Aristotle's theory of, 65–66

Genesis, 198, 219, 249, 262

Geoffrey de Bussero, 348

Geoffrey Plantagenet, 200

Geography, 145, 253–58

Geology, 67

Geometry: Greek, 86–89

Gerard of Brussels, 294; *Book on Motion,* 295

Gerard of Cremona, 204–5, 269, 327

Gerbert of Aurillac (Pope Sylvester II), 188–90, 192, 203, 268

Giles of Corbeil, 335

Giovanni di Casali, 297

God or the gods: as causal agent(s), 23–27, 29, 31, 39–40, 42–44, 111, 113, 115, 119, 200–1, 227, 231, 237, 239, 243, 246, 286, 320; as creator and sustainer of the universe, 198–99, 362; as planetary deities, 202, 275, 277; as prime mover, 251; equated with nature, 140; existence of, proved, 194–95, 233; in Aristotle's thought, 62, 66, 220, 251, 275; in Egypt and Mesopotamia, 16–18; in Galen's thought, 131; in Plato's thought, 39–40, 42–44, 275; in prehistoric cultures, 7–9; Olympian, 21, 23–25, 29, 39; Stoic view of, 82–83, 140. *See also* Allah; Demiurge; Prime mover

Goody, Jack, 12

Grammatical studies, 211

Greek (language): knowledge of, in Rome, 134–35, 147; knowledge of, in the Middle Ages, 186

Gregory of Tours: *A History of the Franks,* 184

Gregory the Great, 184–85

Gregory IX (pope), 217

Grosseteste, Robert, 224, 226–27, 284, 358; cosmology of, 246; on light and vision, 313; skeleton of, 225

Guido de Marchia: on planetary spheres, 272

Guy de Chauliac: *Chirurgia magna,* 341, 343

Gymnastikē, 70

Gynecology, 329

Ḥakam, al-, 180

Haly Abbas. *See* ʿAlī ibn ʿAbbās al-Majūsī

Hall, A. R., 366; on the scientific revolution, 359

Handmaiden ideal, 150–51, 161, 180, 184, 198, 201, 217, 223–24, 226–27, 229, 232–34, 240, 380 n.27

Hārūn ar-Rashīd, 168, 346

Haskins, Charles Homer: on the medieval scientific achievement, 357

Health. *See* Disease

Heart, 122, 128–29, 335; anatomy of, 120–21; Aristotle's theory of, 66; as seat of the passions, 127; as seat of sensation and emotion, 66; as vital center of the body, 128

Heaviness and lightness, 56–57

Heliocentric astronomy: of Aristarchus of Samos, 97

Hellenistic natural philosophy, 69–83; defined, 69

Hellenization: of North Africa, 163; of the Middle East, 163–66; of Islam, 168

Hemorrhoids, 341

Henry II (king of England), 200

Hera, 21, 31

Heraclides of Pontus: on planetary motion, 97; on the diurnal rotation of the earth, 97

Heraclitus of Ephesus, 25–26, 29; on change and stability, 32

Herbals, 348–50

Hermann the Dalmatian, 204

Hernia, 329, 340–41

Hero of Alexandria: on mirrors, 107

Herod the Great, 77

Herodotus, 13, 26

Herophilus of Chalcedon, 120–22, 124–25, 308; influence of, on Galen, 127

Hesiod, 25, 29; on Greek healing practices, 111; *Theogony,* 21, 23, 26; *Works and Days,* 21

Hexis: Stoic theory of, 82

Heytesbury, William, 295

Hipparchus, 144, 146; and observational astronomy, 98; astronomical writings of, 137; on astronomy as a quantitative art, 99; on cosmic dimensions, 97; on the length of the lunar month, 99

Hippocrates of Cos, 113–14. *See also* Hippocratic medicine; Hippocratic oath; Hippocratic writings

Hippocratic medicine, 113, 115–19

Hippocratic oath, 115

Hippocratic writings, 113, 115, 117, 120, 126, 169, 204, 320, 327, 331–32, 339, 364; availability of, in Islam, 323; influence of, on Galen, 125; influence of, on medieval medicine, 318; *On Ancient Medicine,* 118; *On the Heart,* 128; *On the Nature of Man,* 116, 118, 125; *On the Sacred Disease,* 115, 118–19

Hippopede: in Eudoxan planetary theory, 93–94

Homer, 25–26, 29; *Iliad,* 21; *Odyssey,* 21, 23; on Greek healing practices, 111, 113

Honorius of Autun, 200

Horace, 193

Hospitals, 164, 323; medical care in, 347–48; origins of, 345–47

Hospital of Saint John (Jerusalem), 346

Hospital of the Brolo (Milan), 348

Hospitallers, 346

House of Wisdom, 168–69, 263

Ḥubaysh, 169

Hugh of Santalla, 204

Hugh of St. Victor, 196; *Didascalicon,* 277

Humanism, 202

Humors, four primary, 116, 125, 332

Ḥunayn ibn Isḥāq (Johannitius), 169, 263, 323, 331; *Book of the Ten Treatises of the Eye,* 170; medical writings of, 327; on the anatomy and physiology of the eye, 312

Hylē, 372 n.4
Hylomorphism, 372n.4

Ibn al-Haytham (Alhazen), 175, 177, 179–80; influence of, on Western optics, 313; on light and vision, 309–13, 315; on planetary spheres, 264, 266–67; 272; on the equant, 264; *Optics*, 216
Ibn ash-Shāṭir, 178
Ibn Bājja (Avempace): as critic of Ptolemaic planetary models; 267; on motion, 305
Ibn Māsawaih, 169
Ibn Rushd (Averroes), 220, 230, 284; Aristotelian commentaries of, 206, 218–19, 241; as a critic of Ptolemaic planetary models, 267, 271; cosmology of, 251; medical writings of, 342; on chemical combination, 286; on light and vision, 312; on motion, 292, 302, 305–6
Ibn Sīnā (Avicenna), 230, 284; Aristotelian commentaries of, 206, 216, 218; *Canon of Medicine*, 205, 216, 325, 327,. 335; medical writings of, 331, 334, 342; on alchemy, 288; on chemical combination, 286; on light and vision, 312; on motion, 292, 302; *Physics (Sufficientia)*, 219; zoological writings of, 350
Ibn Ṭufayl: as a critic of Ptolemaic planetary models, 267
Immortality (of the human soul), 220–21, 227, 231–32, 234, 237
Impetus, 251, 303–4
Incommensurability: of the side and diagonal of a square, 87
Induction, 50
Inertia, 39, 304
Instruments, scientific, 362. *See also* Astrolabe; Diopter; Quadrant
Ionia, 27
Isḥāq ibn Ḥunayn, 169
Isidore of Seville, 159, 185, 255; *Etymologies*, 158, 184, 254; on astronomy, 268; on medicine, 317; *On the Nature of Things*, 158, 184
Islam, 165; and the classical tradition, 170–71, 173–76; origins of, 166, 168; relationship of, to science, 171, 173–75, 180–82
Ius ubique docendi (right of teaching anywhere), 212

Jābir ibn Ḥayyān (Geber), 288
James of Venice, 205

Johannes of Sacrobosco (John of Holywood): *The Sphere*, 262, 270
Johannitius. *See* Ḥunayn ibn Isḥāq
John of Seville, 204, 269
John Philoponus. *See* Philoponus, John
John XXII (pope), 241
Joseph the Spaniard, 189
Julius Caesar, 133
Jundishapur, 164–65, 168
Jūrjīs ibn Bakhtīshuʿ, 168
Justinian, 73, 164

Kepler, Johannes, 177, 368; on vision, 315
Khusraw I, 164
Khusraw II, 164
Khwārizmī, al-, 268; *Algebra*, 205, 216
Kidney stones, 329
Kilwardby, Robert, 240, 251
Kindī, al-, 176, 230; on light and vision, 310, 312
Kinematics, 296–301, 365, 367–68; defined, 295
Knowledge: prehistoric conceptions of, 10–11. *See also* Epistemology
Koran, 166, 173
Koyré, Alexandre, 366; and the continuity debate, 358–59, 361
Kuba (African tribe), 8, 10
Kuhn, Thomas, 366; on the scientific revolution, 359

Laon: cathedral school of, 192, 195
Law, natural, 7, 200; Stoics on, 80
Legal studies, 193, 206, 209, 211, 241
Lesbos, 62, 74
Leucippus, 26, 30, 34
Lever, law of, 109–10
Liberal arts, 138, 149, 156, 158, 188, 192, 206, 209, 211, 227
Library of Alexandria, 74, 98–99
Light: Aristotle's theory of, 68; Greek theories of, 308–9; in Robert Grosseteste's cosmology, 246; Islamic theories of, 309–11; mathematical theories of, 106–8; radiation of, as a coherent process, 309–10; radiation of, as an incoherent process, 310; reflection of, 106–7; refraction of, 106–8; Strato's theory of, 77; Theophrastus' theory of, 75; theories of, in medieval Europe, 313, 315
Lindisfarne, 322
Liver, 127, 335

Logic, 33, 184–85, 188, 191, 193, 211–12, 223, 230; origins of, 26, 35
Lombard, Peter: *Sentences*, 262
Love and strife: as cosmological principles, 31, 34
Lucan, 193
Lucretius, 79, 158; *On the Nature of Things*, 138, 140
Lungs: Aristotle's theory of, 66; Galen's theory of, 128
Lupitus, 189
Lyceum (Aristotle's), 47, 72–75, 77, 212

Macrobius, 143, 193; *Commentary on the Dream of Scipio*, 145, 197
Madrasahs, 174
Magic, 8, 20, 360; and medicine, 115
Magnetism, 274
Maier, Anneliese: on the medieval scientific achievement, 357–58
Ma'mūn, al-, 168–69, 177
Manicheism, 163
Manṣūr, al-, 168
Manṣūr ibn Isḥāq, 327
Mappaemundi, 255, 353
Maps, 254–58. *See also* Geography
Marcus Aurelius, 74, 125, 148
Marginality thesis, 173–74
Mark of Toledo, 204
Martianus Capella, 158, 193; on astronomy, 268; *The Marriage of Philology and Mercury*, 145–47, 152, 188, 197
Materialism, 29–31; of Epicureans, 80; of Stoics, 80–81
Mathematics: applicability of, to nature, 32, 40–41, 60, 85–86, 110, 202–3, 294–301, 304–7, 361–62, 365, 368, 371n.15; Babylonian, 13–16; early medieval, 189–90; Egyptian, 13–14; Greek, 86–89; origins of, 13; relationship of, to astronomy, 94–95, 104, 261; relationship of, to physics, 94–96, 104, 262
Matter: primary, 283–84, 287–88. *See also* Form and matter
McMullin, Ernan: and the continuity debate, 359
Mean-speed theorem, 300
Mecca, 166
Mechanical philosophy, 362
Mechanism, 31, 79, 81, 83
Mechanization: in physiology, 129
Medical practitioners: Jewish, 329; numbers of, 329, 332; types of, 111, 113, 115, 118, 317–18, 325, 327; women as, 325, 329
Medical schools: Bologna, 330, 332, 341–42; Ferrara, 330; Montpellier, 330, 341; Oxford, 330 Padua, 330; Paris, 330; Tübingen, 332; Turin, 332
Medical studies, 206, 209, 211–12, 241
Medicine: as a school subject, 325–26, 330; as a craft, 317; Byzantine, 162; clerical, 329; domestic, 9, 327; early medieval, 317–25; Egyptian, 18; Greek, 111–31, 317, 320–21, 345; in prehistoric cultures, 9–10; in the universities, 327, 329–32; Islamic, 171, 323–25, 341–42; learned, 115, 118–19, 317, 329–30, 332; medieval European, 317–23, 325–48; Mesopotamian, 20; monastic, 318, 320–23, 325, 346; origins of, 9–10; preventive, 333; relationship of, to astrology, 279–80; relationship of, to philosophy, 115, 124–25, 131, 317, 324, 326–27, 330–31; relationship of, to religion, 18, 20, 113, 115–16, 119, 131, 320–23; Roman, 124–31, 317–18, 320–21; specialization within, 10. *See also* Disease
Mendicants, 223
Menstrual blood: in Aristotle's theory of generation, 65
Mercury: sun-linked motion of, 91, 97, 144, 147, 268
Merton College (Oxford), 210, 295–96, 299, 305
Merton rule, 299–301, 368
Mesopotamian science, 13–18, 20; influence of, on Greek science, 20, 98–99
Metallurgy, 290
Metals, 288
Metaphysics, 361–62, 366–68. *See also* Underlying reality
Meteorology, 224, 252–53, 283: Aristotle on, 67
Methodists (school of physicians), 124
Meton, 89
Metonic cycle, 89
Microcosm/macrocosm, 140, 202, 246, 275
Middle (or mixed) sciences, 86
Middle Ages: defined, 183
Midwifery, 118, 317–18, 327
Miletus, 27, 29
Mineralogy, 75
Minima naturalia, 286–87
Miracle, 113, 198, 200–1, 220, 234–35, 239, 242–43, 320–23

Mirrors, 107
Mixture (*mixtum*), 285–86, 288
Momentum, 304
Monasticism, 153–58, 191
Mondino dei Luzzi, 342; *Anatomia*, 343
Mongols, 181
Monism, 29–31
Monophysites, 163–64
Monopsychism, 221, 227, 233–34, 237
Monte Cassino, 153, 204, 231, 318, 326
Montpellier: medical school at, 330; university of, 341
Moon: distance of, from earth, 97–98; nature of, 54
Morphē, 372n.4
Motion: Aristotle's theory of, 56, 58–60, 301–2, 365; as a quality, 293, 296; celestial, 303; defined, 291; forced, 58–59, 62, 301–4; geometrical representation of, 296–99; in the mechanical philosophy, 362; infinitely swift, 305; mathematical analysis of, 294–301, 304–7; natural, 58–59, 62, 301–2; nature of, 240, 292–93; projectile, 59, 162, 302–4; quantity of, 299; Strato's theory of, 76; uniform, 294, 298, 300; uniformly accelerated, 295, 299–301. *See also* Dynamics; Kinematics
Mousikē, 70
Mover: identification of, in natural motion, 302; identification of, in projectile motion, 302–4; necessity of, in all cases of motion, 58–59, 301
Mozarabs, 204
Muʿāwiyah, 166
Muḥammad, 166, 173
Mumps, 117
Mūsa, sons of, 169
Museum of Alexandria, 74, 99
Muwaqqit, 174
Mythology: Greek, 21–25

Naples: university of, 231
Natural history, 212, 230, 348–53, 368; Aristotle on, 63–64
Natural philosophy: defined, 3–4; relationship of, to ethics, 78, 80, *See also* Science
Natural place: Aristotle's doctrine of, 57–58
Natural spirit, 379n.22
Natural theology, 130
Naturalism, 27, 39, 198, 200–2, 219, 221, 227, 231, 235, 237, 246; in medicine, 116, 118, 320

Nature (*physis*), 39, 40, 45, 224, 281–82; Aristotle on, 52–53; prehistoric attitudes toward, 4–11
Neleus, 76
Neoplatonism. *See* Platonic tradition
Nerves, 121–22, 127, 129; anatomy of, 120
Nestorians, 163–65, 168–69
Newton, Isaac, 4, 361
Nicholas of Damascus, 77; *On Plants*, 349–50
Nicolaus Bertrucius, 343–44
Nisibis, 164–65
Noah, 255
Noah's flood, 231, 234
Non-naturals, 333
North Africa, 163
Nose-bleed, 334
Number: as the ultimate reality, 31–32; theological uses of, 203; theory of, 146, 203
Number systems, 14–16
Nutrition, 122, 127, 129

Observation, 282. *See also* Empiricism; Experience; Experiment; Sense perception
Obstetrics, 121, 328–29
Ockham, William of: on motion, 292–93; on the relationship between philosophy and theology, 242
Odysseus, 23
Old Testament, 169
Omnipotence (divine), 220, 237–39, 242–44, 260, 293, 323
Optics, 368; as an instrument of conversion, 226; Greek, 105–8, 307–9; Islamic, 177, 179–80, 308–12; medieval European, 312–15; of Aristotle, 67–68. *See also* Light; Vision
Oral culture, 5–12
Order: in the animal and human frame, 130–31; in the cosmos, 39–40, 42–44, 63; source of, for Plato, 39–40; source of, for the Stoics, 81
Oresme, Nicole, 239, 241, 248, 258–61, 367; on astrology, 280; on motion, 297–300, 307; on the possible rotation of the earth, 365
Oribasius, 318
Orleans: cathedral school of, 192
Otto the Great, 203
Otto III, 188–89
Ovaries, 121

Ovid, 193
Oxford: as an educational center, 193, 206; university of, 208–10, 216–17, 224, 226, 295, 330, 367. *See also* Merton College

Padua: university of, 228, 330
Paideia, 70
Pantheism, 216, 218
Pantokrator hospital (Constantinople), 346
Parallax, 252, 376n.16; solar, 98; stellar, 97
Paris: as an educational center, 206; cathedral school of, 192, 195, 198; university of, 208–9, 216–18, 223, 226–28, 231, 235, 240–41, 294, 297, 330, 367
Paramenides, 34–35: on change and stability, 32–33, 51
Parthenon, 72
Patronage, 136
Pecham, John, 240; on light and vision, 313–15
Pergamum, 27, 125
Peripatetics: defined, 73
Periplus, 254
Perspectiva: defined, 307
Perspective: visual, 105–6, 307–8
Peter Abelard. *See* Abelard, Peter
Peter Lombard. *See* Lombard, Peter
Peuerbach, Georg: on planetary spheres, 272
Phaedrus, 139
Pharmacology, 18, 113, 129, 318, 329, 333–35, 339, 348–49
Philinus of Cos, 122
Philo of Larissa, 139
Philoponus, John, 77, 162; on motion, 302–3; 305–6
Philosopher's stone, 288
Philosophy: origins of, 25–27
Philosophy of nature. *See* Science
Physicians. *See* Medical practitioners
Physiologus, 351
Physiology: animal, 64; human, 119, 121–22, 124, 127–29
Physis, 82, 281. *See also* Nature
Pia mater, 120
Place: Aristotle's doctrine of, 58, 249; medieval discussions of, 249, 293; 365; Strato on, 76
Plague, 111, 280, 320, 339
Planets: Aristotle on the motion of, 61; Mesopotamian knowledge of, 17; motion of, 91–92, 95–104, 267, 270–71; nature of,

250; Plato on the motion of, 42; retrograde motion of, 91, 94. *See also* Cosmology; Astronomy
Plants, 318
Plato, 31, 35–44, 67, 71, 73, 80, 150, 198, 230; and the theory of forms, 36–39, 48; cosmology of, 39; geometrical atomism of, 85; influence of, 77, 125, 127, 130; on change and stability, 37, 51; on light and vision, 105, 313; on planetary motion, 90–91; on the applicability of mathematics to nature, 86, 294; on the elements, 283; on the requirement of uniform circular motion in the heavens, 92–93; on the three faculties of the soul, 127; *Phaedo*, 38–39; *Republic*, 35–38; *Timaeus*, 39, 42, 105, 130, 136, 138, 147–49, 169, 193, 197, 201–2, 215, 245, 251, 275. *See also* Platonic tradition
Plato of Tivoli, 204
Platonic solids, 41, 88
Platonic tradition, 73, 77, 127, 130, 136, 138–39, 145, 150, 162, 171, 185, 188, 197, 200, 202, 205, 216, 218, 224, 227, 229, 245–46, 277, 284, 302, 313, 379n.9
Pliny the Elder, 142, 146, 159; *Natural History*, 141, 144, 189, 254; on astronomy, 268; on medicine, 317
Pneuma, 81–82, 120–22, 127, 129, 379n.22
Pneumatists (school of physicians), 124
Popularization of science, 136–38, 139–41, 144–47, 149, 158–59
Porphyry, 165; *Introduction to Aristotle's Logic*, 149; logical works of, 188
Portolan charts, 256–57
Poseidon, 21, 23, 26
Posidonius, 138–40
Potentiality. *See* Actuality and potentiality
Praxagoras of Cos, 120
Prehistoric attitudes toward nature, 4–11
Prime mover, 66, 220, 251. *See also* Unmoved mover
Primitive mentality, 10
Primum mobile, 249, 267
Printing: influence of, on anatomical knowledge, 345
Privation, 51
Profatius Judaeus, 269
Prognosis (medical), 117, 125
Proportion: theory of, 146
Providence (divine), 79, 83, 220, 227, 237, 323

Pseudo-Apuleius: *Herbal*, 349
Pseudo-Aristotle: *Mechanical Problems*, 109;
   On Plants, 349–50
Pseudo-Dionysius, 188
Psychē: as source of order for Plato, 40;
   Stoic theory of, 82
Psychology, 212, 220, 224
Ptolemaic dynasty, 74
Ptolemy, Claudius, 99, 146, 368; *Almagest
   (Mathematical Syntaxis)*, 104, 169, 177,
   204–5, 211, 216, 252, 262–64, 269, 272;
   Geography, 138, 258; on light and vision,
   107–8, 308–9, 312; *Optics*, 106; *Tetra-
   biblos*, 275–76, 279; *Planetary Hypotheses*,
   104, 262, 264; planetary models of, 98–
   104, 262, 271
Pulse: as a diagnostic tool, 120–21, 126, 331,
   335, 337–38
Purging: as a medical therapy, 116
Pythagoras, 13, 25. *See also* Pythagoreans
Pythagoreans, 31–32, 35, 41, 87, 134, 371n.15;
   on planetary theory, 97; on the application of mathematics to nature, 294; on the
   ultimate reality, 85

Quadrant (astronomical), 177, 264, 269
Quadrivium, 156, 187, 190–91, 211, 268; defined, 138
Qualities: intensification and remission of,
   296; intensity of, 296–98; quantity of, 296;
   secondary, 362; the four primary, 55–56,
   116, 118, 126, 283–84, 288, 332, 334
Quintessence, 55, 61, 246, 250
Quintilian, 193

Rain, 143, 252
Rainbow: Aristotle's theory of, 67, 309; me-
   dieval theories of, 252–53
Rationalism, 33–34, 37–38, 79, 202, 219,
   221, 231, 235, 237; in theology, 193–97.
   See also Faith and reason
Rationalists (school of physicians), 124
Ratios, 306–7
Raymond of Marseilles, 268
Rāzī, al- (Rhazes): *Almansor*, 325, 327; *Book
   of the Secret of Secrets*, 288; medical writ-
   ings of, 331, 342
Reason and revelation. *See* Faith and reason;
   Rationalism
Reflection of light, 106–7; in the rainbow,
   253
Refraction of light, 106–8, 311, 377n.25; in

the rainbow, 253; law of, 108; within the
   eye, 311
Regimen, 116, 333
Regiomontanus, Johannes, 272
Regular geometrical solids, 40–41, 88
Reichenau, 190, 318
Reims, 188
Religion. *See* Islam; Christianity
Renaissance, 359, 360; conception of, 356
Resistance: in motion, 304–7
Respiration, 121, 128
Rete mirabile, 127, 129
Retinal image, theory of, 368
Retrograde motion: of the planets, 101, 144,
   147, 268, 270
Rhazes. *See* Rāzī, al-
Rhodes, 138
Richard of Middleton, 284
Ripoll, 189–90
Robert de Courçon, 217
Robert of Chester, 204, 268
Roger Frugard: *Surgery*, 341
Roman science, 124–31, 133–51, 317–18,
   320–21; relationship of, to Greek science,
   133–36, 147–49
Rome, 133
Russell, Bertrand, 2

Saints, cult of, 320–21
Salerno: as a medical center, 325–26,
   329–30, 342
Sampson hospital (Constantinople), 346
Schools: Athenian, 72–73, 75–77, 80; Carolingian, 184–85; cathedral, 185, 191–93;
   episcopal, 184: Greek, 70–73, 75–77, 138;
   Islamic, 174; medieval, 145, 155, 158, 184,
   191–93, 197; monastic, 155–56, 158, 185,
   191–92; Roman, 151–53. *See also* Curriculum; Education; Universities
Science: Byzantine, 161–62; defined, 1–4,
   45; eastward diffusion of, 163–66; Islamic,
   170–82, 188–90; place of, in Roman culture, 137
Scientific method, 361–62, 366–68., *See
   also* Demonstration; Empiricism; Experience; Experiment; Observation; Rationalism; Sense perception
Scientific revolution (of the 16th and 17th
   centuries), 358–63, 366–68
Scotus, John Duns, 242, 284
Scriptoria, 155–56
Seasons, 275, 277

Secondary qualities, 34, 79
Semen: in Aristotle's theory of generation, 65
Seminal causes, 198
Seneca, 140, 193, 258; *Natural Questions*, 197
Sense perception, 224. *See also* Empiricism; Experience; Experiment; Observation
Septimius Severus, 125
Seville, 181
Sexagesimal numbers, 14–16
Siger of Brabant, 234–37
Simplicius, 77, 162
Simulacrum, 308
Skeletal system, 127
Skepticism, 139, 242
Smyrna, 27, 125
Socrates, 35, 71
Solstices, 43, 91
Sophists: educational program of, 71
Soranus, 318
Soul: Albert the Great's theory of, 230–31; Aquinas's theory of, 233; Aristotle's theory of, 65, 68, 220, 224, 231, 365; as the form of the body, 220; Avicenna's theory of, 231; of the world, 42, 82, 246; Plato's theory of, 231; Siger of Brabant's theory of, 234; Stoic theory of, 82
Spain, 188; Christian reconquest of, 181
Sphericity of the earth, 42, 58, 145, 158, 245, 253
St. Gall, 318
Stagira, 47
Stars, 98
Stereographic projection, 264, 266
Stoa poikilē, 72, 73, 80, 212
Stoic tradition, 83, 125, 130, 138–40, 150, 197–98, 245, 248, 250, 275
Stoics, 80–83; influence of, 83, 125,130, 248; on cohesion, 81; on conflagration and regeneration of the cosmos, 82–83, 140; on the continuity of matter, 81; on cosmology, 82–83; on divinity, 81; on pneuma, 81; on the soul, 82; on void space, 81
Strabo, 76
Strato of Lampsacus, 74, 76, 121; as head of the Lyceum, 76; corpuscularianism of, 77; on heavy and light bodies, 76; on light, 77; on motion, 76; on place and space, 76; on the acceleration of falling bodies, 76; on void, 77
Substance, 49, 282–83, 285

Sulla, 73, 76
Sulphur-mercury theory, 288
Sun: distance of, from earth, 97–98
Supernatural, the: as distinct from the natural, 8, 27, 44; defined, 27; in Epicurean thought, 78; in medieval thought, 200–1, 221, 231, 235. *See also* Naturalism
Surgeons, 339, 342
Surgery, 18, 324, 329, 339–42
Swerve: in Epicurean philosophy, 79–80
Swineshead, Richard, 295; on motion, 307
Sylvester II (pope), 188. *See also* Gerbert of Aurillac
Symonds, John A., 356–57
Syriac, 165, 169

Technology: medieval, 190; relationship of, to science, 1, 3, 5
Teleology: denunciation of, by Lucretius, 140; in Aristotle, 53–54; 362; in Galen, 130, 131; in Stoic thought, 83; in Theophrastus, 75. *See also* Final cause
Temperature, 296
Tempier, Etienne, 236–40
Tension: Stoic theory of, 82
Terrestrial region, 54–55, 82, 162, 247, 250, 252, 275
Tertullian, 120, 150, 321
Thābit ibn Qurra, 170, 177, 252, 269
Thales, 25, 27, 29
Themistius, 162
Theodoric of Freiberg, 253
Theodoric the Ostrogoth, 148
Theological studies, 193–96, 206, 209, 211, 229, 240–41
Theology: as a science, 4; relationship of, to philosophy, 201, 221, 223, 227, 229–38, 240–42, 261; relationship of, to science, 250, 260–61, 293
Theophrastus, 47, 74–75, 139; *On Stones*, 75
Theorica planetarum (*Theory of the Planets*), 271–72
Theriac, 335
Thierry of Chartres, 198, 203, 246
Thorndike, Lynn: on the medieval scientific achievement, 357
Tides, 274, 277
Time-keeping, 184, 263
Tio (African tribe), 8
Titus, 141
Toledan Tables, 268, 272

Toledo, 180–81, 204; as a center of translation, 327
Translation: from Arabic to Latin, 197, 203–6, 215, 246, 268–69, 277, 279, 288, 303, 312, 326–27, 364; from Greek to Arabic, 165, 168–71, 180, 263, 288, 303, 308, 323; from Greek to Latin, 147–49, 156, 197, 203, 205–6, 215, 246, 269, 277, 279, 312, 317–18, 323, 351, 364; from Greek to Syriac, 165; into European vernacular languages, 332, 341, 351
Transmutation: of the elements, 41, 56, 252, 279, 283, 287–88
Trephining, 342
Trepidation, 177
Trigonometry, 102; spherical, 263
Trinity (divine), 232
Trivium: defined, 138
Trojan War, 21
Tropic of Cancer, 90–91
Tropic of Capricorn, 90–91
Trotula (Trota), 328–29
Tübingen: university of, 332
Turin: university of, 332
Two-sphere model of the cosmos, 42–43, 90–91

Ulugh Beg, 264
Umayyads, 166, 168, 180
Underlying reality: the question of, 27–32, 35–39, 44, 48
Universities: as guilds, 207–8; curriculum of, 211–12, 216–18, 222, 241, 364; enrollments in, 209; internal organization of, 208; origins of, 206–8; relationship of, to the church, 209. *See also under the names of particular universities*
Unmoved mover, 62, 251, 275
Urinalysis: as a diagnostic tool, 126, 331, 335–36
ʿUthman, 166

Vacuum, *See* Void
Varro, 144, 146; *Nine Books of Disciplines*, 138
Veins, 120, 122, 127, 128
Velocity: as a measure of motion, 294–96, 304, 306–7, 367; as a technical term, 60; instantaneous, 295
Vena cava, 128
Venus: sun-linked motion of, 91, 97, 144, 147, 268

Vesalius, Andreas, 345
Vespasian, 141
Vich, 188, 190
Virgil, 193
Vision, 368; Aristotle's theory of, 67–68; atomistic theories of, 105, 308; extramission theories of, 75, 105–6, 308–9, 311, 313; Greek theories of, 308–9; intromission theories of, 105, 308–9, 311–13; Islamic theories of, 309–12; mathematical theories of, 105–6, 308–9, 311, 313; Plato's theory of, 42, 105; Theophrastus's theory of, 75; theories of, in medieval Europe, 313, 315
Vital heat: Aristotle on, 65–66; Galen on, 127–29; Theophrastus on, 75
Vitruvius, 140
Vivarium, 156, 318
Vivisection: human, 120
Void, 30, 78, 80, 129, 238–41, 365; Aristotle on nonexistence of, 56; extracosmic, 248; motion in, 56, 304–5; Stoic theory of, 81; Strato on, 77
Voltaire, François Marie Arouet de, 356
Voyages of exploration, 257

Water: as an element, 31; as the underlying reality, 29
Water wheel, 190
Wearmouth (monastery of), 158
Weights, science of, 108–10, 294
Wheel: invention of, 5
William of Auxerre, 217
William of Conches, 200–1, 221; *Philosophy of the World*, 200
William of Moerbeke, 205
Witelo: on light and vision, 313, 315
Wounds, 118
Writing: alphabetic, 11–13, 27; hieroglyphic, 11; invention of, 11–12; relationship of, to the origins of philosophy and science, 12–13, 27; wide dissemination of, among the Greeks, 13

Yaḥya ibn Barmak, 168
Yates, Frances, 360
York, 185

Zarqālī, al-, 268
Zeno (of Elea), 35; on change and stability, 33
Zeno of Citium, 73, 80

Zeno's paradoxes, 33
Zeus, 21–23, 31
Zodiac, 89: defined, 17; signs of, 279

Zoology: medieval, 350–53; of Albert the
    Great, 230; of Aristotle, 63–66
Zoroastrianism, 163